AF726263

ORGANIZATIONAL INFORMATION SYSTEMS IN THE CONTEXT OF GLOBALIZATION

IFIP – The International Federation for Information Processing

IFIP was founded in 1960 under the auspices of UNESCO, following the First World Computer Congress held in Paris the previous year. An umbrella organization for societies working in information processing, IFIP's aim is two-fold: to support information processing within its member countries and to encourage technology transfer to developing nations. As its mission statement clearly states,

> *IFIP's mission is to be the leading, truly international, apolitical organization which encourages and assists in the development, exploitation and application of information technology for the benefit of all people.*

IFIP is a non-profitmaking organization, run almost solely by 2500 volunteers. It operates through a number of technical committees, which organize events and publications. IFIP's events range from an international congress to local seminars, but the most important are:

- The IFIP World Computer Congress, held every second year;
- Open conferences;
- Working conferences.

The flagship event is the IFIP World Computer Congress, at which both invited and contributed papers are presented. Contributed papers are rigorously refereed and the rejection rate is high.

As with the Congress, participation in the open conferences is open to all and papers may be invited or submitted. Again, submitted papers are stringently refereed.

The working conferences are structured differently. They are usually run by a working group and attendance is small and by invitation only. Their purpose is to create an atmosphere conducive to innovation and development. Refereeing is less rigorous and papers are subjected to extensive group discussion.

Publications arising from IFIP events vary. The papers presented at the IFIP World Computer Congress and at open conferences are published as conference proceedings, while the results of the working conferences are often published as collections of selected and edited papers.

Any national society whose primary activity is in information may apply to become a full member of IFIP, although full membership is restricted to one society per country. Full members are entitled to vote at the annual General Assembly, National societies preferring a less committed involvement may apply for associate or corresponding membership. Associate members enjoy the same benefits as full members, but without voting rights. Corresponding members are not represented in IFIP bodies. Affiliated membership is open to non-national societies, and individual and honorary membership schemes are also offered.

ORGANIZATIONAL INFORMATION SYSTEMS IN THE CONTEXT OF GLOBALIZATION

IFIP TC8 & TC9 / WG8.2 & WG9.4 Working Conference on Information Systems Perspectives and Challenges in the Context of Globalization
June 15–17, 2003, Athens, Greece

Edited by

Mikko Korpela
University of Kuopio
Finland

Ramiro Montealegre
University of Colorado at Boulder
USA

Angeliki Poulymenakou
Athens University of Economics and Business
Greece

KLUWER ACADEMIC PUBLISHERS
BOSTON / DORDRECHT / LONDON

Distributors for North, Central and South America:
Kluwer Academic Publishers
101 Philip Drive
Assinippi Park
Norwell, Massachusetts 02061 USA
Telephone (781) 871-6600
Fax (781) 681-9045
E-Mail <kluwer@wkap.com>

Distributors for all other countries:
Kluwer Academic Publishers Group
Post Office Box 322
3300 AH Dordrecht, THE NETHERLANDS
Telephone 31 78 6576 000
Fax 31 78 6576 254
E-Mail <services@wkap.nl>

Electronic Services <http://www.wkap.nl>

Library of Congress Cataloging-in-Publication Data

A C.I.P. Catalogue record for this book is available from the Library of Congress.

Organizational Information Systems in the Context of Globalization
Edited by Mikko Korpela, Ramiro Montealegre, and Angeliki Poulymenakou
ISBN 1-4020-7488-3

Copyright © 2003 by International Federation for Information Processing.

All rights reserved. No part of this work may be reproduced, stored in a retrieval system, or transmitted in any form or by any means, electronic, mechanical, photocopying, microfilming, recording, or otherwise, without written permission from the Publisher (Kluwer Academic Publishers, 101 Philip Drive, Assinippi Park, Norwell, Massachusetts 02061), with the exception of any material supplied specifically for the purpose of being entered and executed on a computer system, for exclusive use by the purchaser of the work.

Printed on acid-free paper.
Printed in United Kingdom by Biddles/IBT Global

Contents

Part 2: Organizations, Corporations and Institutions

Part 3: ICT Industries and Systems Development

Acknowledgements

We are very grateful for the valuable help and kind support of many people in setting up and in ensuring the success of this conference.

David Avison and Chrisanthi Avgerou IFIP Working Group 8.2 and 9.4 chairs respectively (for 2002) have been instrumental in encouraging us to pursue this endeavour and in offering their ample support, experience and advice along the way. We also extend our thanks to their successors, Julie Kendall and Abiodun Bada. Geoff Walsham, our General Chair, has been a key reference point for us throughout our course of work for the conference.

Special thanks are due to the individuals who served as reviewers for the submissions. It is through their care, attention, and insight that the conference and this book came to fruition and we appreciate their efforts. We thank, of course, all the authors who trusted us with their research and worked hard to accommodate reviewers' comments within a very limited time frame.

The conference organisers are indebted to the IFIP/UNESCO fund organisers Anselmo Del Moral and Plamen Nedkov, and to Dewald Roode IFIP Technical Committee 8 Chair for ensuring financial support for three conference delegates from developing countries. We are also thankful to Commonwealth Secretariat for providing funds for the same purpose.

The members of the local organising committee Spiros Borotis, Kostas Samiotis and Maria Karyda, have carried out the final editing and formatting of this volume in Athens. A very big thanks is owed to them, particularly to Spiros who was overall responsible for this tedious process. Janice DeGross has shared with us her invaluable experience in proceedings editing and we thank her for that. Nikos Korfiatis has been responsible for setting up the conference site, a work commended by many people involved in this conference.

We also acknowledge the support of the Department of Management Science and Technology of the Athens University of Economics and Business for local and administrative arrangements.

Mikko Korpela
Ramiro Montealegre
Angeliki Poulymenakou

Acknowledgements

Conference Chairs

General Chair

Geoff Walsham
University of Cambridge, UK

Program Co-Chairs

Mikko Korpela
University of Kuopio, Finland

Ramiro Montealegre
University of Colorado, Boulder, USA

Organizing Chair

Angeliki Poulymenakou
Athens University of Economics and Business, Greece

Program Committee

David Allen, UK
Chrisanthi Avgerou, UK
Abiodun O. Bada, USA
Richard Baskerville, USA
Dubravka Cecez-Kecmonovic, Australia
Claudio Ciborra, UK
Francois Cooren, USA
Kevin Crowston, USA
Mike Cushman, UK
Enrique Dans, Spain
Robert Davison, Hong Kong
Bill Doolin, New Zealand
Georgios Doukidis, Greece
Brian Fitzerald, Ireland
Neki Frasheri, Albania
Michael Gallivan, USA
Seymour E. Goodman, USA
Judith Gregory, Norway
Ole Hanseth, Norway
Penny Harvey, UK
Richard Heeks, UK
Lisa von Hellens, Australia
Rudy Hirschheim, USA
Evangelos Kiountouzis, Greece
Krisana Kitiyadisai, Thailand
Heinz Klein, USA
S. Krishna, India
Esselina Macome, Mozambique
Shirin Madon, UK
Jonathan Miller, South Africa
Emmanuel Monod, France
Eric Monteiro, Norway
Robert Moreton, UK
Eduardo Morgado, Brazil
Anja Mursu, Finland
Joe Nandhakumar, UK
Ojelanki Ngwenyama, USA
Mayuri Odedra-Straub, Germany
Nancy Pouloudi, Greece
Renata La Rovere, Brazil
Nancy Russo, USA
Sundeep Sahay, Norway
Ann Séror, Canada
Ulrike Shultze, USA
Leiser Silva, USA
Abimbola Soriyan, Nigeria
Donaldo de Souza Dias, Brazil
Lucy Suchman, UK
Natalia Volkow, Mexico
Gert-Jan de Vreede, The Netherlands
Louise Wittaker, South Africa
Eleanor Wynn, USA

Additional Reviewers

In addition to the program committee, the following individuals also undertook reviews for the conference:

Margunn Aanestad, Norway
David Avison, France
Bongsug Chae, USA
Ricardo M. Checchi, USA
Mark Chun, USA
Tony Cornford, UK
Ela Dabrowska, UK
Noornina Dahlan, Malaysia
Klarissa Engelbrecht, South Africa
Helen Hasan, Australia
Marius Janson, USA
Timo Käkölä, Finland
Ken Kendall, USA
Niki Panteli, UK
Angeliki Poulymenakou, Greece
Prem Premkumar, USA
Mehdi Raoufi, Sweden
Helen Richardson, UK
Gamila Shoib, UK
Craig Standing, Australia
Wal Taylor, Australia
Eileen Trauth, Ireland
Geoff Walsham, UK
Edgar Whitley, UK

1

PERSPECTIVES AND CHALLENGES OF ORGANIZATIONAL INFORMATION SYSTEMS IN THE CONTEXT OF GLOBALIZATION

Introduction to the Theme

Mikko Korpela[1], Ramiro Montealegre[2] and Angeliki Poulymenakou[3]
[1]University of Kuopio, Finland; [2]University of Colorado, USA and Instituto de Empresa, Spain; [3]Athens University of Economics and Business, Greece

1 INFORMATION SYSTEMS AND GLOBALIZATION

In this book, the context and process of globalization are understood as a complex web of increasingly deepening interrelationships among countries, organizations, social groups, and individuals. Information and communication technologies (ICTs) are seen as a vital part of the globalization agenda.

Falling trade barriers – the European Union, the North American Free Trade Agreement, Mercosur – global capital markets, global communication, global customers, a changing regulatory environment, technologies that support a span of control transcending national boundaries, and the need to find revenue growth in new markets are all driving corporations to globalize. More importantly, societies, social groups, governments, policymakers, non-profit organizations, and individuals are all increasingly challenged by changes in their professional, social and personal spaces and activities, brought about by connectivity and digitization.

The information systems research agenda has been dealing for a number of years now with these issues. Globalization is a force of unification and conformity – as seen, for example, in the proliferation of global brands such as Nike, Coca-Cola, McDonald's. Globalisation is also a force of divide, diversity, and social struggle for prosperity. Despite dominant policy and managerial rhetoric, not all globalisation messages can be considered as positive, and the multifaceted impacts of the globalisation process on local contexts are far from being well understood, adequately researched, or dealt with in a balanced manner. The multidisciplinary work carried out on the societal and organisational implications of information systems has significant contributions to make in deepening our understanding on these new phenomena and the changes they bring about.

In 1973, the Secretary-General of the United Nations reported to the General Assembly that the number of computers was about 80,000 in the USA, 11,000 in Japan, 7,000 in the UK, 1,200 in Brazil, 200 in India and 30 in Nigeria. The report concluded in the following main points about the role of "the computer" in global socio-economic development (United Nations, 1973):

> *In looking to the computer as an aid to development, three points are particularly noteworthy. First, it must be emphasized that computers are only tools. [...] Their effectiveness depends on those who plan, approve, implement, manage, monitor and evaluate activities in which the computer is involved. The human element, in short, is the sine qua non for the use of this technology. Secondly, computers and their technology are costly, in terms of both the hardware and software involved and the education, training and retraining of personnel at all levels. Thirdly, the computer and its technology require and generate a broad socio-economic environment [...].*
>
> *Finally, [...] the computer is not a panacea for administrative and development problems. Possession of a computer may at first create at least as many problems as it solves. Its use requires highly trained systems designers, analysts, programmers and operators. Without such trained personnel, the computer cannot be used effectively. It also requires reliable data which have to be collected and transcribed suitably. The computer demands a considerable infrastructure, or else it cannot function.*

Today, thirty years later, we speak about information systems instead of "the computer," acknowledging the need to focus on the interaction between technology and its context. Despite the dramatic development in technology, however, the conclusions of the United Nations report are still surprisingly valid. ICTs have an enormous potential for socio-economic and human development, but only when applied meaningfully in organizational and societal activities - as information systems - by knowledgeable people, and supported by appropriate infrastructures.

The Internet has now reached all countries, but the relative difference between countries in the number of Internet users per capita is as high, if not higher, than the difference in mainframes thirty years ago. Optimism is tempered by the awareness that most fundamental and technological progress is still "outside and has to be imported," and that a process of "haves and have-nots" polarization is underway that could exacerbate differences among national groups, aggravating the already existing gaps. This raises an important question about the overall ability of relevant actors to exploit the malleability of these new technologies and to implement applications – nurtured in the proper social, cultural, and technological environment – to capitalize on their full potential.

This was what motivated two working groups of the International Federation for Information Processing (IFIP) to call for a conference to discuss the perspectives and challenges of information systems in the context of globalization in Athens, Greece, in June 2003.

2 BACKGROUND OF THE CONFERENCE

IFIP is a non-governmental, non-profit umbrella organization for national societies working in the field of information processing. It was established in 1960 under the auspices of UNESCO as an aftermath of the first World Computer Congress held in Paris in 1959. IFIP's principal aims were and are to foster international cooperation, to stimulate research, development and applications, and to encourage education, dissemination and exchange of information on all aspects of computing and communication.

In 1977, IFIP established a working group concerned with "the interaction of information systems and the organisation." The working group, WG 8.2, seeks to generate and disseminate knowledge about and improve understanding of the role and impact of information technology across a range of social levels (society, organization, individual) and across a diversity of spheres (marketplace, workplace, home, community). One of the most influential events organized by the WG 8.2 was a conference in Manchester in 1984 on research methods in information systems, which became a breakthrough for non-positivist research approaches (Mumford et al., 1985). WG 8.2 has ever since been the home for interpretative and critical research on information systems as social phenomena. Although some of its conferences have explicitly included a societal dimension (Avison et al., 1993), the focus of the working group has been on the organizational and workgroup levels of analysis. The overwhelming majority of WG 8.2 members are from North America, Europe and Australia. The spirit of the working group has thus been favourable to societal and cultural diversity, but research in its realm has not been geographically global in scope.

Another IFIP working group, WG 9.4 on "social implications of computers in developing countries", emerged from a working conference in New Delhi in 1988 (Bhatnagar and Bjørn-Andersen, 1990). WG 9.4 has always emphasized a social viewpoint on the application of ICTs, and therefore the focus has been on organizational information systems in addition to national issues of ICT policy. The working group has been the main forum for information systems research on and in developing countries, but also for more general discussion on the role of ICTs in development and the need for societal, cultural, and organizational adaptation in information systems research and development (Avgerou and Walsham, 2000).

In the WG 8.2 conference in Aalborg in 2000, Geoff Walsham reviewed the state of information systems research on issues of globalization. He stated that although globalization was a popular buzz word, most of the existing research was conducted in North America and Europe on the organizational and sub-organizational levels, implicitly assuming that the results were universally applicable. He urged information systems researchers around the globe "to study particular individuals, groups, organizations, or societies in detail, and in context" (Walsham, 2000).

Following Walsham's call, the idea of a joint conference by WG 8.2 and WG 9.4 was pushed forward by the chairpersons of the WGs, David Avison and Chrisanthi Avgerou. They searched for possible sites and organizers for the conference. In July 2001, a proposal by the Athens University of Economics and Business was approved, with Angeliki Poulymenakou as the Organizing Chair. Ramiro Montealegre was appointed as Program Co-Chair representing WG 8.2 and Mikko Korpela representing WG 9.4. The three conference chairs worked closely together as a team, supported by Geoff Walsham as the General Chair and the WG 8.2 and 9.4 chairs.

The conference call focused on organizational information systems from a global perspective. Contextually aware studies of information systems from all parts of the world form the basis for comparisons and inferences from a global viewpoint in the conference. "Globalization" is seen as a contradictory process, which may imply increased transnational uniformity but also increased diversity, through increased interconnectedness of local actors.

3 THE REVIEW PROCESS AND THE CONFERENCE

The Call for Papers for this conference attracted 111 paper and panel submissions on a diverse set of topics relating to the conference theme. Each paper was reviewed by two anonymous reviewers, who represented both WG 8.2 and WG 9.4 viewpoints. In case the reviewers' recommendations differed, the three conference chairs provided a further review. At the end of the review process, we had as many as 52 papers and 2 panel proposals that were recommended to be presented in the conference. The increased interest in the conference urged us to accept them. This is an unusually large number of papers for a typical WG 8.2 or 9.4 meeting, but the joint nature of the conference further justified it.

However, that many papers could not be accommodated in the proceedings. The selected papers varied significantly both in terms of topics but also in terms of the 'maturity' of the work, according to our reviewers' comments. We then decided to select 26 papers and 2 panels to be included in this book, and to include the remaining 26 papers in an electronic supplement that is available to the conference participants on CD-ROM. The electronic papers will be submitted to various information systems journals for potential publication afterwards in an edited form. All the papers selected for this book were also edited by the authors according to the reviewers' comments, and the final papers were accepted by the proceedings editors.

In addition to the selected presentations and panels, the conference programme includes keynote presentations by Prof. Saskia Sassen and Prof. Geoff Walsham, as well as an invited panel.

We are very pleased with the originality, diversity, and quality of the papers selected and we hope that the reader will feel equally satisfied. The selected papers reflect the diversity of information systems research with respect to research approaches, theoretical traditions, research subject, and geographical focus. We have papers employing diverse research methods, drawing on multiple analytical theories, and representing the various world continents.

4 OVERVIEW OF THE PAPERS

We provide below an overview of the papers accepted for publication in this volume. The papers exemplify an extremely rich and diverse array of themes, theoretical approaches, local contexts and practical implications currently researched under the conference's general theme.

Notably, through the research presented in this volume, we gain insights in local contexts that are comparably less frequently researched through an information systems perspective. Moreover, even when well-researched contexts such as the UK are the basis for empirical work, the research reported in this volume casts a different eye on them, focusing mainly on issues that transcend the usual managerial and economic welfare maxims.

Empirical research reported in this volume has been conducted in the following national, regional and local contexts (listed in alphabetical order): Australia (including aboriginal communities), Brazil, China, Ecuador, Egypt, India, Ireland, Jordan, Korea, Malaysia, Maldives, Mozambique, New Zealand, Peru, Portugal, Russia, Scandinavia, Spain, South Africa, UK and the USA. The electronic supplement adds Argentina, Bangladesh, Canada,

Caribbean, Chile, Cuba, The Gambia, Ghana, Greece, Kenya, Nepal, Nigeria, Oman, Philippines, Venezuela and Yugoslavia to the list.

The topics of research covered in this volume span the micro (individual), meso (organisational), and macro (society) levels. Despite the obvious risks embedded in any attempt to group the individual pieces of work under any classification scheme, we believe there is merit in eliciting some generic themes under which research documented in this volume can be seen to unfold. We have opted against a theory versus practice divide, as this would be rather artificial – even papers that do not have a particular real life context documented and analysed have a strong element of practice in terms of their specific focus on particular implications for it. We have also opted against a "Western" vs. "non-Western" divide, to highlight the diversity of contexts within each theme.

We see five main domains where the dynamic interplay between the global and the local is studied in relation to information systems in the works included in this volume:

1. Work practices and their transformation
2. Organizations, corporations and institutions
3. ICT industries and systems development
4. Societal dynamics
5. Cultural, philosophical and policy issues

In what follows, we present a brief overview of the papers in this volume according to the above generic categories of topics, followed by the panels.

4.1 Work Practices and their Transformation

Kathy McGrath examines the omni-present theme of organizational modernization through ICT adoption. She is applying a discourse-analytic perspective on the emergency services of the UK National Health Service. Her analysis shows how calls for cultural change quickly converge on narrowly conceived target-meeting efforts both at the ministerial and at the managerial level in her research context. Participants in the local context do not accept framing modernization in such terms, as McGrath shows.

Emilio Mosse and Sundeep Sahay 'unpack' the relationship between communication practices and information flows in their study of community health information systems in Mozambique. They use Castell's metaphor of counter networks to analyze how information processing practices are influenced by communication practices across various levels of administration and how the marginalization of groups and regions can be avoided when implementing such systems.

Jens Pors and Jesper Simonsen's work studies the relationship between work practice and IT artefact in the context of a groupware application adoption by distributed work teams. Their case setting is a financial institution, spanning four countries. They review the fundamental concept of coordination, as this is manifest in the artefact design as well as in the intentions to support collaborative work. Their extensive, multifaceted case study analysis leads them to suggest six generic factors influencing the integration of such technology into a work practice. These factors address management, membership, evaluation potential, reduction of complexity, exemplification of experiences, and active facilitation.

Helen Richardson applies Bourdieu's critical social theory on the study of Customer Relationship Management (CRM) technology utilization. Her context is CRM used in Call Centres serving a variety of client firms in northwest England. Richardson discusses the role of these systems as purveyors of organizational intelligence and identifies internal conflicts inherent in their use: worker control seems to take precedence over worker flexibility, or empowerment. Richardson uses critical social theory for the interpretation of social relations around CRM system use, in her context. Her interpretive lens helps us visualize some of the relevant dynamics of the social situation she examines such as the accommodation and bypassing of formal management arrangements imposed through the technology, the incongruence of 'knowledge management' aspirations through mere passive recording of data, and the aggressive management tactics of domination poorly concealed under standardization claims.

Robert Tucker and Niki Panteli report on research dealing with the issue of trust in virtual teams. They challenge current understandings which claim that trust is created 'swiftly' in such environments and they elaborate on the difficulties created by conflict and power differentials. Their empirical study on eighteen virtual teams stresses the value of shared goals formulation, which they frame as a continuous, multifaceted process throughout a virtual team's 'life'.

In the conference, this theme is augmented by the electronic papers by **Helen Hasan and Kathryn Crawford** on distributed communities of learning and practice, **Dawn Jutla et al.** on information technology innovation, **Adekunle Okunoye** on knowledge management, **Lucy Firth et al.** on community health information systems, as well as **Kostas Samiotis and Angeliki Poulymenakou** on knowledge management interventions.

4.2 Organizations, Corporations and Institutions

Mariyam Adam and Michael Myers interpret the implementation of an Information System for the Maldives Customs Service as a neo-colonialist attempt for domination of the global technological onto the local context. In their case study they illustrate neo-colonialist behaviour in the attempts to streamline local practices with international policy directives issued by the projects' funding body. The study reflects on the potential impacts of such 'globalization facilitated by software'.

Nabeel Al-Qirim studies ICT innovation adoption in small and medium enterprises in the context of three electronic commerce adoption cases in New Zealand. The conceptual framework driving the empirical research is developed through a synthesis of Rogers' diffusion of innovation theory with relevant extant theory on ICT adoption. The empirical work he conducted corroborated some of the factors in the framework, while many others were not supported, partly due to a tendency of the firms to treat electronic commerce initiatives as the ground for experimentation rather than as a vehicle for strategic advantage.

Jose Esteves, Joan Pastor and Joao Carvalho document a case of ERP implementation in a Portuguese medium sized enterprise. They develop a comprehensive factor model for the implementation process in such contexts, which they then populate with data from their case with special emphasis on organizational actions and reactions over the project's duration. Their findings are further interpreted from a cultural context perspective leading the authors to conclude that further research within this line of thinking is capable of shedding light into cases of problematic ERP adoption in local contexts diverging from the common corporate mould.

Wei Liu and Chris Westrup study globalization by looking at how Multi-National Enterprises use ICTs for organizing. They compare a UK with a China based venture to discuss cross-cultural coordination and control. Their main conclusions show that far from introducing managerial uniformity by alleviating time and distance barriers, and by enforcing standard practices, the use of ICTs necessitated human agency for coordination as well as for exploitation of available management facilitation features. In non-Western contexts, the seeming 'closeness' through ICT warrants reinterpretation based on local context characteristics.

Esselina Macome studies IS implementation with an emphasis on the global-local context debate within a case where the system was specifically earmarked as a modernization initiative. The suppliers and implementers of an invoicing system represent 'global' in her study in the Mozambican Electricity Company, while 'local' is represented by the company's IT and management staff and the system's end users. Using Rogers' diffusion of innovation theory coupled by an actor-network theory interpretation, she draws a number of recommendations for implementation team formation and membership, accommodation of changes to work practices, and organizational stabilization efforts.

Garry Pan, Shan Pan, Xin-Xin Chen, and Donal Flyn relate global IT products and local implementation contexts in their study of ERP implementation in China. They provide insights on the experiences of five firms in this rarely researched local context. Their analysis is focusing on alignment issues regarding management and operational process features on the one hand, and package features on the other. Their study, still at the exploratory stage, suggests that the usual difficulties and dilemmas encountered by western economy firms when implementing ERP, are further precipitated in the Chinese context due to an ambient sense of uncertainty with this new technology, as well as cultural 'distance' from it expressed by local players.

The electronic conference papers by **Bongsug Chae** on dynamic institutional theory, **Baltazar Chilundo and Margunn Aanestad** on vertical and integrated health programmes, **Evangelia Kopanaki and Steve Smithson** on inter-organizational systems, **Renata Lèbre La Rovere** on information management networks, **Carmen Joham and Junilee Pradhan** on IT acquisition, **Ann C. Séror** on virtual institutional infrastructures, as well as **Andrew Stein et al.** on ERP implementation contribute to this theme.

4.3 ICT Industries and Systems Development

Aini Aman and Brian Nicholson address in their study the global software industry. They examine offshore software development commissioned by UK companies to Malaysian firms. They wish to explore and discuss a broad range of characteristics of the client – offshore vendor relationship, and to this end they exploit the COCPIT framework. They propose a new contingency framework that can be used to monitor and 'explain' managerial situations in offshore software development outsourcing.

Ping Gao and Kalle Lyytinen study telecommunications transformation in China. They interpret current reform initiatives in this context through Giddens' structuration theory perspective. Their analysis of how signification, domination and legitimation evolve over time is juxtaposed with related telecommunications strategy reform actions. This study investigates a major market change phenomenon from the perspective of the behaviour of decision makers in a path dependent process The insights gained may help us understand the future market reform decisions of this major international player.

Ciara Heavin, Brian Fitzerald, and Eileen Trauth trace the evolution trajectory of Ireland's software industry. They develop a comprehensive analysis framework by synthesizing extant literature on IT industry strategy, and development in the context of developing economies. They apply their conceptual framework on empirical evidence to illustrate how each of their proposed factors has been contributing to the industry's evolution and to recommend strategy directions for further consideration.

Jan Pries-Heje, Richard Baskerville and Galina Hansen examine the adoption of high-speed software development practices in Russia. Their study sheds light in an important economy under development and its local adaptation of global software practices. Using extant theory on the enablers of software practice diffusion, they analyze four Russian software firms in terms of their software development practice trajectories. They apply an organizational development perspective to conclude that barriers to the diffusion of practices are related to language, the 'youth' of Russian software firms, leadership challenges, and the effects of extensive foreign outsourcing.

S.K. Puri and Sundeep Sahay tackle the pervasive issue of participation in IS development in the context of GIS projects aiming to support the implementation of development related policies in India. In two GIS implementation case studies, they contrast phenomena reflecting the centrality of project control with the need to promote substantial user involvement, not least for effective harnessing of local, indigenous knowledge. Using the reflexive lens on institutional structure versus human, participatory agency, they interpret contrasting situations regarding GIS implementation scenaria by focusing on the interplay between 'scientific' and 'indigenous' knowledge dominance, and alternative views of management agency.

This theme is also addressed by the electronic conference papers by **Pamela Y. Abbott** on software outsourcing, **Marisa D'Mello** on global software organizations, **Gaye Kiely and Brian Fitzgerald** on the use of methods in IS development projects, **Anja Mursu et al.** on methods for sustainability in IS development, **Brian Nicholson and Erran Carmel** on offshore software development, as well as **Rajeev Sharma and S. Krishna** on global software work.

4.4 Societal Dynamics

Tony Cornford and Ela Klecun-Dabrowska deal with questions of social (in)exclusion in the context of community information systems that support healthcare needs in a London geographic community. They argue that the relative dearth of solid ideas and debate regarding 'informational responses' to social exclusion is manifest in the domination of 'technology as a solution to access problems' type of thinking. Their extensive case study in the London Borough of Lewisham questions the value of focusing on consequences when examining the impact of ICTs on exclusion. They propose a process driven line of thinking, focusing on the need to embed such ICT interventions in a wider agenda of sustained regeneration and reform of the related institutional framework.

Justine Johnstone examines the evolving character of non-governmental development organizations as active knowledge agents within a variety of policy influencing and development processes. She documents knowledge roles implicit in the current behaviour of such agencies focusing on their informational, epistemic and conceptual elements. Johnstone suggests that these three dimensions can aid NGOs reflect on their approach and practice. She illustrates further her point by suggesting how these three dimensions can help NGOs frame their use of networking technologies to augment their core competencies.

Tony Salvador, John Sherry and Alvaro Urrutia discuss in their work the role of shared models of ICT access (telecenters, cyber-cafes) as means for easing the digital divide. Their ethnographic study of cyber-cafes and traditional cafes in six countries across the globe illustrates the 'social' nature of these places and contrasts this with design maxims of functionality and efficiency dominating the set-up of some of these places. They argue in favour of an evolutionary approach in the set-up and running of these settings more akin to local characteristics of social life than to the technological options available.

Craig Standing, Ian Sims, Rosemary Stockdale, and Arjen Wassenaar are debating the capacity of government sponsored e-marketplaces to act as bridges of the digital divide experienced by small and medium size firms in the Australian context. They raise questions related to the capacity of such firms to reap the benefits of participation in such digital network arrangements and explore them in the context of two cases, a central and a regional government marketplace initiative. Factors emerging through this study include shared ownership structures, the move from transaction to collaboration and information sharing business models, and the need to provide a wider rationale for marketplace development beyond narrow conceptions of direct economic returns.

Prodoromos Tsiavos, Ian Hosein and Edgar Whitley debate the relationship between technology and jurisdiction. They exemplify their discussion through the case of copyright and peer-to-peer technologies in conjunction with the regulatory environments surrounding their control, at various stages of such technologies' development. In their review, the existing schools of thought on the relationship between technology and regulation are either calling for mere replication of current regulatory thinking into digital spaces, or calling for fundamental regulatory reform. The authors move on to make the case for a third option: the co-evolution of technology driven regulatory change and of established legal practices.

The electronic conference papers contributing to this theme include the ones by **Elaine Byrne and Sundeep Sahay** on participation in primary health care information systems, **Shirin Madon** on e-governance, **Alexander Osterwalder** on ICT in developing countries, as well as **Elpida Prasopoulou and Athanasia Pouloudi** on IT as a stabilization mechanism in strategic alliances.

4.5 Cultural, Philosophical and Policy Issues

Chrisanthi Avgerou in her paper reviews the historical trajectory of thinking that has linked ICTs and economic development within the discourse of international development agencies. She examines four recent influential publications on this topic, arguing that they all focus on the reciprocality between the development of conditions that foster ICT innovation, and the creation of an effective market. Avgerou reveals the controversy surrounding economic theories of development that are the theoretical foundations on which such current discourse has been based. She argues in favour of socio-economic theory of development as more appropriate for this discourse as it reflects the dynamics of ICT and development, "a situated, context-specific process that is entangled with indigenous politics and historically-formed institutions".

Claudio Ciborra and Diego Navarra debate dominant policy rhetoric promoting the use of ICTs to improve state governance practices as a lever for development in late developing countries. They look into the case of Jordan's e-government policy through the lens of New Institutional Economics, as the links between enforcement-transition costs and the outcomes of policies and interventions. Their analysis sensitizes us to the complex web of factors intervening between e-government implementation and effective governance, making a clear case against beliefs that effective governance capacity can be substituted for all-encompassing ICT interventions.

Kenneth Kendall, Julie Kendall and Muhammadou Kah approach the issue of ICT policy development for developing countries from a discourse-analytic perspective. Using Web discussion lists for relevant discursive material, they propose a 'dramatistic analysis' method aiming to illustrate the dynamics underlying sustainable policy development.

Mike Metcalfe and Carmen Joham in their paper attempt to break the dominant mould of thinking and research on knowledge sharing. They point our attention to patterns of knowledge exchange away from the exchange of written knowledge, towards the oral and the visual domains. They review the challenges that these forms of knowledge exchange introduce for the design of technical systems and then they augment their position with the cultural and social aspects pertinent to these situations. Their empirical context is a project aiming to support remote aboriginal communities in their exchanges with state government. Their study recommends useful enhancements to knowledge sharing facilities of dominant technological infrastructures, such as the Internet.

Gamila Shoib and Joe Nandhakumar deal with IS implementation in 'non-Western' contexts. Their argument is that 'Western' technologies carry with them a particular rationality that is incongruent with respect to the local context of implementation. They reflect on two large-scale case studies to illustrate the situated nature of rationality regarding IS implementation decisions. They indicate some problematic aspects of dominant rationality in the context of their topic, namely the treatment of technology as uniformly superior, the treatment of success as an 'absolute construct', the shallow understanding of human agency, and the futility of the 'West/non-West' dichotomy itself.

This theme is also studied in the electronic conference papers by **Saheer Al Jaghoub and Christopher Westrup** on building software industry and the role of nation state, **Dimitrios A. Brachos** on digital divide, **Peter Meso et al.** on the diffusion of national IT policies, as well as **Jill Slay et al.** on understanding "culture" in IS security research.

4.6 Panels

Robert Davidson, Ernest Jordan, Carol Hsu and Chrisanthi Avgerou in their panel discuss issues of professionalism in the IS domain. Particularly they wish to explore the role of professional societies in the promotion of a 'global' set of IS professional norms, a stance on which they apply a cautious outlook as they indicate the danger of over-westernization of the global community's perspective on IS professions.

Joseph Feller, Brian Fitzgerald, Jan Ljunberg, and Magnus Berquist's panel on open source and free software casts a wide net over this hot topic for economies under development. They wish to explore the ethos characterizing these development communities, the economic, business and societal rationales underlying their existence, forms of collaboration and knowledge sharing developed in them. The panelists also promise to cast their attention on the philosophical foundations of the debate on open source and free software as a public good.

5 CONCLUSION

As technology becomes more pervasive and capable, there is a danger that the debate will remain anchored on techno-centric issues as to what to use the technology for, or how to make it work. This book provides rich insights on how global technology in action must be considered in view of local expectations and constraints. This is particularly relevant in the current discourse relating technology to development – the theme permeating most of the works included in this volume. Research reported in this book achieves in our view a number of important objectives. It illustrates local ICT utilisation and policy development contexts that are not as frequently examined as North American and Western European corporate contexts. The research papers included exemplify the role of social theory in approaching ICT utilisation challenges in a globalisation context. The debates raised on implementation, policy, organisations and organising, and on social dynamics, increase our awareness on the diversity of perspectives we need to delve upon when framing the role of ICTs in the globalisation agenda. The equal representation of managerial and non-managerial decision making contexts alerts us to the fact that ICTs should not be considered only as a corporate wealth creation prerogative.

Social actors, managers, and researchers are faced with great challenges and opportunities in the development and use of information systems. However, making fashionable commitments and investments for their countries and organizations without systematically analyzing needs and individual and environment characteristics is irresponsible. If relevant stakeholders fail to understand the role of ICTs in accordance with their situations and idiosyncrasies, it will only create new barriers, limitations, and foreign necessities in their societies. We have to identify the extent to which countries and organisations must follow existing models or develop new ones.

We hope that the collection of papers in this book provide a deeper understanding and appreciation of this area, and thereby stimulate further research and discussions. This book has important implications for both research and practice. It is significant in that it represents one of the few books to consider information systems in the context of globalization. For practitioners, it provides useful insights. There is a qualitative difference in the approaches that have been followed. Given that there are no universal sources of advantage, social welfare, or individual well being, and that there are no two countries alike, additional research holds the promise of helping us unravel the dynamics of the global versus local, and of uniformity versus diversity regarding ICT adoption.

REFERENCES

Avgerou, C., and Walsham, G. (eds.) *Information Technology in Context: Implementing Systems in the Developing World*, Aldershot: Ashgate, 2000.

Avison, D., Kendall, J. E., and DeGross, J. I. (eds.) *Human, Organizational, and Social Dimensions of Information Systems Development*, Amsterdam: Elsevier, 1993.

Bhatnagar, S. C., and Bjørn-Andersen, N. (eds.) *Information Technology in Developing Countries*, Amsterdam: Elsevier, 1990.

Mumford, E., Hirscheim, R., Fitzgerald, G., and Wood-Harper, T. (eds.) *Research Methods in Information Systems*, Amsterdam: Elsevier, 1985.

United Nations, *The Application of Computer Technology for Development: Second Report of the Secretary-General,* UN Publication E.73.II.A.12, 1973.

Walsham, G. "Globalization and IT: Agenda for Research", in *Organizational and Social Perspectives on Information Technology*, R. Baskerville, J. Stage, and J. I. DeGross (eds.), Boston: Kluwer Academic Publishers, 2000, pp. 195-210.

About the Authors

Mikko Korpela is Research Director of the Healthcare Information Systems R&D Unit, University of Kuopio, Finland, and Docent at the Department of Computer Science at the same university. He was born in Finland and achieved his DTech in Information Systems at the Helsinki University of Technology in 1994. He was a Visiting Researcher at the Obafemi Awolowo University, Nigeria, for 1989. Later he was founding Secretary of IFIP WG 9.4 and an initiator of the Health Informatics in Africa (HELINA) conferences. He is member of IFIP WG 8.2 and AIS as well as Chair of Finnish Social and Health Informatics Association. His research interests include West African political history, activity development and information systems development particularly in healthcare. Mikko can be reached by e-mail at mikko.korpela@uku.fi or through his home page at http://www.uku.fi/~korpela.

Ramiro Montealegre is an Associate Professor of Information Systems at the University of Colorado, Boulder, and a Visiting Professor at Instituto de Empresa in Madrid. He was born in Guatemala, and achieved his Bachelor in Engineering degree from Francisco Marroquín University, Guatemala. His master's degree in Computer Science is from Carleton University, Canada. He received his Doctorate in Business Administration from the Harvard Business School in Management Information Systems. His research focuses on the interplay between information technology and organization transformation in highly uncertain environments. Ramiro can be reached by e-mail at Ramiro.Montealegre@colorado.edu or through his home page at http://leeds.colorado.edu/interactive /employees / profiles / show.cfm?id=1323

Angeliki Poulymenakou is Assistant Professor in Information Systems in the Department of Management Science and Technology of the Athens University of Economics and Business. Prior to that she has served as lecturer in the Information Systems Department of the London School of Economics and Political Science. She is an academic member of ELTRUN, the e-business research centre of the AUEB. She holds a first degree in Mathematics (Athens), and MSc and PhD degrees in Information Systems (London School of Economics). Her current research interests focus on the study of technological intervention in the areas of organisational learning and knowledge management, and on the study of dynamic organisational networks from an information systems perspective. Angeliki can be reached by e-mail at akp@aueb.gr. More information on her activities can be found at http://www.eltrun.aueb.gr.

Part 1:

Work Practices and Their Transformation

2

ICTS SUPPORTING TARGETMANIA[1]

How the UK Health Sector is Trying to Modernise

Kathy McGrath
Brunel University, UK

Abstract: This research is concerned with a globalising discourse of modernisation, in which information and communication technologies (ICTs) are deeply implicated. It examines efforts to modernise the emergency arm of the UK National Health Service (NHS) so that it is more responsive to contemporary public demands. Pervasive in managerial and ministerial discourses about modernising the NHS are perceived joint needs to change its culture and to mobilise the potential of information technology. Drawing from critical social theory, the main argument of this paper is that the way the ambulance services are regulated and monitored is giving rise to a modernisation programme in which cultural change and ICT innovation is being narrowly conceived in terms of achieving response time targets. So in this case the process of modernisation is acting to deny autonomy locally and reinforce a 1950s image of ambulance workers that they would reject as not at all modern.

Keywords: Modernisation, information systems, critical social theory, cultural change, health sector, performance targets.

1 INTRODUCTION

Whether one subscribes to the view that we are now in an age of high modernity (Giddens, 1991), a postmodern era (Clegg, 1990), or that we have never been modern at all (Latour, 1993), organisations still strive to show

---

[1] Targetmania is a neologism reflecting a popular usage of -mania as the eager pursuit of something. So targetmania indicates the eager pursuit of targets. Without presupposing that eagerly pursuing any phenomenon is essentially good or bad, this paper reveals that the government's discourse of modernising the UK ambulance services focuses attention on pursuing performance targets, and it explores what is being achieved by this pursuit

that they are modern. In these discourses, they discuss the efficiency of their operations, the quality of service they provide, and the benefits for staff and consumers of their adoption of perceived leading edge technologies. Relative to many nations the UK is one where many such modern organisations exist, but just as modernisation programmes deriving benefits from ICTs are unevenly experienced across the world, so they are within nations. Many UK organisations still strive to be modern, and this discourse is perhaps nowhere more prevalent than within the health sector. Literature published by the Department of Health is replete with references to modernising the National Health Service (NHS) and government ministers and health service managers reinforce these messages.

This research is concerned with these modernisation efforts, in which technological rationalisation and change in UK health care organisations is aiming to make the NHS more responsive to contemporary public demands. It focuses on the ambulance services, and examines the outcomes they are achieving in efforts to exploit the potential of ICTs to improve the service they provide. The main argument of this paper is that the modernisation message for ambulance services focuses attention on meeting targets that are disjoint from the context that would give them meaning. Not only does such disjunction act to constrain local innovations that might give rise to cultural change, but also, against these homogenising measures of performance supported by ICTs, outcomes achieved indicate that parts of the service are getting worse. Following this introductory section, I present the theoretical basis of the study, followed by a description of the research setting and method used to conduct the investigation. Then I present and discuss the research findings and draw conclusions from the study.

2 THEORETICAL PERSPECTIVE

Pervasive in managerial and ministerial discourses about modernising the NHS are perceived joint needs to change its culture and to mobilise the potential of information technology. Although this research takes up both themes, it eschews the rationalistic perspective evident in the popular management literature that seems to inspire several government publications and ministerial speeches. In this way, the views that organisational culture can be changed in line with managerial objectives (Schein, 1985), and that particular technologies have properties that produce such effects, are seen as problematic. Attempts to change ways of organising or to implement technology to create organisational effectiveness may give rise to unintended consequences, which in turn may challenge the original objectives.

2.1 Power-Knowledge Relations

This research explores the above view by adopting a Foucauldian concept of power-knowledge relations, which allows movement between a loose coupling of forms of organising and technological developments in the broader environment to an increasingly tighter coupling of the individual and technology in a specific change effort. These moves may be understood as circular movements between poles of bio-power and disciplinary power through a number of intermediate positions in which regulative methods and disciplinary techniques are fused (Foucault, 1977).

Bio-power is concerned with human life and bodies in general, in terms of how they are supervised and regulated. Issues such as 'propagation, births and mortality, the level of health, life expectancy and longevity, with all the conditions that can cause these to vary' (Foucault, 1979: p.139) are the concerns of bio-power. It is focused on specifying and controlling a whole population and therefore should be understood as having a global dimension.

Disciplinary power is a supervisory mechanism targeted at groups and individuals, and hence on their functioning and performance within an institutional setting. It targets the body by addressing issues such as 'its disciplining, the optimization of its capabilities, the extortion of its forces, the parallel increase of its usefulness and its docility, its integration into systems of efficient and economic controls' (ibid.). It is an individualising mechanism for subjugating bodies in a specific context through the calculated and meticulous control of time and space.

The techniques of bio-power and disciplinary power are not in opposition. Rather their evolution constituted 'two poles of development linked together by a whole intermediary cluster of relations' (ibid.). Among these linkage mechanisms, Foucault focuses on the deployment of sexuality as one of the most important arrangements in which bio-power was joined with disciplinary power. The deployment of ICTs in organisations constitutes another form in which concerns to specify and discipline bodies may be fused together.

2.2 A Regime of Truth

Foucault links power and knowledge in a circular relationship through discourse. He refers to this relationship as a *regime of truth*, arguing that:

Each society has its regime of truth, its "general politics" of truth: that is, the types of discourse which it accepts and makes function as true; the mechanisms and instances which enable one to distinguish true and false statements, the means by which each is sanctioned; the techniques and procedures accorded value in the acquisition of truth; the states of those who are charged with saying what counts as true. (1980: p. 131).

In this context, truth is not a fundamental, universally accepted, scientific truth but rather should be understood as knowledge that is deemed to be true by a given community. Interpretations of this knowledge emerge in social discourse and enable and constrain the courses of action available to members of the community. Action (conceived as the exercise of power) may reproduce or change the domain of knowledge, which in turn may reconfigure the power relations. Discourse, then, has three elements: a community in which the discourse emerges, institutions with knowledge and authority which delimit the discourse, and grids of intelligibility, or *dispositifs*, which enable particular practices to be related to each other in a specific type of discourse.

So, the way that the UK government finances, regulates, and monitors the performance of the NHS can be linked with specific change efforts within the ambulance services in which the discourse of modernisation is being enacted. Closing the circle, the disciplinary mechanisms employed to categorise particular groups within this discourse are implemented in the name of patient care and therefore these local efforts may be understood as accomplishments in which a particular service discharges its responsibility to foster the health of the nation.

3 RESEARCH SETTING AND METHOD

There are 35 UK ambulance services, 32 regional services in England and a single national service in each of Wales, Scotland and Northern Ireland. Although plans to modernise the NHS affect all UK services, variations apply in Scotland, Wales and Northern Ireland. This research focuses on ICT implementation and related cultural change within the 32 English services.

UK governments, past and present, have been prime movers in these developments. Through a diverse mix of consultative groups, professional advisors, surveys and opinion polls, successive governments claim to have understood contemporary demands on the health service, and they have attempted to enact various strategies for responding to these demands. In the last 20 years, we have seen moves to introduce more professional managers into the NHS (Department of Health, 1983), the establishment of an internal market (Department of Health, 1989) and, more recently, a proposal to replace the internal market with a system of integrated care (Department of Health, 1997). Each move has expressed a concern with patient care, and increasingly information technology has been articulated as having a central role. After 20 years of effort, then, we may ask why it is that the NHS, and specifically the ambulance service, is finding it so difficult to modernise. What characterises the discourse of modernisation and the discursive practices of ambulance workers that makes them reluctant partners?

3.1 Ambulance Performance Standards

For nearly 30 years, the official measure for assessing whether UK ambulance services are responding to the demands of the public has been based on how frequently they get to emergency scenes within predefined target times set at national level. Between 1974 and 1996, these performance measures, then called ORCON (Operational Research Consultancy) standards, required that:

- 50% of emergency calls should receive a response within 8 minutes, and
- 95% of emergency calls should receive a response within 14/19 minutes in urban/rural areas respectively.

Only the 14/19-minute requirement was incorporated in the Patient's Charter (Department of Health, 1996), and services focused upon this target despite clinical evidence that some conditions require a faster response. In 1996, a government-initiated review of ambulance performance standards (Chapman, 1996) identified a requirement to respond to patients according to their clinical need, and hence to prioritise emergency calls so that those with the greatest need receive the fastest responses. The new ambulance performance standards were thus established. They reflect the requirements of the national performance assessment framework for coronary heart disease (Department of Health, 2000), in which the aim is to achieve a 40% reduction in deaths from heart disease and stroke in the under 75s by 2010. Some services started to move towards the new standards in 1997, but since March 2001 all services have had targets of responding to:

- 75% of category A (life-threatened patient) calls within 8 minutes, and
- 95% of categories B (serious, not life-threatened) and C (neither serious nor life-threatened) calls within 14/19 minutes in urban/rural areas.

Both the 8-minute and the 14/19-minute standards are included in *Your Guide to the NHS* (Department of Health, 2001), the current replacement for the Patient's Charter, but early evidence shows that the 8-minute standard is now the focus of attention (Department of Health web site, 2002).

3.2 Computerisation Within the Ambulance Services

Ambulance services have adopted particular types of ICTs in their drive to enhance performance. From the late 1970s to the mid-1990s, performance to ORCON standards became an increasing focus of attention, especially when the requirements of the internal market in the NHS started to emerge in 1989. In line with these standards, ICTs addressed speed of response assuming all calls had equal priority. During this period, these developments took the form of implementing computer aided dispatch (CAD) systems.

CAD systems were concerned with taking and responding to emergency calls more quickly based on up to date information about the position and status of suitable resources. However, by incorporating monitoring mechanisms, in particular tracking the locations and status of resources, they challenged the task orientation of vehicle crews, who previously had relative freedom to manage the cycle time involved from receiving an emergency call to notifying availability to attend another incident. Typically this cycle may be from 50 to 80 minutes, and includes the times to get to the incident, to assess the patient at the scene, to convey the patient to hospital, to comply with hospital procedures regarding handover of the patient, and to notify availability following patient handover. Both ORCON standards and the new ambulance performance standards address the first task in this cycle only.

The move, in the late 1990s, to the new ambulance performance standards changed the basis on which emergency responses are provided and service performance is measured. At this time, services modified their CAD systems to embed within them applications referred to as priority dispatch (PDS) systems. Through a structured series of question and answer protocols and embedded rules, PDS systems guide call takers through a process that suggests a clinical determinant indicating the severity of a patient's condition. Thus they enable the priority of each call to be assessed so that vehicles can be dispatched to patients according to the priorities assigned.

Although the determinant attached to each medical condition is determined at organisational level, ambulance services have limited freedom to decide on the category of response provided for each complaint. Indeed, in some cases the priority category is mandated at national level, as in the government imperative that suspected cardiac arrests receive a category A response (Department of Health, 2000). In this way, a deviation from the discipline can have serious consequences for the ambulance service and the control room staff concerned, who will be accountable for the outcomes of such action. This eventuality may be seen as giving rise to a situation in which the suggested determinant is accepted uncritically, rather than risk the consequences of overriding it.

3.3 Research Method

The data for this research were drawn from three main sources: literature produced by the Department of Health, including statistics about ambulance service performance; seminars and exhibitions held at the UK national conference for ambulance services (Ambex) in July 2000; and my empirical work with a UK ambulance service. The literature outlines the government's strategy for modernising the NHS, the Ambex conference addresses the way that ambulance services are responding to it, and my fieldwork gave me some detailed insights on the type of initiatives at the heart of these responses. At the conference, I heard presentations from the Health Minister, several ambulance service senior managers and chief executives, and a number of researchers working in the field. I also saw demonstrations of the ICTs and other resources being used by ambulance services to address the modernisation programme. My fieldwork with one service involved a longitudinal study of 1-2 days per week for 12 months during which time I observed a change programme to implement the priority dispatch initiative. I use the theoretical framework outlined earlier to analyse these efforts, and specifically to address the research question posed at the start of this section about how the discourse of modernisation challenges existing practices.

4 CASE ANALYSIS

The ambulance services are a community in the sense that acting as providers of pre-hospital patient care grounds the identities of each of them. All are subject to the same government monitoring mechanisms and each has access to public funds and other resources relative, in principle, to the demands of the population it cares for. Moreover, as part of responding to the needs and expectations of those they serve, each has adopted particular types of ICTs (perceived as performance enhancing) called CAD systems and PDS systems. In Foucault's (1980) sense, then, these mechanisms are the basis of a regime of truth for the ambulance service community, and modernisation in line with government imperatives is a discourse to which they are being applied.

Bio-power circulates in the arrangement in which government through the various arms of the NHS addresses the health needs of the population. The funding bodies, patients' charters, ambulance performance standards, and other regulatory methods constitute grids of intelligibility, which aim to address patient care by matching resources to demand in systems of categorisation, measurement, and normalisation. The specific aim to reduce premature deaths from coronary heart disease is an example of particular importance. This goal involves a regulatory mechanism with four main aspects. Calls to the ambulance services are categorised based on clinical need, targets are set for emergency response times to each category of call, the performance of the services against target is continually measured, and funding for subsequent periods of operation is based on an evaluation of previous performance. The norms of performance for ambulance services, effective since March 2001, are enshrined in the new ambulance performance standards, and the implementation of information systems was seen as 'essential to the proposed changes' (Chapman, 1996: p. 6).

Disciplinary power operates at the level of individuals and groups. The implementation of local PDS systems introduces a disciplinary mechanism within working practices designed to optimise staff members' individual performances with the aim of improving the overall performance of the service. It aims to standardise the process of call taking so that the priority assigned to each call is a function of the call taker's compliance with the discipline rather than a function of his or her individual human judgement of the clinical priorities. Furthermore, by limiting the dialogue with the caller to a predetermined exchange of questions, answers and instructions, it aims to provide a response to the patient in the fastest time possible.

Moreover, local quality assurance units have been introduced within ambulance services to monitor compliance with the new discipline. These units evaluate a small percentage (typically 1-3%) of the emergency calls received each month. Call takers receive an evaluation record and compliance rating which feeds into their periodic assessments by managers. Thus call takers are subjected to a disciplining mechanism and may be subjectified by it in the sense that, by reflecting on their evaluations and addressing the issues raised, they participate in their own monitoring.

So the implementation of PDS systems embedded within CAD systems, which in combination are inscribed with the values of the new ambulance performance standards, attempts to fuse together the government's plans for modernising the NHS and the ambulance services responses to them. However, there are some contradictions within this regime of truth and some unintended outcomes. For example, members of the ambulance services note that framing the performance targets in terms of response times means that:

> *If you get there in 8 minutes and the patient dies you've succeeded, but if you get there in 9 minutes and the patient lives you've failed (Director of Corporate Resources, Ambulance Service, quoted in Nicholl et al. (2001: p.16).*

In this way, many argue:

> *They are not real targets. The real targets are the ones that have clinical outcomes (Priority dispatch manager, Ambulance Service)*

The view that clinical outcomes are the real measure of success is an increasingly pervasive discourse within the ambulance service community. The new ambulance performance standards may be seen as medium for and an outcome of this discourse. On the one hand, the new standards emphasise the need to provide an 8-minute response to *life-threatened* cases, which the previous standards did not; so current assessments of service performance take into account what the outcomes for patients *might be*. On the other hand, the outcomes *achieved* (for example, the number of patients that leave hospital alive following out-of-hospital cardiac arrest) still do not form part of the performance measures for ambulance services. Indeed, hospitals do not pass information to ambulance services about the clinical outcomes achieved by emergency patients, so that ambulance performance figures are de-rooted from the context that would give them meaning. Moreover, such one-way traffic sustains the 1950s image of ambulance workers as 'glorified porters' (Radio 4, 2002a), and reinforces the sense among them that they occupy a marginal position in NHS plans (Ambex, 2000).

4.1 A Discourse with the Health Minister

At the Ambex conference in July 2000, the Health Minister advised the ambulance services of what they need to do to modernise:

> *We need to change the way the NHS works ... status quo is unacceptable ... public expectations have changed ... technology has advanced ... what we are about is treatment based on clinical need. ... We are not trying to create league tables ... however we do need to learn from each other. ... In the past there may have been a culture where one trust competed with another ... but as far as this government is concerned, this is not the way. ... We need to teach the public to help themselves ... to make the service the same wherever you are ... to work together. ... Targets need to be met.*

At this time, a number of ambulance services had already installed a PDS system, and several of them were already getting close to or regularly meeting the new standards. Indeed, the Minister said 'I cannot but pick out Staffordshire Ambulance Service' in this regard, while declaring that the services yet to achieve their targets would be expected to show considerable progress by

March 2001. The chairman of the London Ambulance Service asked the Minister how much she believed could be achieved by the new practice of triaging (prioritising emergency calls) over the telephone, by comparison with the former practice of triaging when the patient was examined. She suggested that 'there are a number of days you get through on a wing and a prayer', but she held fast to the aims of telephone triage so that resources could be dispatched according to patient need – a need assessed using a PDS system. Still, members of the ambulance services are concerned about the efficacy of the new regime because they feel that without a medical examination much uncertainty is involved in assessing the urgency of clinical need. Indeed, they cite cases where prioritisation at the point of call receipt does not achieve the desired outcome – that those in the greatest need receive the fastest responses.

For example, when a member of the public calls the ambulance service to report that someone has collapsed in the street, the caller's responses to standard questions may result in an automated assessment that the patient is life-threatened and needs a category A 8-minute response. Yet when a caller describes the condition of a person who, say, has fallen off a building site, the assessment may indicate a serious and very painful back injury, but not one that is life-threatening, hence a category B 14/19-minute response. However, when ambulance staff arrive on the scene, they may assess the first patient as a substance abuser, who lost consciousness under the influence of a substance and does not require an *emergency* response at all. The second patient invariably will be conveyed to hospital, hence had the greater clinical need, but the seriousness of these conditions could not be gauged until the patients were examined. So there is a concern that the ICTs in use for call prioritisation are not yet sophisticated enough to pick out reliably and safely calls requiring different levels of response (Nicholl et al., 2001).

Responding to a question from a senior executive at the East Anglian Ambulance Service, the Minister said:

Any notion that we are prepared to lower the standards – no.

This exchange reflects concerns within the ambulance services about the demands of meeting 'a far tougher standard' (IT director, Ambulance Service). In effect, priority dispatch requires that fewer calls receive an 8-minute response than did ORCON, but the new standard is perceived as more demanding because ORCON allowed *any 50%* of calls reached in 8 minutes to be counted. In this way, the ORCON regime created pressure to dispatch resources as quickly as possible to each call regardless of how urgent it seemed, while deemphasising the implications of having no resources available when, say, a "suspended" patient – one who has stopped breathing – was reported. Under a priority dispatch regime, services must reach *75% (during 2003/04, 90%) of category A* calls in 8 minutes. These calls may be only 20-30% of their work, but they occur all of the time. In this way, services have to manage their resources much more dynamically than before to ensure that their vehicle responses match a changing profile of call priorities in the control room. In the above scenario, this would mean diverting an ambulance on its way to a lower priority call to respond to the "suspended" patient.

A further comment from the Ambex 2000 conference is interesting in this regard. The chief executive of Staffordshire Ambulance Service argued that staff should dispatch an ambulance as quickly as possible and worry about the priority later – when the patient is examined. Delegates were well aware that Staffordshire had been responding to over 80% of *all* calls within 8 minutes for some time, and therefore had not only been exceeding the new 8-minute target before it took effect, but exceeding the then current ORCON 8-minute target by over 60%. Nevertheless, none of the other trust board directors suggested they were sanguine enough to attempt a deferred prioritisation strategy in their services, lest in delaying triage they failed to achieve a sufficiently high number of 8-minute responses to be sure of including the required number of life-threatened cases. Indeed, the chief executive of the Essex service, an advisor to the Department of Health, argued that prioritisation is essential *when the call is received*, especially for the larger services like Essex and London – the latter has about 9 times the call volume of Staffordshire. Still, it is interesting to note that the service that is out-performing all others, including the 50% that are smaller than it, is one that has chosen not to *dispatch* by priority.

4.2 Outcomes Achieved

The following table summarises the latest available (Department of Health web site, 2002) yearly performance figures for ambulance services.

Table 1. Summary of ambulance service performance figures 1999-2002

	99/ 00	00/ 01	01/ 02
a. No of services operating to the new performance standards	20	31	32
b. No of (a) achieving the category A 8-minute standard	1	3	14
c. No of (a) achieving the category B/C 14/19-minute standard	12	13	16
d. No of (a) where cat. A performance dropped from previous year	*	1	1
e. No of (d) where performance dropped more than 5 percentage points	*	0	0
f. No of (d) where cat. A performance dropped in 2 successive years	*	*	0
g. No of (a) where cat. B/C performance dropped from previous year	*	11	3
h. No of (g) where performance dropped more than 5 percentage points	*	0	3
i. No of (g) where cat. B/C performance dropped in 2 successive years	*	*	3

* Indicates no performance figures available

Results in Table 1 show a mixed performance against government targets. Against the category A target, services tend to improves their performance year on year, an increasing number are reaching the 75% target, nevertheless in 2001/02 only 14 (of 32) achieved it. Category B/C responses are more variable. Performance fluctuates year on year, with 40% of services showing a drop in performance in 2001/02, and 10% of cases showing a significant drop of more than 5 percentage points. Moreover, 10% of services have a declining category B/C performance over 2 years, and only 50% of services are yet achieving the 95% target. The current trend, then, seems to be for ambulance services to focus resources on meeting the 8-minute standard, with the result that category B and C patients in some areas are experiencing a less prompt service. Clearly, one compelling issue is how this performance is affecting clinical outcomes, but while hospitals and ambulance services continue to report their performance separately there is no way of comparing patient outcomes according to wait times for an ambulance. However, one can question the extent to which comparing ambulance services using the current standards of performance gives rise to meaningful results.

Although the national framework for coronary heart disease (Department of Health, 2000) requires that suspected cardiac arrests receive a category A response, not all patient complaints are specified as clearly. Services have some scope to decide how they allocate other complaints to the categories A, B and C, which affects the response targets they must meet. Moreover, services vary in the degree of accuracy with which their systems record response times to calls. For example, some services require ambulance crews to press a button in their cabs when they arrive at an emergency scene so that their computer systems can record the time precisely. Others require crews to fill in a paper form, which may not be completed until patient handover at hospital up to one hour later, when the arrival time may not be remembered so precisely. When the clock starts ticking is also variable. A Department of Health (1998) directive states that this should be when the caller's telephone number, the incident location and the chief complaint have been logged by the call taker, but not all services time this process so precisely. Moreover, some services use automated Caller Line Identity (CLI) to identify the caller's address, which may be the incident location, so the time between call connection and the start of timing is variable, and this time is not reflected in current performance figures even though patient outcomes depend on it.

So efforts 'to make the service the same wherever you are' is a discourse of an ideal type, the more so when we consider how the clinical skills of ambulance crews may vary, and how geographical area affects the time needed to make an emergency journey. This argument is not to suggest that we should accept a postcode or some other form of "lottery" for the ambulance service, but that a focus on national standards may be stifling local innovation and change which might find other ways of improving the service. Indeed, the history of the microcomputer, the Internet, and several strategic information systems has shown us how local innovations can radically change the way that people and communities work.

5 DISCUSSION

This research has focused on how the UK ambulance services are trying to modernise. Although themes of cultural change and ICT innovation are articulated in the messages from health service managers and government ministers, they are still firmly rooted in an instrumental discourse that sees these processes linked directly with organisational performance. Moreover, that performance is measured in terms of a narrow response time target that seems meaningless to many required achieving it. So the government's rationale for modernising the NHS reinforces an identity of ambulance workers that they reject – that their value should be measured by their ability to provide rapid transport rather than contribute to patient outcomes.

Within the ambulance service that was the focus of my fieldwork effort, staff respond to current modernisation initiatives in ways that reflect a number of the concerns raised at the Ambex conference. Questioning the efficacy of a computerised call prioritisation regime, some control room staff will bypass or override certain PDS protocols, drawing upon their own knowledge and experience to assess the clinical priorities. Most, though, will not risk this course of action even when they feel a need for it, lest they find themselves accountable to a coroner's court and subject to disciplinary action. Many ambulance crews will respond especially rapidly to calls that they judge may provide an opportunity to demonstrate their clinical skills, irrespective of the priority assigned by control room staff using the PDS system. In this way, a road traffic accident will usually receive a very rapid response, even when information from the caller suggests there are only minor injuries because the accident happened at very low speed in congested traffic conditions. So the working practices of staff indicate a desire to contribute to clinical outcomes, a recognition of the need for prioritised responses, but a lack of trust in PDS systems and a challenge to the way that being responsive is conceived within the new performance standards.

So, although recent moves within the ambulance services endeavour to differentiate between categories of patient so that those in the greatest need get the fastest responses, three major problems with targets are evident in these developments. First, the emphasis in the new standards on providing the most rapid responses to category A cases means that this target is focused upon at the expense of category B and C cases, some of whom now receive a slower and hence (on a response time measure) a worse service. Second, national targets deny autonomy locally. Only one service claims to reject the telephone triage message, but Staffordshire achieved autonomy by exceeding the targets without accepting the message. Surely we should now ask if there are ways of improving the service without focusing narrowly on response time targets. Third, target outcomes are constructed. Ambulance services time their responses in different ways, some of which are more open to construction than others. Moreover, when scrutiny is intense and targets seem meaningless how the construction is achieved may be open to abuse. Indeed, in a recent poll of health service managers (Radio 4, 2002b), 1 in 12 admitted to 'massaging their figures'. Although this poll was not specifically targeted at ambulance services, it reveals that the truths NHS performance figures purport to tell are not accepted by all within the service, far less that they constitute an objective truth that all of us should accept.

6 CONCLUSION

Modernisation efforts may be accepted because they embody desirable goals – to be seen as abreast of the latest developments in a field, as adopting contemporary technologies, and as being responsive to the world in which we live. Alternatively, they may challenge existing ways of organising and established ideas of providing a service. Both such manifestations are evident in current modernisation programmes within the UK NHS, but the difficulties with implementing these programmes does not reduce to a straight conflict between those advocating new ways of being responsive and those wishing to retain existing ways. Being modern is narrowly conceived in terms of performance targets so that innovation and change means getting faster. Moreover, since targets are framed by response times they are disjoint from the patient outcome context they are alleged to serve. The in-order-to aspect of modernisation is being addressed, but a link with the because-of motives (Ciborra, 1999) that would give it meaning has not been established.

Modernisation is a globalising discourse in which the use of ICTs is deeply implicated. However, in the UK ambulance services those advocating modernisation and those who need to accept and enact it are at most loosely aligned and the way that ICTs are being adopted creates limited scope to fuse these interests together. Perhaps, though, a different type of fusion should be attempted. Rather than just trying to make conversations with callers so fine-grained that a patient's condition can be determined over the telephone, the potential of ICTs to connect different communities should be exploited. If ambulance services knew more from hospitals about the outcomes achieved by their patients, the need for innovation and change might seem more compelling and then ICTs would achieve more than supporting targetmania.

REFERENCES

Ambex. *Ambulance Services Association UK National Conference*, Harrogate, UK, 1-3 July 2000.

Chapman, R. *Review of Ambulance Performance Standards*, Department of Health, UK, July 1996.

Ciborra, C. "A theory of Information Systems Based on Improvisation", in *Rethinking Management Information Systems*, W. Currie and R. Galliers (eds.), Oxford: Oxford University Press, 1999.

Clegg, S. *Modern Organizations: Organization Studies in the Postmodern World*, London: Sage, 1990.

Department of Health. "Report of the NHS Management Inquiry", *Department of Health and Social Security*, London, 1983.

Department of Health. *Working for Patients*, London: HMSO, 1989.

Department of Health. *The Patient's Charter*, London, November 1996.

Department of Health. *The New NHS: Modern, Dependable*, London: HMSO, December 1997.

Department of Health. "Measuring When the Clock Starts and Stops", *MISC* (98:23), London, 1998.

Department of Health. "Modernisation of Ambulance Services", *HSC* 1999/091, London, April 1999.

Department of Health. "National Service Framework for Coronary Heart Disease", *HSC* 2000/012, London, April 2000.

Department of Health. *Your Guide to the NHS*, London, September 2001.

Department of Health Web Site. http://www.doh.gov.uk/public/sb0213.htm. Statistics Division entries, accessed 7 October 2002.

Foucault, M. *Discipline and Punish*, Harmondsworth: Penguin, 1977.

Foucault, M. *The History of Sexuality, Volume 1: An Introduction*, Harmondsworth: Penguin, 1979.

Foucault, M. *Power/Knowledge: Selected Interviews and Other Writings 1972-1977*, C. Gordon (ed.), Brighton, Sussex: Harvester Press, 1980.

Giddens, A. *Modernity and Self-Identity*, Cambridge: Polity Press, 1991.

Latour, B. *We Have Never Been Modern*, Harvester Wheatsheaf, Hemel Hempstead, 1993.

Nicholl, J., Turner, J. and Martin, D. *The Future of Ambulance Services in the United Kingdom: A Strategic Review of Options for the Future of Ambulance Services*, London, Ambulance Service Association, 2001.

Radio 4. "Emergency", July 15, 2002a.

Radio 4. "Today Programme", October 7, 2002b.

Schein, E. *Organizational Culture and Leadership*, San Francisco: Jossey-Bass, 1985.

About the Author

Kathy McGrath is a lecturer in Information Systems at Brunel University in the UK. She was born and educated in England, and holds an MSc and PhD in Information Systems from the London School of Economics. Her research focuses on the development, management and use of ICTs, and their organisational and social consequences. Kathy can be reached by e-mail at Kathy.McGrath@brunel.ac.uk.

3

COUNTER NETWORKS, COMMUNICATION AND HEALTH INFORMATION SYSTEMS

A Case Study from Mozambique

Emílio Luís Mosse and Sundeep Sahay
University of Oslo, Department of Informatics, Oslo, Norway

Abstract: Mozambique, actively trying to deploy ICTs like many developing countries, faces critical problems in providing health care services to its population, especially those living in rural areas. These attempts are extremely complex due to a variety of constraints ranging from inadequate infrastructure to manpower shortages, to a culture that does not yet value the "efficient use of information." In this paper, we unpack the relationship between communication practices and health information flows within the context of the primary health care system of Mozambique, and argue that understanding this relationship is crucial for developing context-sensitive approaches to implementing computer-based health information systems. Through an empirical analysis of ongoing efforts in a health district within the Health Information System Program (HISP), we seek to understand the communication practices which surround processes of collection, analysis and transmission of health data up and down the various levels of the health administration hierarchy. The nature of these communication practices has a visible impact on the quality, timeliness and usage of health data. The paper describes the functioning of the health information system as a network comprising multiple human and non-human actors within deep-rooted socio-cultural structures. Inspired by Castells, we examine how understanding the reciprocal relationships between communication practices and health information can help to develop "counter networks" to provide hitherto marginalized groups and regions in the potential access to "network society."

Keywords: Information and communication technologies, Health information systems, Mozambique, Actor-Network Theory, Network Society, work practices.

1 INTRODUCTION

Mozambique, like many other developing countries, has in recent years been attempting to introduce information and communication technologies (ICTs) in various sectors to promote socio-economic development. One important sector is health, especially in the domain of primary health. These ICT initiatives take place in a context that is historically and culturally shaped; the socio-cultural structures are reflected in patterns of how work is currently done. The work practices and existing communication practices that surround the collection, storage, transmission and analysis of routine health data, particularly within the district health level, are often in tension with those assumptions inscribed in the technology and proposed management methods of work. Understanding the nature of these tensions is crucial for the planning of new ICT implementations, and this requires understanding local improvisations and work practices, and how these can be more effectively integrated (see e.g. Sahay and Avgerou, 2002). The informal and locally improvised ways in which communication takes place is often in direct tension with the conditions and implicit assumptions made for the implementation of ICT initiatives like the Health Information System Program (HISP) that is described in this paper.

The paper is organized as follows. In the next section, we outline our conceptual approach to the study of communication, and analyze its role in the creation of "counter networks". Following this, we provide some details of the context of the Mozambique health system and the ongoing HISP, which is the basis of the empirical study. In Section 4, based on the empirical work within the health district of Xai-Xai, we describe the communication practices that surround the health information flows. In Section 5, we analyze the case material based on key concepts from Actor-Network Theory and the concept of counter networks. This analysis helps us to develop some implications for the implementation of HISP. Finally, we draw some concluding remarks arising from the study.

2 COMMUNICATION PRACTICES AND "COUNTER NETWORKS"

In discussion below, we outline our conceptual approach to the study of communication, and relate it to the metaphor of "counter networks." Drawing from Castells (1996; 2000), "counter networks" is an analytical concept that we propose to help analyze the nature of communication practices at district level, and relate it to the HISP implementation.

Our interest is to understand how communication practices both constitute and are constituted in the flows of health information. By health information flows, we refer to the flow of health data from the community to the district, to the province and to national levels of the health administrative hierarchy in Mozambique.

We take an interpretive research approach to analyze health information flows in the context of the socio-cultural processes that surround their construction and use, seen from a social systems perspective. Interpretive research is concerned with developing a deep and contextual understanding of the phenomenon. We focus on the subjective processes that surround the social construction of communication practices understood through the use of qualitative research methods (Walsham, 1993; Martin and Nakayama, 1999). The aim of such interpretive analysis is to understand rather than to predict behavior. We seek to understand "communication practices in context", where the context is a health district of Mozambique. These practices relate to how health data are collected, their storage, analysis, and transmission up and down the different levels of the administrative hierarchy. These practices involve people, their meetings and conversations with others, their physical movements to circulate information, their use of various forms of artifacts, such as book registers and forms, and current attempts to apply ICTs. These social interactions are regulated by the formal requirements of the health information systems (for example, to submit monthly reports) and are enacted in the everyday life of health staff often in informal and locally improvised ways. The interpretive perspective helps to focus on these formal and informal communication practices that surround, and are constituted in the flow of health information in the particular context of a health district. Such an approach implies an epistemology that assumes the social construction of knowledge and reality (Walsham, 1995).

Communication practices are conceptualized as being mutually constituting and constituted by health information flows, implying that they shape and are also shaped by health information flows. For example, the ways data are collected, stored and transmitted reflect how people communicate with each other, how they use artifacts, and how these information flows are shaped through these communication practices. This reciprocal relationship implies a structurational process (Giddens, 1984), which Yates and Orlikowski (1992) draw upon in their study of "genres of communication." Our emphasis on the socio-cultural context within which communication is shaped implies that health information flows are intricately tied up with questions of identity (Amant, 2002), the context of communication (Hall, 1983), and the norms and meanings that shape social interaction (Yates and Orlikowski, 1992).

Our empirical analysis aims at unpacking the mutual relationship between health information flows and communication in a Mozambican health district, with a view to understand the nature of communication practices and their relation with HISP. Developing such a subtle understanding of this relationship, we believe, creates the potential to develop meaningful implications for a more context sensitive implementation approach. We see the unfolding of these communication practices as crucial to the development of the notion of "counter networks", as a metaphor to emphasize those people, regions and sectors who for reasons of history, geography, economic resources and knowledge have been excluded from formal networks, as reflected in notions of the "global village" and "information society".

Our usage of the term "counter networks" draws its inspiration from Castells' (1996) discussion of globalization processes in the "network society". Castells argues that groups and also regions that have been historically excluded and marginalized, for example the primary health care (PHC) sector in a developing country like Mozambique, run the risk of being systematically marginalized in the future if they are not able to "link up" (using the power of new ICTs) with ongoing processes of globalization that provide them greater potential access to funds, new knowledge, and a broader visibility. However, this networking is problematic to achieve in practice due to historical reasons such as the inadequate infrastructure and capacity of people to engage with new technologies and forms of knowledge. Thus building these "counter networks" can be viewed metaphorically as a deliberate attempt to try and transcend existing inhibiting conditions, in order to include marginalized entities, such as the primary health care sector, in a broader "network of flows". Castells sees ICTs as a fundamental tool to enable these networking processes. An important aspect of "counter networking", we argue, rests in the understanding of locally constructed communication practices, and analyzing approaches by which they can be seriously "taken into account" by ICT initiatives like HISP.

HISP is an initiative aimed at computerization of PHC processes in developing countries, including Mozambique, in selected health districts. HISP was initiated in South Africa in 1995, and now represents a collaborative research and development program between University of the Western Cape and University of Cape Town (South Africa), University of Oslo (Norway), Universidade Eduardo Mondlane (Mozambique) and the ministries of health of South Africa and Mozambique. Within this research program, HISP has developed a free open source software application – District Health Information System (DHIS) database, based on MS Access and designed for data collection, analysis and reporting. The aim of HISP is to develop health information systems to support the emerging decentralized health administrative structures in various developing countries (Braa and Hedberg, 2002). The broader agenda of HISP is to enable local control of health information at the district and sub-district levels and thus to empower local users to have greater control of their work processes. Typically in most developing countries, the existing situation of collected health data is that it is sent to higher levels for analysis and reporting rather than being used to support local action (Wilson et al., 1989; Opit, 1987). HISP seeks to introduce computers at the district levels as contrasted to the existing focus of computerization at the province and national levels. By placing computers at the district and through training of the district staff, HISP seeks to redress this current imbalance in focus and encourage the local analysis and use of information to support health care in peripheral areas. However, bringing about this change in practice is an extremely complex and long-term task. One of the reasons for this is the nature of existing systems of communication, which we argue are in tension with the assumptions inscribed in efforts like HISP. We discuss some of these issues in the case description below.

3 RESEARCH SETTING

The research is set in Mozambique, a country situated in Southern Africa, with a total area of 801,590 square kilometer, and divided into eleven provinces, which are further subdivided into a total of 131 districts. The current population of the country is estimated at 19 million, of which more than half are under 20 years old and more than 60% are illiterate. The growth rate of Mozambique is estimated at 1.1% (Census, 1997) with a population density of 20 persons per square kilometer. The population is predominantly rural; the urban population is estimated at only 13%. The extensive rural spread of the population, a condition magnified by nearly two decades of devastating civil war (1976 - 1992) that destroyed large portions of the communication and physical infrastructure (including roads, telephone lines, and health facilities) has meant that access to health care services for the majority of people is problematic. For all these reasons many of the rural communities have existed in "isolation", devoid of active contact with the formal governmental systems. This isolation has seriously affected all aspects of socio-economic development processes, effects of which are still being felt, for example in the provision of health care. The low levels of literacy and nearly non-existent computer literacy especially in the rural areas, make the introduction of any form of ICT-based systems a herculean task. To enable these historically excluded and marginalized regions to become part of larger globalization processes is not a task of "simply technically plugging in the computers", but requires a strategy for the systematic construction of "counter networks", that takes into serious consideration the existing socio-cultural conditions.

We discuss one such attempt of counter networking in HISP within the PHC domain, a sector that has attracted insignificant amounts of funding and attention as compared to the hospital system or the private health care sector. We first discuss the overall structure of health organization of Mozambique. Set within this broader context, we provide details of our empirical work at the micro level of the health districts.

3.1 Health System Organization

Mozambique has adopted the Primary Health Care (PHC) approach of the World Health Organization (WHO, 1978) in an attempt to have a direct and positive impact on the communities through the expansion of health services. This PHC approach has been argued to be effective for the provision and organization of health care services in developing countries (Amonoo et al., 1984; Newell, 1989). After Mozambique became independent in 1975, the number of health facilities was greatly expanded in all regions, a trend which still continues (Roemer, 1992). The health information system (HIS) of Mozambique has been described as being primarily top-down, implying that while planning and implementation of health systems start at the level of the central Ministry of Health, MISAU[1], health data are collected at the local community level and sent up to the ministry (Braa et al., 2001). Feedback from MISAU to the local levels is limited, and data collection and its upward transmission are seen to reflect the existing hierarchical organizational structure (Jeremias and Sitói, 1997) across four levels: community, district, province and national (see Figure 1). The Mozambique health system employs a total of about 17.000 people, of which approximately 400 are medical doctors, 9500 are health technicians, and the remaining are administrative staff.

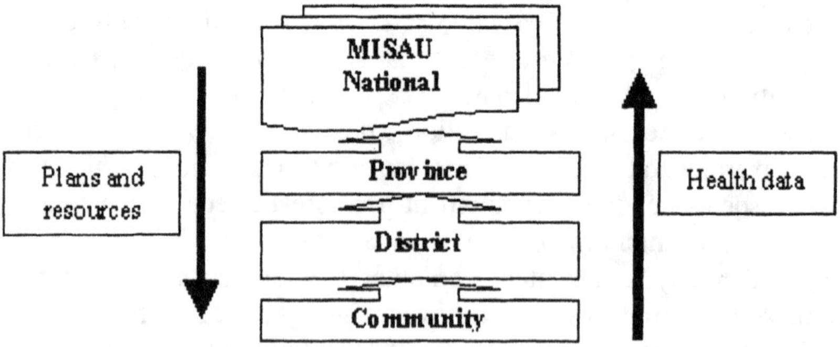

Figure 1. Health system organization in Mozambique

[1] Ministério de Saúde

The community level is important for providing basic care, and there have been positive experiences in the participation and involvement of community groups such as non-governmental organizations (NGO) and Agentes Polivalentes Elementares (APEs[2]) in the management of services and health programs. Within the community level, health centers and health posts (health units) are responsible for the provision of PHC services, and also for the collection of health data in various forms and its transmission to the next level, the district. At the community level, data related to a patient (for example, a pregnant woman) are first entered in register books and tick registers, and at the end of the month are summarized and sent to the district, indicating the number of patients seen in each health unit, and categorized by disease. The people involved in this process of health data transactions are mainly health personnel with limited formal medical training. Among factors inhibiting the effective transmission of data to the higher levels are physical distance, poor transportation connections, and time pressures experienced by health workers. The poor communication that results from these conditions contributes to incomplete, untimely and often incorrect transmission of health data.

The next level is the district, which consists of a network of health units, responsible for the management of health care in its jurisdiction. The most important authority in this network is the district health office, DDS[3] that is responsible for the overall administration, planning, and implementation of the health services. With respect to health data, the DDS is responsible for the aggregation of all the data received from health units within the district, and its transmission to the province. In addition to the data from health units, the DDS also works with NGOs, such as World Vision and Save the Children, which operate within the district. The province is responsible for sending monthly report to MISAU, based on further aggregation of the health data from the provincial districts.

For this discussion the community and the district levels are the most important. We now describe the case study at the micro level of analysis of communication practices.

4 THE CASE STUDY

The study was carried out in the health district of Xai-Xai. In Table 1 below, we provide an overview of the number and qualifications of personnel that work in the health district of Xai-Xai. A quick glance at the

[2] An APE is a person trained for a few months to provide first aid in the community, usually where there is no health unit.

[3] Direcção Distrital de Saúde

table indicates how understaffed the facility is and also the low educational skill level. The level of health personnel in the district varies from Medical Doctor to "Servant". Besides the doctor, the education level of the people running the health units is mostly limited to elementary or primary school.

Table 1. Number and educational level of personnel in Xai-Xai health district[4].

Position	HC Chicumbane and RH Chicumbane	HC Maciene	HP Siaia	HP Chongoene	HP Nhancutse	HP Chilaulene	HP Zonguene	HP V. Lenine	HP Chipenhe	HP J. Nyerere	HP Banhine	HP Nhamavila
Medical Doctors	1											
Pré-Universitário	5	1										
Básico	25	1	2	1	1		1	2	2	2		1
Elementar	4	3		1		2	1				1	
Servant	15	7		2	1	1	1	1	1	1	1	1
Other	9	3	1		1							
Total	**59**	**15**	**3**	**4**	**3**	**3**	**3**	**3**	**3**	**3**	**2**	**2**

The rural hospital in Xai-Xai has 93 beds, caring for approximately 200 inpatients per month, and provides several forms of primary health care. The health centre is responsible primarily for the provision of the Extended Immunization Program (PAV) and Mother and Child (SMI) programs. The personnel responsible for these programs comprise the core planning team, NEP[5] responsible for both the entire district and the health center.

Field research was conducted through detailed participant observation and interviews with key actors during the period of April to June 2002, with follow up research during 2003. The interviews were conducted with 13 key people engaged with processes of health information systems across three levels, specifically the community, district (DDS) and the province. Two of the DDS staff was interviewed three times each. The aim of the field research efforts was to understand the functioning of the HIS in this particular setting and to determine constraints to the processes of health data collection, storage, transmission and analysis. The interpretive research approach adopted helped us in understanding how people create and associate their subjective and inter-subjective meanings in their interactions with the world around them during the course of their everyday work practices (Walsham, 1993). In addition to the interviews, we were also

---

4 Pré-Universitário refers to completion of "high school" (12 years). Básico refers to 10 years of education, while elementar refers to 7 years of education completed.

5 Núcleo de estatística e planificação

actively engaged in providing training to the NEP staff on the use of the HISP DHIS software and how their existing work could be supported by ICTs.

In this section, we first describe the location, people and infrastructure in the district. We then describe the organization and practice of health information flows in the district.

4.1 Location, People and Infrastructure

Xai-Xai district health office is located 15 kilometers from the provincial capital of Gaza, and has an area of 1749 square kilometers, with a population of 200.896 habitants. This health district comprise of 13 health units including 1 rural hospital, 1 health center and 11 health posts. The rural hospital and the health center which are located in the same geographical boundary serve a population of 74.357 inhabitants. A medical doctor, medical technicians, nurses and midwives staff both the hospital and the health center. However, these entities perform different activities relating to health care, and consequently they have different relations to the HIS. The health center is located in a compound, which consists of many buildings (Figure 2), each typically consisting of one room with a thatched roof (Figure 3), and associated with a particular health program or disease. For example, the three people from NEP sit in the same one-room building, as is the case with the staff responsible for ELAT/ELAL (Tuberculosis and Leprosy programs). There is another room that is the venue for conducting training programs and seminars for health technicians. This room is better equipped than the other rooms as it has a computer, a photocopy machine, a classroom, and a small library. This infrastructure has been donated by the Portuguese Co-operation and Save the Children, and during our stay there we often found this room to be securely locked.

Figure 2. Xai-Xai compound.

Figure 3. The room housing the NEP.

The health posts are located at a relatively large distance from the district office. At the health posts, we asked the patients how long it took them to reach the post and the replies ranged from a half hour to two-hour walk. There was no vehicle available at the health post, and during times of emergency (for example, child birth), a car was borrowed from someone in the community or a public transport (called "shapa") was used. The health posts consist of one or two small rooms, and sometimes have facilities for conducting deliveries, staffed by one trained medical agent and/or nurse. The health posts serve as the first point of entry of the patient into the health system. A card is made out for the patient, which is then supposed to serve as covering the history of the patient, for example starting from the detection of pregnancy to antenatal checkups to delivery and the various immunization activities that are expected to culminate in the child being "fully immunized". These cards are not maintained by the health posts but are given to the patients. Often these cards are misplaced, and a new card is then prepared, and the patient history "reconstructed".

4.2 Organization and Flows of Health Information

Within the hierarchal structure of the district office, the group of three people that constitute the NEP are in charge of the collection, aggregation, and analysis of all the health data from all the health units of the entire district, and its transmission to the provincial level. The team is composed of the chief of HIS (also the chief of NEP), and the persons responsible for SMI and PAV.

In addition to the monthly data sent by health units, the NEP also receives the weekly epidemiological bulletin (BES) which is the basis of the weekly surveillance system for a number of important infectious diseases.

Health data come from the health units and other sources, for example there are seven APEs in Xai-Xai district that send data directly to the NEP. APEs record their health care activities (for example, outpatients seen) on a form that is created specially for this specific district. There are also twenty six Matronas (traditional midwives) in the district who send in separate data documents to the NEP staff. The data first come to the head of the NEP who then distributes the forms to the staff responsible for the PAV and SMI programs respectively. These different people constitute a network of flows and are jointly responsible for compiling the monthly report to be sent from the district to the province.

On an average, health personnel meet more then 100 patients a day leaving them without much time and energy to perform administrative tasks relating to health information. They are also constrained by the fact that they often have to travel to provide supervision to other health units within the district. During the course of our field research, we followed the person responsible for the SMI program. We found that in the morning she met many patients, basically women coming for antenatal observation, delivery, post-natal and family planning issues. Each interaction with a patient was a rather personal and customized one, taking 5-10 minutes on an average. After working with the patients, she went to the NEP room to fill forms about what she did, for example how many patients she met and the number of medicines and condoms distributed during the day. Often, the actual numbers were forgotten, and what was documented was what she thought she remembered. As a result, there were mismatches between patients actually seen and medicines distributed. These "incorrect" numbers became part of the artifacts as they were summarized in the forms sent on to the province. As a result of the "flows", given the lack of any form of quality control and supervision, these data became part of the official archives.

We found communication practices between people responsible for data collection relating to the various programs taking place in a very informal and improvised manner. Being located in the same room, many steps of the data flows took place by people going across to the other table and taking a form or leaving one on the table if the other person was not around. The servants who carried the data from the health posts to the DDS would slip the reports under the door of the office if the door to the NEP room was locked. These reports would often lie unseen on people's tables because they were too busy or not available. As a result, although the BES was supposed to be a weekly report, it was often compiled and sent with two weeks delay or more. The tables of the staff of the NEP were littered with various forms, unstapled and stapled, both from the previous and current months. Forms were often mixed up, misplaced, or just not tended to. When asked why the forms were lying in such a mixed up state, one of the NEP team members responded "there are not enough folders available to neatly group forms by months or diseases". However, the close proximity of three NEP members and the camaraderie and good will that seemed to exist helped to ensure that despite the tremendous work pressure, somehow the monthly routines around data collection and its transmission were performed. Though the NEP staff told us that due to time constraints it was not possible for them to perform any systematic analysis of data, we found many tables and graphs pinned to the walls. This display could be seen as a reflection of "orders being obeyed", since the province had instructed the districts to make greater use of graphs and analysis.

Also interesting were the different levels of trust ascribed to the data by the NEP staff depending on whether the forms were carried by the servants or by the medical nurse. Needless to say, forms carried by the servant were trusted less than if carried by the nurse. The carrier of the data influenced the perceived quality of data since clarification over discrepancies in data could be provided by the nurse and not so by the servant.

The health information flows in the NEP were rather convoluted. Data would come to the head of the health information system in a stapled bunch. He/she would unstaple them and pass the respective forms to concerned individuals who needed to fill in and return them back to the head of NEP. The completed forms would then be restapled and sent to the DDS Director who was then responsible for signing them and transmitting them to the province. We could not detect any formal system of quality control and supervision, and the signing appeared to be a routine formality. Though the reports were supposed to be sent to the province on a monthly basis, there were often significant delays in their transmission. Contributing to these delays was the distance between the district and province, poor transportation means, and the fact that the district would try to combine the sending of reports with other tasks that needed to be done in the province, for example, the collection of medicines. In cases of delays, rarely did the province take action and demand that the reports be sent on time the next time around. As a result, the poor communication between the different levels of the health network within the district, contributed to the poor quality and untimely flows of health information.

At the provincial level, the person responsible for statistics entered the paper-based forms received from the district into a computer. The person entering the data has completed 10 years of education (básico) and is currently studying for the pré-universitário qualifications (12 years), but lacks any statistical and medical training. The lack of formal advanced training has potentially important implications for ensuring the quality of health information, and the level of statistical analysis that could be carried out. After entering the data into the computer database, print-outs were taken and the paper forms were circulated to the province staff responsible for different programs. These people performed their individual analyses, and the Head of provincial statistics also conducted a similar analysis. Reports were sent to the central level by fax or were carried in person in paper form or in floppy disks. We were told that often faxes were sent to the wrong person in MISAU and not the person responsible for national health statistics. A phone call would then come from MISAU a few weeks later to remind the province that the data were not received, leading it to be re-sent more carefully. Similar stories were also told when reports were sent in person; normally the reports were sent with any person going towards Maputo on other errands (personal or official). This person would often leave the floppy disk or paper forms with anyone in MISAU, since he would be scared to go directly to the Director; sending the reports in person did not necessarily imply that the data would reach the concerned person in MISAU on time. When received by floppy disk, sometimes the data would be entered into the MISAU database successfully, but sometimes the disk was just put away in the cupboard and sometimes the floppy disk were unusable. The effect of these inconsistencies was clearly evident when we analyzed the national database and detected tremendous gaps in the data. For example, sometimes 4 or 5 months in a particular year did not have data, and other times many facilities had blank entries against their name.

In addition to the monthly reports, both the district and the province are responsible for producing three-, six-, nine-monthly and annual reports describing the health status of the district and province during these time periods, and comparing it with corresponding periods of the previous year. This analysis was also seen to have many inconsistencies and gaps.

5 DISCUSSION AND CONCLUSIONS

Through the case discussion and analysis, we have analyzed the mutual linkages between communication practices and health information flows within and across the four different levels of the health administrative structure (community, district, province and national levels). In general, communication practices were seen to be informal, locally specific, and taking place in an improvised manner. For example, forms were slipped under the door with no real accountability on who left the forms and when. Who carried the data influenced the perceptions of data quality, and who carried it was less by design and more based on the chance event of who was traveling when in a particular direction. While these local practices influenced the overall quality, timeliness and usage of formal data, at another and more informal level, to outsiders like us, the work seemed to be going on with a degree of harmony that obscured the seeming chaos within. The routines of the bureaucracy were being followed within "acceptable" constraints of timeliness, quality and inconsistencies. In general, it seemed to be accepted that data collection was seen as a task that had to be performed primarily to meet the needs of the bureaucracy, and was thus subservient to the primary and mammoth task of providing health care to vast numbers of people with very limited resources.

The sense of apparent harmony within the DDS in Xai-Xai could suggest that the existing networks built around these local communication practices are reasonably well aligned and robust, tied up with certain of the existing inscriptions embedded in the national HIS. These inscriptions are embedded in the "node" of the NEP in Xai-Xai based on social networks of the surrounding health information flows which are deep rooted in the local identity culture expressed in ways of working. It is this "deep rootedness" which comes into tensions with new ICT-based initiatives and management methods that require certain systematization of processes (for example, how and where data are to be filled or not) and the use of rules for the aggregation, analysis and transmission of data.

The ongoing attempts to introduce HISP in Xai Xai have made little impact to date. The HISP facilitators installed the software in the district computer and provided training to potential users in 2001. However, after many months when we went back in 2002, we discovered that the software had never been opened after that training session. The question that arises then is, "why was that the case?" One of the reasons could be the inappropriate facilitation support provided by the HISP trainers who had not followed up or returned to the site for many months. But is that the only reason? We tried to rectify this issue of inadequate support by being based in the district for two full weeks to ensure more active and continuous facilitation. During these two weeks we provided the NEP staff with training on how to use the software, enter their routine data and perform some basic analysis. We also educated them about the aims and objectives of HISP. The basic principle of HISP is to work with health personnel who are responsible for health data to encourage their capabilities to use the collected data as meaningful information in providing PHC.

Despite their apparent enthusiasm and interest to learn something novel, the NEP staff could not translate their interests and routines to those demanded by the HISP approach and software. For example, they could not find the time to dedicate to learning the DHIS software and to practice data entry. The existing system of communication that was based on co-location and face-to-face interaction in the NEP room was disrupted by their extended presence in the computer room that was a few buildings away. During the course of our training efforts, we found that the NEP staff were constantly being called by the hospital staff to do other jobs, for example to a patient or go to the pharmacy and collect more condoms. The whole exercise of HISP and the training was seen as something that was being added on and burdening them beyond what they already needed to do. They thus found it difficult to understand the meaning of ICT in their everyday work and ongoing systems of communication and work structures. HISP, in its present form, thus represents a network that is only partially aligned, with the interests of the local actors only partially translated into that of the system. Many reasons have been pointed out to contribute to this state including existing communication patterns, infrastructure constraints, educational levels, and a culture that does not yet value the "efficient use of information". However, our analysis suggests that there is a counter network in Xai-Xai deeply rooted in local communication practices and social network.

The existing experiences thus raise the question, how might such "counter networks" be extended to address these ongoing challenges? At a more fundamental level are these counter networks desirable? Taken from a normative perspective, we can argue that these counter networks are desirable as a number of existing challenges around the health care system could be addressed with more effective information management. A simple example could be that a "proper" report on drug consumption in the district could help the province authorities to send the right medicines to the correct place and in a timely manner. As one doctor told us, better information can at least help them to make better claims of what they need, and how currently these needs are not being met. Whether the province authorities will act on the information they receive is another question, but at least the availability of relevant information is an important starting point for advocacy.

If we agree that counter networks are a desirable and effective means of trying to make the situation of previously excluded groups and regions visible over the network, and with it create the potential of appropriate action (for example, the better deployment of resources), the challenging question concerns how can they be further constructed? We have argued that a starting point and fundamental step in this process is to understand in depth the existing communication practices and how these are intricately tied up with the flows of health information. Such an understanding can help develop closer insights into how these practices can both support and be supported by initiatives like HISP. Training and facilitation efforts need to take into account broader infrastructure constraints, the skill level of people, the time and timing of when and how training is conducted, and the personal motivations of the staff. For example, our training efforts could have been more effective if the computer was placed in the NEP room rather than the present situation where it is in a different building. This sensitive placement of the computer would lead to smaller disruptions in the existing work practices. Communication practices are intricately tied up with deeper questions of identity of the health staff, and understanding these relationships is crucial for developing context-sensitive approaches for implementation of computer-based information systems. Indeed, to create counter networks linking the marginalized and excluded to the network society (following Castells) requires first to recognize the existence of counter network in the local context. These are the issues which we will continue to pursue in the future stages of our empirical and theoretical explorations.

ACKNOWLEDGEMENTS

We would like to thank Dr. Judith Gregory, from IFI, UiO for her valuable comments on earlier versions of this paper. Comments received from the reviewers have also been very useful in improving the quality and focus of the paper.

REFERENCES

Amant, St. K. "When Culture and Computers Collide Rethinking Computer-Mediated Communication According to International and Intercultural Communication Expectations", *Journal of Business and Technical Communication*, (16:2), 2002, pp. 196-214.

Amonoo-Larsen, R., Ebrahim, G., Lovel, H. and Ranken, J. District Health Care. *Challenges for Planning, Organization and Evaluation in Developing Countries*, London: Macmillian Press, 1984.

Braa, J. and Hedberg, C. "The Struggle for District-based Health Information Systems in South Africa", *The Information Society*, (18:2), 2002, pp. 113-128.

Braa, J., Macome, E., Costa, J., Mavimbe, J., Nhampossa, J., José, B., Manave, A., and Sitói, A. "A Study of Actual and Potential Usage of Information and Communication Technology at District and Provincial Levels in Mozambique with a Focus on the Health Sector", *Electronic Journal of Information Systems in Developing Countries*, (5:2), 2001, pp. 1-29.

Castells, M. *The Rise of the Network Society, The Information Age: Economy Society and Culture*, Vol. I, Oxford: Blackwell, 1996.

Castells, M. "Globalization in the Network", *Prometheus Journal*, (4), 2000, pp. 109-123.

Census, in *INE – Instituto Nacional de Estatística* (1997), in http://www.ine.gov.mz

Giddens, A. The Constitution *of Society: Outline of the Theory of Structuration*, Cambridge: Polity Press, 1984.

Hall, E. T. *The Dance of Life: The Other Dimension of Time*, Garden City, NY: Anchor, 1983.

Jeremias, L. and Sitói, A. "Moçambique Perfil do Sistema Nacional de Saúde (2ª Versão Corrigida)", *MISAU*, Maputo, 1997.

Martin, J. and Nakayama, T. "Thinking Dialectically about Culture and Communication", *Communication Theory*, (9:1), 1999, pp. 1-25.

Newell, K. "The Way Ahead for District Health Systems", *World Health Forum*, (10), 1989, pp. 80-87.

Opit, L. J. "How Should Information on Health Care Be Generated and Used?", *World Health Forum*, (8), 1987, pp. 409-417.

Roemer, M. *National Health Systems of the World: The Issues*, Oxford: Oxford University Press, 1992.

Sahay, S. and Avgerou, C. "Information and Communication Technologies in Developing Countries" (Guest Editorial for Special Issue), *The Information Society*, (18:2), 2002.

Walsham, G. *Interpreting Information Systems in Organizations*, Chichester: Wiley, 1993.

Walsham, G. "Interpretive Case Studies in IS Research: Nature and Method", *European Journal of Information Systems*, (4), 1995, pp. 74-81.

WHO. "Report of the International Conference held at Alma Ata", *URSS*, 6-12 September 1978, Health for All Series, Nº 1, WHO, Geneva.

Wilson, R. G., Echols, B. E., Bryant, J. H. and Abrantes, A. "Management Information Systems and Microcomputers in Primary Health Care", *Aga Kahn Foundation*, Geneva, 1989.

Yates, J. and Orlikowski, W. J. "Genres of Organizational Communication: A Structurational Approach to Studying Communication and Media", *Academy of Management Review*, (17), 1992, pp. 299-326.

About the Authors

Emílio Mosse is a Ph.D. candidate at University of Oslo (UiO), Department of Informatics at the Information Systems Group, Oslo, Norway. He is a lecturer in the Department of Mathematics and Informatics at Eduardo Mondlane University (UEM) in Maputo. He was born in Mozambique. His research interest is health information systems. He can be reached by e-mail at: emosse@ifi.uio.no (UiO) or mosse@nambu.uem.mz (UEM).

Sundeep Sahay is a Professor in the Department of Informatics at the University of Oslo (UiO), Norway. He was born in India. His interests are in understanding the inter-relationships between globalization processes, the use of ICTs and their implications on work practices in different organizational settings. He has published extensively both in IS and organization studies. He can be reached by e-mail at: sundeeps@ifi.uio.no (UiO).

4

COORDINATING WORK WITH GROUPWARE

The Challenge of Integrating Protocol and Artefact

Jens Kaaber Pors and Jesper Simonsen
Computer Science, Roskilde University, Denmark

Abstract: One important goal of employing groupware is to make possible complex collaboration between geographically distributed groups. This requires a dual transformation of both technology and work practice. The challenge is to reduce the complexity of the coordination work by successfully integrating the protocol stipulating the collaboration and the artefact, in form of the groupware application, mediating the collaboration. This paper analyses a generic groupware application that was deployed in a large financial organisation in order to support working groups distributed throughout four countries. Using the CSCW framework of coordination mechanisms, we have elicited six general factors influencing the integration of the groupware application in two situations.

Keywords: Groupware, Distributed Work, Collaboration, Coordination, Coordination Mechanism, Protocol, Artefact, Integration.

1 INTRODUCTION

For a number of years it has been acknowledged that introducing groupware in an organisation is indeed a difficult matter that requires attention to the implementation of the technology as well as to a range of organisational and social factors (Bullen and Bennett, 1990; Orlikowski, 1993; Grudin, 1994). In order for a groupware application to be successfully adopted, the introduction and the implementation of the application have to comprise a number of organisational and social factors (Grudin, 1994): It has to address a real need from the group members; they should have a clear understanding of a mature use of the application and the positive impacts on their daily work; training and facilitation might be necessary, as well as perhaps even preparations to prevent premature rejection. The IS community in general and the Computer Supported Cooperative Work (CSCW) research community in particular have addressed this issue. This has led to an acknowledgement of ethnographically-informed systems design approaches (Plowman et al., 1995; Harper, 2000), while others have discussed employing groupware by means of tailoring (Mørch and Mehandjiev, 2000; Weigang and Haake, 2000). Studies of the consequences of introducing new technologies in organisations emphasise the importance of the practical use situation and the mutual agreements on how to use a groupware application, and this often entails changes in work organisation and practice (Orlikowski, 1993; 2000, Mark, 2002). The CSCW community is now becoming aware that both focusing on supporting existing work practices (as in ethnographically-informed design) and focusing on flexible and transparent technologies (by means of tailoring or customising software) represent too narrow of a perspective (Bowker et al., 1997).

Our starting point corresponds with Berg (1999). We aim for a "reconsideration of the notions of 'supporting' work and 'transparent' technologies" (p.373). We strive for "an empirically grounded grasp of the functioning of information technologies in work practices" (ibid.: p.374), where we "want to work towards an account that emphasizes the *active mediating* roles of these artifacts. That is to say: an account that, in pointing to the activities of information technologies in [...] practices equally highlights the transformations of the work-activities that are mediated through them" (ibid.: p.376). The challenge is to grasp the intricate relation between this duality of changes when trying to understand how technological artefacts are employed in work practices. We refer to this transformation of both technology and work practice in general as the *integration* of groupware and work practice. When focusing on coordinating work with groupware, we use the CSCW framework related to coordination mechanisms conceptualised as protocol(s) of coordination embedded in a computational artefact (Schmidt and Bannon, 1992; Carstensen, 1996; Schmidt and Simone, 1996). Thus the general question we follow in this paper is "when employing groupware in order to coordinate work, what conditions influence the integration of groupware and work practice?" or – to be more specific – "*what are the factors that actively influence the integration of protocol and artefact?*"

This paper presents an empirically based contribution to the above stated question by investigating how coordination of collaboration in geographically distributed groups is achieved with a generic groupware application. We have analysed the deployment and use of the web-based groupware application Lotus QuickPlace™ (in the following referred to as QP) in a large financial distributed organisation (in the following referred to as "Beta"), that had just emerged as a result of a major merger. In Beta more than 100 different QPs comprising in total about 3000 active users and more than 20 Gb of documents have emerged in less than 2 years. In this paper, we report on two differing cases with respect to the character of work and degree of integration. Both cases represent situations where coordinating work with groupware was attempted. The aim was to enhance collaboration in a distributed development project and in a distributed organisational section in charge of a recurrent task. To provide an answer to our research question, we have elicited six overall factors that have influenced the integration of QP in both cases.

In the following, we first provide a brief description of the conceptual framework of coordination mechanisms, the research method applied, and the overall starting point for deploying QP in Beta. Then, we give a more detailed account of each of the two cases: A project evaluating the possibility of creating a single customer security architecture across four different countries, and the translation section in charge of producing corporate financial reports. From these two accounts, we derive the factors that have influenced the integration of protocol and artefact in each situation, and the differences between the two situations are discussed in the light of these factors. In the concluding section we outline the implications of our research approach with regards to groupware employment and to integration of groupware with work practices.

2 COORDINATION MECHANISMS AS PROTOCOL AND ARTEFACT

The CSCW framework related to coordination mechanisms is especially relevant for our research aim since it explicitly captures the duality of changes related to the integration of groupware and work practice. The framework depicted in Figure 1 is derived from Schmidt and Bannon (1992), Carstensen (1996), and Schmidt and Simone (1996): Collaboration is analytical distinguished as comprising 'real' work (e.g. what the individual actor is doing) and the coordination of work (i.e. the articulated coordination involved in distributed activities needed in order for many actors to perform a cooperate effort). Coordination of work is sometimes referred to as articulation work (Schmidt and Bannon, 1992; Star and Strauss, 1999).

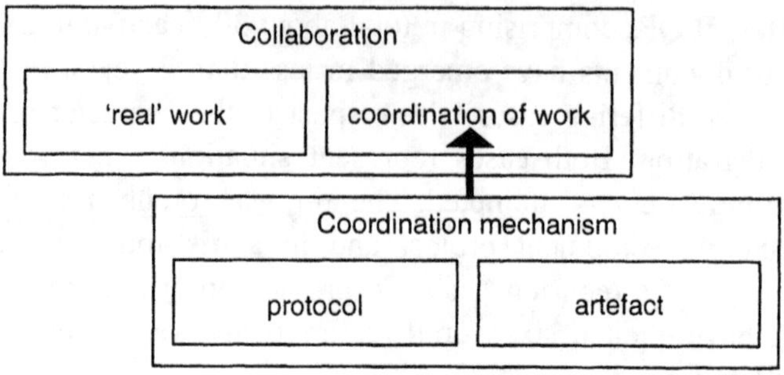

Figure 1. The conceptual framework of 'coordination mechanism'

A collaborative work task is distinguished from other kinds of work by presupposing mutual dependencies between multiple actors. In the instances where the use of groupware mediates mutual dependencies of collaborative work, a coordination mechanism is established. This duality is captured by the definition of a coordination mechanism as "a construct consisting of a coordinative protocol (an integrated set of procedures and conventions stipulating the articulation of interdependent distributed activities) on the one hand and on the other hand an artefact (a permanent symbolic construct) in which the protocol is objectified" (Schmidt and Simone, 1996: p.165). The protocol stipulates how the collaboration is carried out and the artefact is the material component that mediates the protocol. In our case the agreements on how to collaborate constitute the protocol while the groupware is the artefact of the coordination mechanisms. Introducing a groupware application thus involves developing a new protocol as well as taking the use of the artefact into account. At the same time the artefact (a generic groupware application) has to be tailored in order to objectify, reflect and support this specific protocol thus together forming an integrated coordination mechanism.

The motivation for employing groupware while aiming at establishing coordination mechanisms is to make complex collaboration possible. Conditions that enforce complex collaboration exist when geographically distributed actors must collaborate in order to accomplish a work task. A coordination mechanism supports the reduction of the complexity of coordination for each actor, thus "affording an increase in complexity of the work practice without a simultaneous increase in complexity in individual interactions" (Berg, 1999: p.391). With a successful coordination mechanism in place, the members of a distributed working group are able to cope with such complex conditions for coordination.

We thus define the criteria for establishing such a computational coordination mechanism as the combination of three objectives: Attribution to meet a need (for a complex and problematic collaboration), integration of the coordination mechanism with work practice (integration of protocol and artefact), and the resulting reduction of complexity involved in the coordination (hereby supporting the collaboration).

Our aim is to close in on an understanding of the duality of changes involved when generic groupware applications are to be integrated in work practices. Our means to meet this aim is to derive the factors influencing the integration of protocol and artefact in order to reduce the complexity of collaboration by leading to stable and integrated coordination mechanisms.

3 RESEARCH METHOD

The study is part of the DIWA research program (www.diwa.dk) investigating the design and use of web-based applications supporting geographically distributed work practices. The research has been conducted with a grounded and interpretative approach (Golden-Biddle and Locke, 1997) drawing on experiences from several initiatives in Beta during 2000-2002:

- An initial investigation of needs and strategies for intranet applications (6 interviews);
- An analysis of strategies and practices for Beta's organisational change support and special interest groups (6 interviews);
- An analysis of specific use of QP in newly established organisational units, projects, and teams handling recurrent tasks (7 interviews);
- Analyses of three development projects (based on interviews and observations);
- Document analysis of 90 requests for a QP from managers to IT operations stating the intended aim of using QP;
- A survey reporting from 53 QPs;
- An analysis based on a log of all http transactions to and from the QP server during a 10 month period.

The interviews lasted between 1 and 2 hours, structured by an interview guide, which was sent to the informant in advance. Interviews were tape-recorded, and later transcribed verbatim. All our analyses of this multifaceted material were reported and discussed with management and other informants from Beta and constitute the basis of the research presented in this paper. The two cases presented here have been chosen from this material to illustrate situations, where groupware is employed with the explicit aim of facilitating the coordination work of a geographically distributed group. The analyses of the two situations show how the different circumstances for achieving the integration of QP actively contribute to this process as described by the identified factors.

4 INTRODUCING GROUPWARE IN BETA

Beta is a large international financial corporation and as a result of a recent merger it comprises several financial companies based in four countries. The organisations of the former companies have been merged into corporate sections and a number of corporate projects have been initiated. This created an instant need for a platform independent tool to support inter-organisational communication, since Beta at the beginning had no secure mail infrastructure, no local area network to exchanges files on, and no corporate intranet. The groupware application: Lotus QuickPlace™ release 2 (www.lotus.com/quickplace) was chosen by the communications department of Beta as the standard application to support communication within geographically distributed corporate projects and sections of Beta.

Setting up the central QP server requires virtually no integration with the existing IT infrastructure and offers secure Web-based workspaces accessible through web browsers. In addition, IT operations had good experiences with Lotus products. QP is a generic system (Bansler and Havn, 1994), which means that it needs to be tailored to the specific need of the group of users. The standard configuration of a QP offers some basic facilities for discussion, calendar, user administration, index, search tool, and a tutorial. The person(s) with manager rights of a newly installed QP must start by designing an initial structure setting up a home page and creating and naming folders and rooms accessible in the QP and invite other users, granting each of them access rights as either manager, author or reader. Thus a QP is understood as a single instance of the generic application, and acts as a separate groupware tool for a specified group. Similar browser-based groupware applications are e.g. eRoom (www.eroom.com), Projectplace (www.projectplace.com), and BSCW (bscw.gmd.de), the latter especially familiar within academia (Bentley et al., 1997).

Apart from a customised logo appearing on every page of the application, QP has been deployed in the generic standard configuration leaving it up to the projects and organisational sections to decide how to integrate the groupware in their work practices. A specific QP is initiated on the server by request from the manager of a distributed group to IT operations. The manager is fully in charge of how to utilise QP. Users initially learn to use QP from the build-in tutorial and Beta offers no further formal training in using, tailoring, or integrating QP.

4.1 Employing Groupware for a Recurrent Task

The production of the financial reports of Beta involves translation of an English master into the four languages of the different countries. The completed financial reports are to be released simultaneously to the stock exchanges and the press. The translation of the English master is initiated about one week before the release deadline. At this time the master is not in its final state and corrections occur several times up to the deadline. This requires new versions of the English master to be distributed during the translation process. These changes have to be coordinated very tightly within the group of translators. During the preparation and translation of the final documents, the information is highly confidential. Emailing drafts by the Internet was considered insecure and prior to the introduction of QP, fax transmissions were used to exchange drafts that often were more than 50 pages long. This involved a very complex coordination process with only a cumbersome infrastructure to support it. The original incentive to use QP for the coordination of the translation process was thus the complex security measures.

QP was originally chosen and set up by the manager of the translation section. He initiates the translation process upon receiving the almost final English master for the quarterly or yearly financial reports. The master document is distributed to the translators via a new room created for that purpose in the QP. The translators work in parallel on the texts and usually in different geographic locations. The frequent changes to the master document right up until the deadline have to be propagated throughout the section. When each translator has completed a part of the documents, he or she uploads it to a language specific folder in QP with a specifically versioned name. It then becomes available to all others for their work and in this way the progression of the work becomes visible in QP. The manager can track the status instantly and the members can see how the others are performing thus revealing whether they e.g. are behind schedule. The way QP is used when the translation of financial reports takes place is an illustrative example of how complex coordination work is carried out using QP as part of a coordination mechanism.

The coordination involved in the collaboration of the translators is complex, while the work of e.g. translating individual documents might be less difficult even though it can be complicated work in a linguistic sense. QP is the artefact in a computational coordination mechanism supporting the coordinating work by mediating mutual dependencies (Schmidt and Simone, 1996). In addition QP provides an overview of the process as well as performing some of the tedious footwork that the collaboration entails. Using QP in this way has lead to a substantial reduction in the complexity of coordination compared to using fax transmissions – both among the translators and for other involved actors responsible for publication of the final financial report.

The manager of the translation section has put a serious effort into integrating the technology to the work practices of the translators. An important motivation stemmed from the problems experienced using fax transmissions and problems foreseen if exchanging documents as attachments to e-mail - as he stated: “E-mail is a mess for this purpose!” Even if the issue of security was resolved, the versioning of the translated documents is complex and problematic when mediated via e-mails: This would require all involved actors in the translation process to build and maintain private archives storing the frequently distributed new versions of the documents. Using these archives (typically stressed by time pressure) would inevitably lead to errors working on a wrong version of a document.

With the introduction of QP the manager produced guidelines for the proper use and on occasions he phoned up people to help and/or persuade them to use the QP correctly. In our understanding this has proven instrumental in promoting the use of the QP, since in his position, ranking at least one level above the translators, he can act as personnel manager, section manager, and manager of QP as well as facilitator. Integrating QP as part of a coordination mechanism required restructuring of work practices and explicit agreements on how to use the application alongside tailoring of the artefact with different rooms, folders, templates etc. The recurrence of the translation of the financial reports every three months creates a ‘naturally occurring’ opportunity for reconsidering the use of QP, since it provides occasions for evaluation and re-design. In addition to this the character of work is well suited for the emergence of a coordination mechanism, since the protocol of the coordination work is well defined. The context for carrying out the recurrent task is relatively well known and stable surroundings make it easier to focus the efforts to integrate the groupware.

4.2 Employing Groupware in a Development Project

In Beta an elaborate way of organising distributed projects for change processes has been developed and implemented. All projects are organised aiming at an overall 6 months time box. Corporate development projects present a highly complex work setting, both geographically distributed and managerially heterogeneous as they involve IT-departments that were ‘independent’ before the merger. The conditions for performing coordination work within a development project are thus relatively diverse and shifting compared to the recurrent task, since the tasks and members change from one project to the next. Thus collaborating on the subject matter of the project requires a great deal of coordination work including negotiations of the means and goals of the project itself. This was indeed the case for the project evaluating the possibility of creating a single customer security architecture.

Establishing meetings and workshops for a group of IT-specialists and IT-managers distributed throughout four countries is not an easy task. They are all very busy people and they have to travel several hours to meet. The project members got together at meetings as often as practically possible and the main body of interaction took place in these occasions. These meetings also provided occasions for clarifications and agreements on how to proceed with the project work.

The project manager experienced difficulties getting trough the agenda he had planned for the meetings: Informing each other in order to establish the current status often took up most of the time leaving little time for discussing possible solutions based on the investigations and analyses done so far. He wanted to use QP in order to enable this coordination to happen even between the meetings. Then the meetings could focus more on the subject matter of the project in terms of discussing problems and alternative designs of security solutions. One meeting concluded by identifying questions about eight issues on the IT infrastructure for each national IT-section in Beta. The manager asked the individual members to examine these questions in their respective countries and report back to the project via a response in QP. He created the questions and documents in the QP and expected the members to post their answers as responses to the document. The idea was that when a member had posted his initial answer, other members could take this into account concerning their answers. In the QP they could see each other's answers, and they would also have the possibility to respond to individual answers and e.g. ask clarifying questions or make comments. This would have enabled the project manager to present a final answer at the next meeting. However, such an openness to information can also be a barrier for using the artefact. How to go about presenting the local practices can be a delicate matter of strategic disclosure and nondisclosure. It can be difficult or impossible to foresee the consequence of sharing information uploaded to the QP considering an issue to be negotiated. As opposed to face-to-face meetings the control of other members immediate interpretations and subsequent use of the material is lost. Using QP to mediate negotiations might thus have unforeseen impacts on the agenda and how an issue will be discussed. It might entail misunderstandings and "sidetracking" of the negotiation as well as weakening the negotiating position of individual actors. As a result, the members did not respond via the QP. Instead they continued to bring their answers to the next meeting and the coordination of the solutions to the particular questions was carried out during the meeting.

The observed use of the QP in the development project resembles a project archive where the results of the project were developed and maintained. The QP was, as expressed by the project manager "used as a place for documentation. Here is the collection of documents that are the result of our work." In this way QP primarily supported the post hoc documentation of the project work. QP was a 'nice-to-have' for the project members in order to get their work done, since other means for coordinating work such as e-mail and phone are more immediately gratifying. However the project manager viewed the QP as a 'need-to-have' when managing the issues and deliverables. The case demonstrates how difficult it can be to integrate a coordination mechanism in the work practices of a development project. Agreeing on a new protocol that relates to a new artefact meets several barriers, while the familiar protocol of preparing for meetings works fine. The tight timeframe of projects also puts a limit on how much is invested in the deployment and tailoring of QP.

5 DISCUSSION

Table 1 presents a summarising characterisation of the factors for the two situations described in the previous sections. Based on these characteristics we identify and discuss lessons learned in terms of factors that influence the integration of groupware in each situation.

Table 1. Six factors actively influencing the integration of protocol and artefact

Situation / Factor	Recurrent task	Development project
Management position	Within section (status as personnel manager)	Within project (among peer experts)
Membership	Continuous	Transient
Evaluation and re-design of coordination mechanisms	In between iterative tasks	Difficult due to short life cycle
Reduction of complexity	Substantial	None
Identified need	Need (for all)	Need (mgr.), nice (others)
Facilitator	Present and active	None

The character of work is very different in the two situations. The translation section is responsible for a time critical recurrent task with predefined procedures that is performed intensively over a short period of time. A high degree of systematisation is required to cope with the strong mutual dependencies among the members in the section. The ways of coordinating work are well defined and the protocol and artefact are integrated in a coordination mechanism effectively supporting the collaboration. Any changes to the protocol have to be carefully prepared in advance enabling the necessary coordination to happen smoothly avoiding misunderstandings or other disruptions that may cause sudden halts in the translation process. In the project the development of solutions involves relations to a range of issues. The character of the work – especially in the early phases of a project as in this situation – is highly focused upon negotiation even of the goals and means of the project itself. Negotiation involves implicit protocols (e.g. with regard to how to prepare and present a case for the meeting's agenda) that might be challenged dramatically when integrated with an artefact. This situation has a great need for support but there are several barriers to overcome before the emergence of coordination mechanisms will occur.

5.1 Management Position

Establishment of a new protocol of coordination as well as getting acquainted with the artefact requires all involved actors to adjust current work practices. An influential factor is management's position when promoting the use of groupware for coordinative purposes. In the case of the translation section the manager had a rank as a personnel manager one level above all involved users. The project manager was managing the project while the project members were managers for their respective IT departments. Thus the relationship among all project members is a collegial peer relationship among experts sharing the same profession. This means that changes in work practices in order to establish new coordination mechanisms – in terms of a new protocol that requires using QP – cannot be expected to happen solely because the manager believes this is a good idea. He must also convince his fellow project members that there are obvious benefits from investing resources in this in addition to following up on the agreements.

5.2 Membership

Beta is a relatively new organisation, still there is a difference in how well the members in the translation section and the project know each other. The translation section was established shortly after the merger in order to take care of the upcoming quarterly financial report. The members in this section form a continuous group that share the same aim. They often perform the task under stress and this contributes to a congenial relationship among them. Projects are transient and the project members form a temporary group that meet by virtue of the project over a short period of time. Continuous membership supports a situation of mutual trust and confidence in roles, competencies, power relations, and responsibilities etc., which support coordination work. Collaboration within a newly formed group faces a number of challenges with regard to the establishment of coordination involving mutual agreements and social accountability, considering who is doing what and when. The short lifecycle of a project results in transient memberships that make it difficult to harvest the benefits from investing in establishing new protocols and using a specific technology to mediate them.

5.3 Evaluation and Re-Design of Coordination Mechanisms

The generic nature of groupware such as QP along with the continuous changes in the organisation, necessitate a periodical evaluation and re-design of QP and on the agreed protocols stipulating its use (Schmidt and Simone, 1996). Even though the work practice of a development project has room for experiments with different ways of coordinating work, evaluation and re-design are difficult simply due to the short life cycle of the projects. The initial tailored 'setup' of QP is thus usually a one-shot trial. This might suggest that new coordination mechanisms should be addressing issues that are repeated from project to project, such as the overall project concept or method, use of mandatory or general techniques and tools, etc. The section handling recurrent tasks has a naturally occurring opportunity for reconsidering the use of QP where former experiences can be evaluated and subsequent adjustments incorporated in the coordination mechanism. The recurrent task has an obvious advantage in this respect, since it provides such frequent occasions for evaluation and re-design, and because the character of work is well defined and has been tried several times before.

5.4 Reduction of Complexity of Coordination

The purpose of establishing a coordination mechanism is to reduce the complexity of the coordination work. In the case of the section performing the recurrent task the work practice was transformed to accommodate a new protocol along with QP leading to a significant reduction of the complexity of coordination for all involved: The problems handling fax machines were eliminated and an instant monitoring of the status of the overall process was established. In the project it was the manager that had an immediate need for making the meetings more effective. The project members did not expect a reduction in the complexity of coordination qua the artefact. On the contrary the visibility of the documents, that the project manager encouraged them to upload, might potentially have lead to an increase in the complexity of coordination: The meetings could have become even more complicated to conduct with the addition of a QP based coordination mechanism, due to the member's unfamiliarity with a new protocol dealing with such delicate negotiations. Thus the protocol of coordination was unchanged and the complexity of the coordination remained at status quo.

5.5 Identified Need

Possibly the strongest driver for using QP in the translation section was the earlier experienced problems of coordinating using fax transmissions as the artefact. This created an obvious need for new coordination mechanisms that would reduce complexity. The solution to the problem of coordinating translations entailed a completely new protocol and established QP as a highly critical application. The coordination work had to be completely redesigned but the complexity was foreseen to be drastically reduced. The result is an immediate and sufficient benefit for all and thus a good return of the investment, and in addition a return that can be harvested repeatedly in the quarterly translation processes to come. The project experienced a different situation in this respect, since using QP was suggested as a means to coordinate a process of negotiation that otherwise still could be conducted without serious problems – at least not serious problems related to coordination work. In a single project running for 6 months the need for investing resources in changing work practices must be evident and the return of investment must be promising within the remaining project period. In the reported situation this might only have been the case regarding the project manager resulting in disparity in work and benefit (Grudin, 1994) and a potentially conflicting situation for the other project members.

5.6 Facilitation

The change of the articulation work in the translation section by introducing QP was both very extensive and did also entail a strong commitment to use QP. The manager took the role as an active facilitator to ensure that all involved adopted the new work process. He conducted well-prepared introductions explaining the new protocol and artefact, and he actively contacted the translators to follow up when the translation process was initiated. The involved actors were not IT specialists and since the manager also was the manager of QP he became the person to contact when questions arose regarding using the QP. The continuous membership in the section along with an open and confident atmosphere significantly contributed to the establishment of an effective facilitation. Such a facilitator role was apparently more difficult to establish in the project where the members themselves were IT-specialists and IT-managers. The project manager left it to the members alone to get acquainted with QP and he did not overcome the obstacles of taking an active facilitating role – neither with regard to following up on the new protocol he suggested nor using QP for this purpose.

6 CONCLUSION

The study shows integration as not only involving the question of tailoring the artefact and involving the question of agreeing on and establishing social accountabilities for revised or new protocols, but as also involving complementary questions of integrating protocol and artefact.

We have demonstrated the applicability of the CSCW framework of coordination mechanisms as an analytic tool eliciting six general factors that actively influence this integration: 1) Authoritative management or obvious benefits for all to invest needed resources; 2) Continuous membership or other means of securing confidence in the coordination mechanism in question; 3) Conditions in the work organisation that support continuous evaluation and re-design; 4) Substantial reduction of complexity of coordination, bearing in mind that such a reduction for some might be seen as an increase in work or complexity for others; 5) Identified need for example based on prior experiences with problematic coordination (nice is not enough); and 6) Conditions enabling the establishment of an active facilitating role.

The achievement of complex collaboration involving geographically distributed actors requires considerable amounts of coordination work. An understanding of how coordination mechanisms unfold over time with the gradual integration of protocol and artefact, might further these endeavours in research and practice.

Our enquiries uncover convoluted interdependencies of collaborative work in a way that allows for discussion and comparison of individual cases. The elicitation of the general factors from the case studies is a way of examining the relations between organisational and technological aspects in order to understand the character of distributed work and the felicitous conditions for coordinating work with groupware.

ACKNOWLEDGEMENTS

We would like to thank the informants in Beta, who provided opportunities for studying groupware in practice, our colleagues, Keld Bødker and Kristian Billeskov Bøving, for collaborating on this study, and the anonymous reviewers for useful comments. The DIWA research programme is sponsored by the Danish Research Councils and the IT-University of Copenhagen has also partly sponsored the research.

REFERENCES

Bansler, J., and Havn, E. "Information Systems Development with Generic Systems" *Proceedings of ECIS*, 30-31 May, Breukelen, The Netherlands, Nijenrode University Press, 1994, pp. 707-715.

Bentley, R., Horstmann, T., and Trevor J. "The World Wide Web as Enabling Technology for CSCW: The Case of BSCW", *CSCW*, (6:2-3), 1997, pp. 111-134.

Berg, M. "Accumulating and Coordinating: Occasions for Information Technologies in Medical Work", *CSCW*, (8:4), 1999, pp. 373-401.

Bowker, G. C., Star, S. L., Turner, W., and Gasser, L. (eds.) *Social Science, Technical Systems, and Cooperative Work: Beyond the Great Divide*, New Jersey: Lawrence Erlbaum Associates, 1997.

Bullen, C. V., and Bennett, J. L. "Learning from User Experience with Groupware" *Proceedings of the Conference on CSCW*, October 7-10, 1990, Los Angeles, California, ACM, New York, 1990, pp. 291-302.

Carstensen, P. "Computer Supported Coordination", *Ph.D. Thesis*, Writings on Computer Science, No. 61, Roskilde University, Denmark, 1996.

Golden-Biddle, K., and Locke, K. *Composing Qualitative Research*, London, UK: Sage Publications, 1997.

Grudin, J. "Groupware and Social Dynamics: Eight Challenges for Developers", *Communications of the ACM*, (37:1), 1994, pp. 92-105.

Harper, R. "The Organisation in Ethnography: A Discussion of Ethnographic Fieldwork Programs in CSCW", *CSCW*, (9:2), 2000, pp. 239-264.

Mark, G. "Conventions for Coordinating Electronic Distributed Work: A Longtitudinal Study of Groupware Use", Hinds, P. and Kiesler, S. (eds.). *Distributed Work*, Cambridge, MA: MIT Press, 2002, pp. 259-282.

Mørch, A. and Mehandjiev, N. "Tailoring as Collaboration: The Mediating Role of Multiple Representations and Application Units", *CSCW*, (9:1), 2000, pp.75-100.

Orlikowski, W. "Learning from Notes: Organisational Issues in Groupware Implementation", *Information Society*, (9:3), 1993, pp. 237-250.

Orlikowski, W. "Using Technology and Constituting Structures: A Practice Lens for Studying Technology in Organisations", *Organisation Science*, (11:4), 2000, pp. 404-428.

Plowman, L., Rogers, Y. and Ramage, M. "What Are Workplace Studies For?", *Proceedings of the Fourth ECSCW*, September 10-14, Stockholm, Sweden, Kluwer, 1995, pp. 309-324.

Schmidt, K. and Bannon, L. "Taking CSCW Seriously: Supporting Articulation Work", *CSCW*, (1:1-2), Kluwer, 1992, pp. 7-40.

Schmidt, K. and Simone, C. "Coordination Mechanisms: Towards a Conceptual Foundation of CSCW Systems Design", *CSCW*, (5:2-3), 1996, pp. 155–200.

Star, S. L., and Strauss, A. "Layers of Silence, Areas of Voice: The Ecology of Visible and Invisible Work", *CSCW*, (8:1-2), 1999, pp. 9-30.

Weigang, W. and Haake, J. "Tailoring Groupware: The Cooperative Hypermedia Approach", *CSCW*, (9:1), 2000, pp. 123-146.

About the Authors

Jens Kaaber Pors is a Ph.D. candidate from Computer Science, Roskilde University, Denmark. He was born in Copenhagen and holds a M.A. in Information Studies from The University of Århus, Denmark. His research interests are science, technology and society studies (STS) with regard to work practices employing standard-based infrastructures for cooperation and other distributed activities. He can be reached by e-mail (pors@ruc.dk) or his homepage at www.ruc.dk/~pors.

Jesper Simonsen is an associate professor, Ph.D., from Computer Science, Roskilde University, Denmark. His research interest is the study of work practices of users and designers for the purpose of offering theories and methods for systems design in an organisational context. He is publishing in journals like CACM, HCI, CSCW, and SJIS, along with a number of international conferences. He is also co-authoring a textbook forthcoming on MIT-press about the 'MUST-method' – a participatory design method integrating ethnographically inspired techniques. Hecan be reached by e-mail (simonsen@ruc.dk) or his homepage at www.ruc.dk/~simonsen.

5

CRM IN CALL CENTRES

The Logic Of Practice

Helen J. Richardson
Information Systems Research Centre, University of Salford, UK

Abstract: We are said to be living in a new and global 'knowledge economy' where individuals and organisations are urged to change their way of working, learning and living. Nevertheless there are limitations and contradictions when it comes to managing knowledge, let alone making sense of knowledge management in the context of organisational practice. This paper focuses on research analysing Customer Relationship Management (CRM) systems, their relationship to knowledge management and their use in call centres in the UK. Assuming a critical research approach, case analysis shows significant contradictions between system objectives and outcomes in practice. By referring to the work of Pierre Bourdieu, a sociologist and critical social theorist, Information Systems (IS) researchers are provided with tools of analysis. In this paper I shall show how key concepts of field, habitus, logic of practice and symbolic violence gave fresh insights into the study of CRM systems in call centres.

Keywords: CRM systems, call centres, knowledge management, critical research, field, habitus, logic of practice, symbolic violence.

1 INTRODUCTION

In the process of analysing CRM system use in call centres, this paper stresses the relevance of applying social theory to the theory and practice of IS. Social theory builds our 'intellectual scaffolding' (Walsham, 2001) and by using the conceptual tools provided by the critical social theorist Pierre Bourdieu, this gives us fresh insights into CRM system use in call centres and greater understanding of knowledge management and change. The distinctive issue confronting contemporary business we are told is not the centrality of knowledge but rather the opportunities there are to intensify its production and utilisation (Castells, 1996). Yet there are many contradictions with IT-driven knowledge management. Some stress the need to see production of knowledge in the context of its use (Gibbons et al., 1994). Others view knowledge management as a whole as 'hype' and 'to suppose that the use of computers can suddenly provide us with the ability to manage something we don't understand is humbug' (Bentley, 1999). What is also often overlooked is that 'power structures play a large part in the making and accepting of knowledge' (Adam and Richardson, 2001) and how technologists so often lack understanding of the situated work practices of the systems user communities (Kvasny and Truex, 2000).

One way organisations try to manage knowledge is through the use of CRM systems. This paper presents an analysis of CRM and knowledge management in organisational practice and the rich social interactions at work. In these terms the critical social theory of Bourdieu provides intellectual foundations useful in this critique of IS and organisational research where individuals are often marginalised (Alvesson and Deetz, 2000). Bourdieu shows appreciation of the dialectical relationship between the individual and the world they inhabit. He discusses the unity and regularity of systems and their practical coherence but on the other hand their 'fuzziness' and irregularities both being equally necessary and inscribed into what he terms the logic of practice (Bourdieu, 1990). This is important when looking at CRM system use embedded as it is in notions of transformation – of organisations in a global knowledge economy, of new ways of working, of virtuality central to communication amongst 'stakeholders'.

The paper shall proceed by firstly justifying the critical approach to organisational research. I shall then introduce the problem areas of CRM, its links with knowledge management and CRM use in call centres. I then briefly discuss the key concepts of field, habitus, logic of practice and symbolic violence and their roles in the 'circuit of reproduction' (Bourdieu, 1990). These concepts are applied to the case material resulting from what I have described as a mobile and interpretative field study. In conclusion I stress the important contribution Bourdieu and critical research has made to the study of organisational practice in the context of the 'so-called' knowledge economy.

2 RESEARCH OBJECTIVES

The objective of this research is to introduce a critical theory perspective to IS research specifically drawing on the conceptual tools provided by Pierre Bourdieu. Bourdieu was a 'resister' appearing on various anti-globalisation platforms before his death in 2002. He also challenged the orthodoxy and what is legitimate or not in terms of research and research findings. In this context, this study attempts to give a wider political picture and to question the assumptions that perhaps just focus on the efficaciousness of software or the constraints or otherwise of IS.

The process of conducting critical research means disrupting ongoing social reality in order to question what is often ignored or taken for granted and gain a critical and richer insight into issues raised. It has been noted that social research – particularly management studies – has tended to conform or reproduce dominating institutions and interests (Alvesson and Deetz, 2000). This paper in contrast aims to throw the spotlight on the individuals' working on the call centre 'front line' and tell the stories often left untold in studies of IS and change (Bannister, 2002; Alvesson and Deetz, 2000). Call centre organisations are seen as 'social and historical creations' and the critical research approach aims to 'recognise the influence of history, culture and social positions on beliefs and actions' (Alvesson and Deetz, 2000). Such is the foundation of Bourdieu's critical social theory too. Taking this approach demands critical reflection, imagination, commitment and application of what has been termed a reflexive methodology - an 'interpretive, open, language sensitive, identity conscious, historical, political, local, non-authoritative and textually aware understanding of social research' (Alvesson and Skõlberg, 1999).

What critical research and Bourdieu's analysis in particular shows us, is that there are wider institutional and ideological issues to be discussed when studying IS and organisational change (Alvesson and Deetz, 2000; Bourdieu, 1990). Furthermore this paper follows the established traditions of conceptualising the social and political as well as technical issues of IS in organisational practice (for example see Franz and Robey, 1984 and Markus, 1983).

3 CRM AND KNOWLEDGE MANAGEMENT IN CALL CENTRES

CRM system use in call centres is about strategic knowledge management utilising tools and applications to enhance the accessibility of knowledge, to automate and manage knowledge and organisational intelligence. CRM systems attempts to codify knowledge and so control customer and other relationships.

The case material in this paper focuses on CRM system use in telephone call centres in the UK. They operate to handle the 'frontline' communication with customers, control and to various degrees automate the gate-keeping roles, capture and standardise sales knowledge and oversee customer service. This front line work is vital and significant to an organisation. It is often strategically important work. CRM system use is aimed at streamlining and shortening the key business processes that define a global organisation's relationship to the markets and customers – it is a strategic business issue that requires technology support (Ciborra, 2000). A number of IT tools can be utilised including automated call distribution (ACD) systems using computer telephony integration, voice and speech recognition and response software, integrated volume response and use of the World Wide Web.

CRM systems mean codifying intellectual capital (Light, 2001) with software providing 'scripts' enabling monitoring and call analysis. This metaphorically, if not physically, welds the worker and machine into a streamlined and controlled knowledge system. Architects of CRM systems, for example, want to 'predict the future world of the user and the machine' but Wilson and Howcroft (2000) stress the important focus of contradiction and resistance, always apparent and possible with human-computer system interaction in an organisational context. This is central to Bourdieu's view of the logic of practice also. Although CRM system use in call centres aims to formalise, standardise and rationalise responses with the use of scripts, nevertheless there is also resistance from non-compliant users actively or otherwise failing to act out the 'script'.

Contradictions appear in work practices too. CRM system use in call centres goes hand in hand with the adoption of teamwork and empowerment management methods. However it is clear from this study and others that team-based management is there to control workers in an organisation rather than to empower them or liberate them from control (Truex and Ngwenyama, 1998; Richardson and Richardson, 2002). Recognising potential contradictions of course 'enables understanding of points of conflict and instability in organisations and how these may interact to change and transform organisations' (Orlikowski, 2001).

There are startling contrasts in the image of CRM in call centres. For the consultants and employers, CRM use in call centres means knowledge intensive, strategic use of technology, flexible working and utility of all those new ways of working like flattened organisations, teamwork and empowerment. For others, they are the 'sweatshops of the 21st century' (Belt et al., 2000).

4 CONTRIBUTION OF BOURDIEU

In this section I will discuss the key concepts of field, habitus, logic of practice and symbolic violence in the critical social theory of Pierre Bourdieu in order to show how these concepts are relevant to IS research.

4.1 The Generation of Practice

Bourdieu tries to convey that the social space and the individuals that occupy it are a result of historical struggles; individuals produce the social space they live in and they are in turn produced by it. Therefore they both incorporate and objectify social structures that they inhabit (Wolfreys, 2000). The relationship between the individual and society, or structures and agencies, is expressed in an analogy: compare social activity to an individual sense of play - people are free to act but they can only do so within the constraints of the game that they are playing. The game or social activity allows for improvisation and manipulation of rules and coming to terms with the game is called the logic of practice. The logic of practice involves not just coming to terms with the rules, nor is it a wholly unconscious experience, nor is it purely as a result of rational calculation. It is that people develop strategies of behaviour but these are shaped by their objective situation (Wolfreys, 2000). In a game a field is clearly seen for what it is – 'an arbitrary social construct underlined by everything that maintains its autonomy – explicit rules, delimited time and space and playing involves a quasi-contract' e.g. for 'fair play'. Social fields on the other hand are the products of a long, slow process of absorption and development of unconscious autonomy and are games 'in themselves' not 'for themselves' (Bourdieu, 1990). Multiple fields define the objectified social structures that identify a society. An actor does not consciously embark on the game – they are born into it and actors need habitus to make it work.

Habitus is a system of generative schemes that are durable – inscribed in the social construction of the self and transposable – from one field to another. So Bourdieu can analyse agents as objectively co-ordinated without being the product of rules on the one hand or conscious rationality on the other (Calhoun et al., 1993). The feel for the social game becomes an instinctive part of the make up of individuals via the habitus that becomes a way of behaving based on a sense of what might be achieved (Wolfreys, 2000).

In what Bourdieu describes as the dominant circular path – a causal loop of generation and reproduction, actors internalise the structure of the field as habitus. Habitus in turn generates practice and practices serve to reproduce the structure of the field. Practices are the recognisable patterned actions in which both individuals and groups engage. They are not a mechanical reaction to rules, norms, models but a strategic yet regulated improvisation responding to a dialectical relationship between a specific situation in a field and habitus (Bourdieu, 1973). Practices are generated by dynamically combining past experience, present situation and implicit anticipation of the future consequences of these very actions. Being determined partly by past conditions through habitus, they tend to reproduce the regularities and objective structures of which they are both products. Through the circuit of reproduction the objective relations of the field are produced and reproduced to both reinforce and change the field's objective structure such as class distinctions and schemes of classification (Kvasny and Truex, 2000). Figure 1 serves as an illustration of these points:

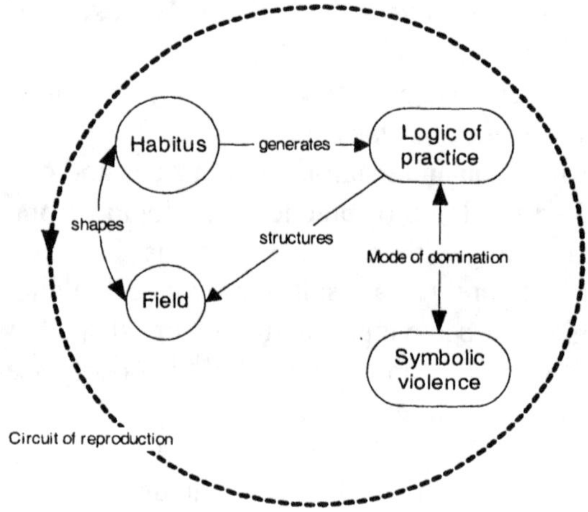

Figure 1: Generation of Practice

Also there is the question of how power relations exist and are maintained alongside the habitus. Domination is maintained in society by means other than direct repression. Society, as stated, is made up of different fields e.g. fields of education, politics, and economics and so on. Within each field people compete for 'capital'. This could, for example, be cultural capital or monetary capital – any capital that is at stake within that particular field. Different forms of capital can be converted to other forms. Once the credentials of this capital becomes generally acknowledged and legitimised then power relations no longer exist between individuals but become objective mechanisms and social institutions that reproduce relations of domination without the need of direct intervention by the dominant group in society (Wolfreys, 2000). The term symbolic violence is coined, being a legitimate call for deference to authority.

Schultze and Boland (2000) have noted three functions of symbolic violence in Bourdieu's analysis - knowledge integration, communication and political domination. Formal controls for example drive behaviours and outcomes through bureaucratic measures. This relates to the knowledge integration function of symbolic violence where written policies, procedures and methodologies are employed to provide a universal ordering and understanding in the social arena (Kvasny and Truex, 2000). Informal control on the other hand involves norms or self-regulation within social classes, for example. This corresponds to communication because controls are embedded deep within codes and shared class meanings. They result from the collective habitus of actors within a social group. The political function means that indirect cultural mechanisms rather than direct coercive social control provide order and social restraint. Symbolic violence is most powerful when it hides the interested nature of managerial actions and when acceptance of an IS by the users serves to perpetuate their domination (Kvasny and Truex, 2000).

5 CASE STUDY

The case material describes call centres in the NW of England. Ironically these are often housed in the old mills and engineering shops once foremost in the industrial revolution but long since idle. That is until the 'virtual' age and it is interesting to consider whether this is indeed the new economic era, or history repeating itself. The area of study has the largest number of call centre jobs in the UK (TUC, 2001). There are many types of call centres including new technology companies with public service centres using 'help desks'; business to business intranets; outsourced services and portal models that rely on attracting maximum users to a web site to drive advertising, negotiate bulk purchasing deals or charge 3rd parties to access customers.

5.1 Mobile Field Study

Rather than taking an organisational or case study model representing examples from each call centre type, this research instead aimed to follow the fortunes of call centre workers as an interpretative and what is described as a mobile field study. In these terms, four call centre workers were involved with the researchers in a two-year study. As the four changed workplaces for a variety of reasons so this enabled aspects of CRM system use in nine different call centres to be analysed (Richardson and Richardson, 2002). In these terms the researchers were mobile, moving as 'shadows' rather than fixing themselves in the organisations and taking snap-shot views.

In practice, we all met up regularly, sometimes engaged in-group discussions, sometimes in one-to-one interviews. Consent was sought for the content and direction of the research not just at the outset but as an on-going concern. Participants supported the research and wanted their voices to be heard and stories told. The powerful tales are of stress, burn out, sacking, job changes and frustration but also of community, solidarity and collective action. Interviews were transcribed and reflexive stories written but always shown and discussed with the individuals concerned.

Call centre work is characterised by temporary, short-term contracts, agency working and casual employment in the UK. The North-West call centre labour force has a large proportion of gay and lesbian workers, women and students reflecting the City Centre locations, low pay and reliance on a bulk of temporary transient staff. Readers may be concerned that a degree of personal detail has been included. However these personal profiles are essential to understand the contribution of culture and history – habitus – to the study of the social relations of IS use. This is one of the strengths of using the conceptual tools of analysis provided by Pierre Bourdieu. Taking a critical research approach means not restricting study to a one-dimensional view of 'the organisation' or 'stakeholders'. Rather the contribution of Bourdieu enables holistic reflection.

The profiles of the interviewees are as follows - names have been changed for reasons of anonymity:

Angela: A woman in her 50's Angela has been working in call centres for nearly 10 years. Most of this has been for a telecommunications organisation in her hometown. She became an example of a highly valued agent, with intensive knowledge of the area and customers (Frenkel et al., 1999) and could give a highly personalised and rich service. Redundancy followed as the organisation transferred to a large, impersonal and centralised service utilising CRM systems to the full. She was alarmed at the change in working conditions finding teamwork and supervision intimidating and the work pressurised and impersonal.

Bill: A young gay man and a dreamer longing for travel and then going back to college. Bill works from call centre agencies but is intolerant of bullying and intimidation from management. If he doesn't like the work or atmosphere, he walks down the road to the next call centre.

Colin: A male supervisor in his 30's in a difficult role at work at one supporting the staff against the more extreme acts of discipline and control and yet under pressure to carry out higher management instructions regarding teamwork, productivity and sanctions.

Denise: A part-time worker and lesbian woman in her early 30's, Denise at the time of the research was also a student on a Business IT postgraduate diploma. She was angered by the 'management claptrap' on the course she studied – high on IT for competitive advantage and low on the consequences for call centre staff.

5.2 Bourdieu and the Logic of Practice in Call Centres

In this section examples from the case material are applied. These are presented under the headings of field, habitus, logic of practice and symbolic violence but as Figure 1 has shown in reality these aspects are strongly interlinked.

5.2.1 Field

Many fields spring to mind when considering the logic of practice in this case study. However the field in focus here is that of CRM use for knowledge management in UK telephone call centres. It is a field fraught with contradictions. Problem areas of CRM system use lie in the limitations of software that assumes business processes are in place. With CRM implementation operational efficiency is also often mistaken for competitiveness and the competing interests of different groups is not recognised or is neglected (Light, 2001).

In the workplaces we studied labour turnover was high. In one centre labour turnover was expected to be 100% in their first year. Scripts and monitoring aspects of CRM software were utilised. Practices like 'hot-desking'[1] were imposed to de-personalise work stations. Workers were generally organised in teams with a team supervisor receiving very little enhanced pay, put there to disseminate orders from the top down.

Management tried various techniques to try to establish control of teams, enhance productivity and instil peer pressure on 'the weakest link'[2]. These included incentives like awarding of food vouchers for improved sales

---

[1] This is the practice whereby workers are not allocated their own work station or work space but sit at the nearest available desk. Individualisation of work space is not permitted.

[2] During team meetings, supervisors encourage analysis of the team performance and identification of individuals who appear to be the weakest link in terms of productivity and so on. Teams were encouraged to assert peer pressure on the individuals identified in order to improve team performance.

figures to the use of punitive actions like imposing star charts to encourage behaviour modification – a technique I have failed to use effectively on many occasions when my children were toddlers!

5.2.2 Habitus

The habitus of our call centre interviewees reflected these aspects of the field. Often in teamwork sessions however orders were side-stepped or ridiculed and had to be dropped. Hot-desking for example failed miserably. Unwritten rules were kept almost unconsciously about who sat where and call centre staff persisted in individualising their work stations with toys and pictures. Yet at the same time the monitoring aspects of CRM system use means that no aspect of the call centre workers day was unaccounted for. They were profiled, listened in to, their opening remarks analysed and wrap-up time and phrases used. The sales content of their conversations, achievement of call numbers and sales targets were scrutinised. Call times were strictly monitored with pressure on for quick 'closure'. Their off-line frequency, intimate toilet habits were considered and discussed at teamwork supervision sessions as were perceived attitudes displayed during conversations with customers. Such is the reality of new ways of working expressed in teamwork and empowerment. As Bill said one day:

*"You can't be early, can't be late, can't go for a sh*t"*

Call centre work can be very stressful and alienating. Calls stream in with always another one waiting. Many workers suffer 'burn-out' and stress driven by the ACD systems, lack of control, remote listening, abusive customers and low pay. Sometimes symptoms would be the inability to stop crying, yet such is the completeness of the CRM system monitoring and control that statistics would be kept of this too. Denise laughed when she remembered:

"There was even a code for running to the restroom to cry"

The habitus of the call centre worker also means coming to terms with uncertain and casual employment. Bourdieu has commented that a feature of work today is generalised and permanent conditions of insecurity and how a constant threat of unemployment, for example, shapes a new type of domination aimed at compelling workers to submission and acceptance of exploitation such as is experienced in call centre teams (Bourdieu, 1999).

Despite facing frequent and sustained verbal abuse from customers call centre staff had to control their responses. This has been described as an aspect of 'emotional labour' where employees have to publicly display an emotion not necessarily felt (Hochschild, 1983). As Denise said:

"I hated being happy, happy all the time – the voice with a smile"

Our interviewees though did enjoy the opportunities for rich communication and dealing with complex issues presented by the customers. Here is when scripts were often bypassed and ignored. Often advice about debt and so on was given beyond that stipulated by the script.

The habitus of the call centre worker is also a story of solidarity and community. Someone was burgled and a collection held to help replace items. Cakes were baked for birthdays. Non-verbal communication was used if a customer became abusive. Then others would listen in and be able to provide comfort after the ordeal. If someone was 'strapped for cash' then their food vouchers won as bonuses were exchanged for money. At one time Bill won £400 worth of vouchers in a national sales competition, although he admitted that this had been achieved by adding items to customer bills and waiting for complaints later! By exchanging these for cash he managed to buy a plane ticket to Jamaica.

Management methods often intimidated, yet solidarity helped in many cases. During the study there was a major official strike against bullying in one of the call centre organisations. This precipitated the Trades Union Congress and call centre organisations to consider good practice (TUC, 2001).

5.2.3 Logic of practice

The importance of the logic of practice as a tool of analysis, as identified earlier, is how lessons can be learned from the past, present and future anticipation of outcomes. Specifically in this case study the logic of practice is the manifestation of the contradictions in the intentions and outcomes of CRM system use in practice.

The logic of practice from CRM system gurus is a story where 'CRM systems provide competitive advantage', 'knowledge can be managed and codified for strategic benefit', 'team work empowers', 'hot-desking will eliminate the garbage of individualisation', 'standardising responses through scripts is effective communication' and 'customer service is an unskilled job, therefore high labour turnover is acceptable, and will not build up future problems'. This study has uncovered a different story. CRM systems are exploitative and work against communication richness; they increase bureaucracy and control rather than embracing empowerment.

CRM and knowledge management needs co-operative skills but of course this development can backfire with a challenging of employers control over the means of production. Knowledge is about experience and as such history is important not just for defining the knowledge held by individuals or organisations but also for understanding the impacts that new IS and change will have – lessons that can be learnt from history. Here is where the habitus comes in 'the habitus, a product of history provides individual and collective practices – more history – in accordance with schemes generated by history' (Bourdieu, 1990). Those who make knowledge a passive recording forget that 'all knowledge is an act of construction and agents far from reacting mechanically to mechanical stimulations, respond to the invitations or threats of a world whose meaning they have helped to produce' (Bourdieu, 1990).

In these terms Bourdieu offers tools of analysis that go beyond the IS constrains or liberates debate and enables a richer understanding and identification of the many organisational and other issues that have intended or unintended consequences.

5.2.4 Symbolic violence

The mode of domination involves the practice of teamwork to exert peer pressure as well as senior management pressure via the team supervisor on the perceived 'weakest link'. Uncertainty and stress leads to an expectation that the job will be short-lived. This habitus results in putting up with the job, responding to supervision or walking down the road to the next call centre rather than resistance. However more crude acts of management oppression spills over into small disputes or even strikes at times. Taking the framework of understanding symbolic violence in organisations outlined earlier by Shultze and Boland (2000) this case study can be analysed as follows:

Knowledge integration aspects involve the use of scripts, rules, codes that even result in codifying responses to stress and also the way team working is used to self-regulate. CRM requires that exact phraseology is imposed for monitoring and analysis purposes. Angela explained that calls were dissected into 4 parts: #1 a welcome and introduction; #2 offering products and services; #3 recap call; #4 positive close. Angela faced difficulties with management for failing to use the exact words "and to recap". Supervisors made her put labels on her monitor with the words on and she had to tick a 'star chart' when she used the words. Her calls were monitored closely and listened in to. Saturdays were colloquially called the 'sacking day'. It was easier and less obvious for management to dismiss someone on a Saturday. One Saturday, Angela was sacked for again using words similar to, but not actually "and to recap".

Communication aspects reveal themselves in the contradictions between the solidarity and community generated by the call centre worker habitus but also the fragile and uncertain nature of the work.

In terms of political domination, management practices were not well concealed. This can be linked to a crude application of CRM software and also a feeling that endless numbers of replacement workers were available. Issues like customer revolts against scripting and automated customer service and a realisation that 100% labour turnover per year is not feasible or profitable may well have impacts on this aspect of domination in the not too distant future.

6 CONCLUSIONS

In this paper I have introduced the critical IS research perspective, specifically drawing on the conceptual tools of analysis of Pierre Bourdieu. In these terms I am following in the footsteps of Schultze and Boland (2000) who applied Bourdieu's key concepts to their study of 'information gatekeepers' and also that of Kvasny and Truex (2000) who established a research agenda using the work of Pierre Bourdieu.

Critical theory has often been criticised for being too theoretical and esoteric (Boudreau, 1997). Moreover critical social theory has had limited exposure in IS research and therefore there is a lack of field experience. This interpretive study examining the social relations of CRM system use in telephone call centres in the UK hopes to make a small contribution to addressing this criticism.

The contribution of Pierre Bourdieu to this study has to be seen in the context of knowledge management in organisational practice. Bourdieu helps us understand the complex historical and cultural factors involved in the social relations of IS use. This comes specifically through analysis of the competing fields, understanding of the role of habitus, contradictions apparent through the application of symbolic violence as a mode of domination and how the relationships between structures and agencies involved manifest themselves in the logic of practice.

Critical social theory suggests that people can change their world (Boudreau, 1997) – there is nothing inevitable about what technology we have and how it is used. Moreover critical social theorists seek to emancipate people and are concerned to find alternatives to existing social conditions as well as challenging taken-for-granted conditions (Alvesson and

Deetz, 2000). In these terms a clear motivation for this research has been to highlight the contradictions between CRM rhetoric and the reality for call centre workers on the 'frontline'. It aims to let their stories be heard.

There are obvious limitations to such a small study in a particular area of the North-West of the UK. CRM use across the organisations concerned would have given further insights. Call centre work is increasingly being outsourced and a particular favourite is setting up centres in developing countries. The dynamics of this social reality would provide a further dimension to this research. As intellectual scaffolding however it is clear that Pierre Bourdieu has provided IS researchers with powerful tools of analysis.

REFERENCES

Adam, A. and Richardson, H. "Feminist Philosophy and IS", *IS Frontiers*, (3:2), 2001, pp. 143-154.

Alvesson, M. and Skõlberg, K. *Reflexive Methodology*, London: SAGE, 1999.

Alvesson, M. and Deetz, S. *Doing Critical Management Research*, London: SAGE, 2000.

Bannister, F. "The Dimension of Time: Interpretive Historiography in Information Systems Research", *Electronic Journal of Business Research Methods*, (1:1), Nov. 2002, pp. 1-10.

Belt, V., Richardson R., Webster, J. "Women's Work in the Information Economy: The Case of Telephone Call Centres", *Information Communication and Society*, (3:3), 2000, pp. 366-385.

Bentley, T. "Knowledge Management? Humbug!", *Management Accounting*, April 1999.

Boudreau, M.-C. "Report on the Panel Discussion Assessing Critical Social Theory Research in IS", *IFIP WG 8.2 Working Conference in Philadelphia*, May 31 to June 3, 1997.

Bourdieu, P. "Contre-Faux" Polity Cambridge cited in Callinicos, A. "Social Theory Put to the Test of Politics: Pierre Bourdieu and Anthony Giddens", *New Left Review*, (I/236), July-August 1999, pp. 77-102.

Bourdieu, P. *The Logic of Practice*, Polity Press, 1990.

Bourdieu, P. "Three Forms of Theoretical Knowledge", *Social Science Information*, (12), 1973, pp. 53-80, cited in Schultz, U. Boland, R.J. "Knowledge Management Technology and the Reproduction of Knowledge Work Practices", *Journal of Strategic IS*, (9), 2000, pp. 193-212.

Calhoun, C., LiPuma, E., Postone, M. (eds.) *Bourdieu Critical Perspectives*, Polity Press, 1993.

Castells, M. *The Rise of the Network Society*, Oxford: Blackwell, 1996.

Ciborra C. U. and Associates (eds.) *From Control to Drift. The Dynamics of Corporate Information Infrastructures*, Oxford: University Press, 2000.

Frenkel, S., Korczynski, M., Shire, K., Tam, M. *On the Front Line: Organisation of Work in the Information Economy*, Cornel University Press, 1999.

Franz, C. R. and Robey, D. "An Investigation of User-Led Systems Design: Rational and Political Perspectives", *Communications of the ACM*, (27:12), 1984, pp. 1202-1209.

Gibbons, M. Limoges, C., Nowotny, H. Schwartzman, S., Scott, P., and Trow, M. *The New Production of Knowledge: The Dynamics of Science and Research in Contemporary Societies*, London: SAGE, 1994.

Hochschild, A. *The Managed Heart: Commercialization of Human Feeling*. University of California Press, 1983. Cited in Belt, V., Richardson R., Webster, J. "Women's Work in the Information Economy: The Case of Telephone Call Centres", *Information Communication and Society*, (3:3), 2000, pp. 366-385.

Kvasny, L. and Truex, D. "Information Technology and the Cultural Reproduction of Social order: a Research Paradigm" in R. Baskerville, J. Stage, J. DeGross (eds.) *Organisational and Social perspectives on IS*, Boston, Mass.: Kluwer Academic Publishers, 2000.

Light, B. "A Review of the Issues Associated with Customer Relationship Management Systems", in *The Proceedings of the 9th European Conference on IS*, 27-29 June, Bled, Slovenia, 2001.

Markus, M. L. "Power, Politics and MIS implementation", *Communications of the ACM*, (26:6), 1983, pp. 430-444.

Orlikowski, W. J. "The Duality of Technology: Rethinking the Concept of Technology in Organizations", in Bryant, C.G.A. and Jary, D. (eds.) *The Contemporary Giddens. Social Theory in a Globalizing Age.* Palgrave Macmillan, 2001.

Richardson, H. and Richardson, K. "Customer Relationship Management Systems and Information Ethics in Call Centres – 'You are the Weakest Link. Goodbye!'", *Australian Journal of Information Systems*, May 2002.

Schultze, U. and Boland, R.J. "Knowledge Management Technology and the Reproduction of Knowledge Work Practices", *Journal of Strategic IS*, (9), 2000, pp. 193-212.

Truex, D. and Ngwenyama, O.K. "Unpacking the Ideology of Postindustrial Team-Based Management", in the *Proceedings of the Conference on 25 years after Braverman's Labor*, Capital and Monopoly, Binghamton, New York, 1998.

TUC "It's your Call. TUC Call Centre Workers Campaign", *Trades Union Congress*, February 2001.

Walsham, G. *Making a World of Difference: IT in a Global Context*, Wiley, 2001.

Wilson M. and Howcroft D. "The Role of Gender in User Resistance and IS Failure", in Baskerville R., Stage J., and DeGross J. (eds.) *Organisational and Social Perspectives on Information Technology*, Boston, Mass.: Kluwer Academic Publishers, 2000, pp. 453-472.

Wolfreys, J. "The Radicalism of Pierre Bourdieu", *SWRecordings 235 Bookmarks*, 2000.

About the Authors

Helen Richardson joined the University of Salford in 1998 after a varied career including working in the field of Social Care and running a Research and Training Unit promoting Positive Action for Women at Work. Her research interests lie in the area of Critical Research in IS particularly looking at Cultures of Consumption and Gender issues in IS. Helen can be reached by e-mail in H.Richardson@salford.ac.uk

6

BACK TO BASICS

Sharing Goals And Developing Trust In Global Virtual Teams

Robert Tucker and Niki Panteli
School of Management, University of Bath, UK

Abstract: There is a widely recognised view that trust is a vital requirement and a 'need to have' quality for effective virtual teams. Despite, however, this acknowledged importance of trust, only limited empirical research exists to date that explores the challenges of creating and developing trust relationships in the global business environment. This paper develops a theoretical framework for conceptualising trust development in this context by taking into account the role of shared goals and power dynamics. Based on data collected on eighteen global virtual teams, we challenge the prevailing assumption that global virtual team members experience swift trust. Within a business environment where conflict and power differentials prevail, building trust is not always a swift process. We find that the process of jointly constructing team goals holds significant value as it may provide the 'glue' to hold team members together long enough to enable trust development.

Keywords: Trust, goals, power, global virtual teams, computer-mediated communication.

1 INTRODUCTION

"The HQ people are making all of the decisions and their goals may be different to ours in Singapore. For example, their goals may be for profit, but they did not discuss this with us. Therefore, when they did certain things it felt as though they were just inflicting their power on us.... Maybe our management in Singapore have told us that our [own] objectives are for cost savings, and this can result in conflict and mistrust within the team" (Global Virtual Team 8, Interviewee G).

The significance of trust as a state of a positive, confident expectation (Cook and Wall, 1980; Currall and Judge, 1995) has been widely recognised in the virtual organization literature. Handy (1995: p.44), for example, has put the point quite succinctly: "Virtuality requires trust to make it work: Technology on its own is not enough". As Lipnack and Stamps also put it (1997: p.225): "in the networks and virtual teams of the Information Age, trust is a 'need to have' quality in productive relationships", whilst according to Platt (1999), trust is essential to any virtual team because these teams do not have everyday interaction and the potential for losing trust is much higher. However, while trust has been identified as a defining feature for virtual organizations (Davidow and Malone, 1993), and while several suggestions have been made for strengthening trust relations within virtual teams, empirical research in this area has remained limited.

In this paper, we aim to cover some of this gap by taking a focus on two factors that we believe are important to our understanding of trust development within the context of global virtual teams: shared goals and power. Even though both shared goals and power dynamics have been recognised as important in developing trust in a virtual team context, these bodies of literature have largely evolved independently of one another.

In what follows, we provide the conceptual foundations of the study and develop a framework that identifies the inter-relationships between shared goals, power and trust. Then the research project conducted for the purpose of the study is described and the methods used for data collection are justified. The results are analysed and discussed using the theoretical framework and their theoretical and practical relevance is identified.

2 CONCEPTUAL FOUNDATIONS

Shared goals are often highlighted as being a key element in the establishment of effective teams. In fact, the literature seems to take for granted that shared goals exist in virtual teams. For example, Lipnack and Stamps (1997: p.7) define virtual teams as "a group of people who interact through inter-dependent tasks guided by a common purpose...". Similarly, Jarvenpaa and Leidner (1999) define a global virtual team as a group of geographically dispersed individuals who work on a joint project or common task. Through the creation of shared goals, groups of people have an inter-relatedness, a shared commitment (Handy, 1995) and a common motivation that adds up to more than just a "bunch of individuals" (Lipnack and Stamps, 2000). These teams exist for some task-oriented purpose, and therefore orientation to task is what distinguishes them from other types of small groups (Lipnack and Stamps, 2000). Shared goals or objectives provide a link between relational and cognitive dimensions (Tsai and Ghoshal, 1998) and can help to provide a sense of shared meaning (Kasper-Fuehrer and Ashkanasy, 2001) or a common business understanding. These goals articulate what the team stands for and the outcomes that they expect or "their shared vision", which can be an important part of a "relational contract" (Kasper-Fuehrer and Ashkanasy, 2001) and an important element of a cooperative relationship management strategy (Allen et al., 2000). It has even been suggested that the best predictors of a virtual team's success are the clarity of its purpose and group's participation in achieving it (Lipnack and Stamps, 2000).

Jarvenpaa and Leidner (1999) have conducted one of the most detailed research projects into studies on trust and virtual teams thus far. Their eight-week study of seventy five teams of university students each consisting of four to six members, highlighted significant differences in the behaviors and strategies between high and low trust teams and supported the existence of swift trust; this type of trust presumes that roles are clear and that each member has a good understanding of others' roles and responsibilities (Meyerson et al, 1996). Their study was initiated with three artificially created, well-articulated tasks for the teams to complete, which, in effect, created shared goals in terms of both the purpose and objectives of these teams. However, whilst this study is comprehensive and provides useful insights, it does have limitations when attempting to apply its findings in a business context where goals are often neither pre-set nor clearly stated.

Furthermore, power differentials, which could influence the degree of inter-dependence among members, are not significant in the case of university students. In business environments, however, power differentials prevail. Power, defined as the capability of one party to exert an influence on another to act in a prescribed manner, is often a function of both dependence and the use of that dependence as leverage (Rassingham, 1999). Indeed, power is an important contextual factor that affects trust (Hart and Saunders, 1997) in that it suggests the existence of a unilateral dependency or an imbalanced relationship (Allen et al., 2000).

Power can take the form of either coercion or persuasion. Coercive power (Allen et al., 2000; Rassingham, 2000) is often apparent when one party possesses a punishment ability. Whilst short-term gains are sometimes available, coercion very much reflects a short-term perspective, which can result in the weaker, more vulnerable, party yielding begrudgingly and engaging in defensive co-operation. This in turn encourages opportunism (Rassingham, 1999; Van der Smagt, 2000) and degrades the relationship (Allen et al., 2000) often into a downward spiral (Rassingham, 1999). Coercion often results in distrust (Allen et al., 2000) and a resultant evasion, deception, and distortion of information. It therefore becomes a significant constraint to relationships, which prevent improvements in coordination, and often results in an attempt by the weaker player to try to escape (Rassingham, 1999). Therefore, whilst coercion can force cooperation (Rousseau at al., 1998), true collaboration requires trust (Kanter, 1994) and as a result coercion is often self-defeating in the long term (Rassingham, 1999). Persuasive power provides a better alternative for enhancing the satisfaction of less powerful partners (Allen et al., 2000; Hart and Saunders, 1997; Rassingham, 1999). It seeks to build trust and helps with the tight coupling of actors with often economic, symbolic and personal benefits (Rassingham, 1999). Whilst persuasion often requires the adoption of a long-term perspective, it is often more expensive and takes time. It also significantly increases the probability of building trust, which occurs when a trading partner is informed about the fullest potential of the relationship (Rassingham, 1999; Hart and Saunders, 1997).

Following from these, we argue that evidence of swift trust has been identified in global virtual teams as this was based on a study that took place within university environments where tasks were well articulated and power differentials were insignificant. In our study, we aim to examine trust development within a global virtual team context taking into account the role of shared goals and the use of coercive and persuasive power. Figure 1 presents a model that identifies the interactions between shared goals, power and trust in this context.

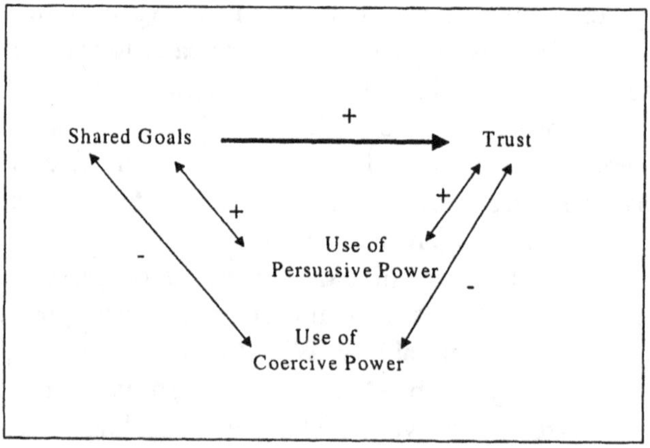

Figure 1. A Model of Shared Goals, Power and Trust

For the purpose of this paper, we have chosen to utilise empirical research that is targeted specifically at exploring the role of shared goals in managing power and influencing trust towards improving the sharing of knowledge within global virtual teams.

3 RESEARCH SITE AND APPROACH

Kappa (a pseudonym) is a company that operates in the high-tech industry and was ranked among the top 20 in Fortune 500 top companies in 2001. It currently serves more than one billion customers worldwide and has an employee base of 150,000 people. Kappa operates as a globally dispersed corporation and makes extensive use of global teams at both intra- and inter-organizational levels. Due to its global characteristics, Kappa was a suitable organization to participate in our study.

In view of the above, the research was specifically focused on gathering information about the effectiveness of Kappa's global virtual teams. Data were collected by interviewing global virtual team members. The criterion sampling tactics (Shaw, 1999), which helped to make decisions on selecting interviewees, were defined as interviewing individuals who: a) were Kappa's employees; b) satisfied the grounded definition of global virtual teams; namely teams that were culturally diverse, geographically dispersed and technology enabled (Jarvenpaa and Leidner, 1999, Lipnack and Stamps, 2000) and c) had worked within a global virtual team for more than two months – thus allowing some exploration of the changes within the team over time.

Whilst many of the interview candidates were from the UK, an attempt was made to vary as many of the aspects of the same as possible. As a result, candidates were selected from across the world and interviews were conducted with employees in Germany, United States, Singapore as well as UK. Similarly, interview candidates were from a range of roles, levels of seniority, cultures, gender, facilitators and involved managers from various business functions across the organization. A total of twelve interviews were conducted and eighteen global virtual team scenarios were explored. The UK interviews were conducted face to face and the remote interviews were conducted via telephone. All the interviews were recorded with the interviewee's agreement and the duration of the interviews ranged from thirty minutes to one hour.

Interviews were organized in a way that encouraged interviewees to recall and reflect on their experiences from working in a global virtual team that they judged to have 'worked well', and conversely their experiences from working in a team that 'did not work well'. This approach resulted in a free flow of data and often produced two case scenarios from each interview. The interviews were guided by prompt questions aimed at exploring the background to the team, the amount of time that the interviewee had worked within the team, the purpose of the team, the frequency of face to face interactions, the distribution of power amongst the team members, the level of trust within the team, and whether and how trust changed over time.

4 RESULTS

The collected data provided information on the experiences and effectiveness of eighteen virtual teams all of which operated at a global level. The majority of these teams were intra-organizational, but four were inter-organizational. All interviewees expected that their teams would continue collaboration in the future. Six of the teams did not work well whilst two other teams improved their effectiveness over time. Table 1 presents a synopsis of the characteristics and experiences of Kappa's virtual teams.

Table 1. Synopsis of Findings: Team Characteristics and Effectiveness

Team	Interviewee	Interviewee location	Inter/Intra organizational teams	Worked well?	Shared goals?	Power differentials?	Trust?	Collaborative?	Swift trust?
1	A	UK	Intra	Y	Y	Y	N-Y	N-Y	N
	B	US		Y	Y	Y	N-Y	N-Y	N
	C	GER		Y	Y	Y	Y	Y	Y
2	D	SGP	Intra	N-Y	N-Y	Y	N-Y	N-Y	N
3	D	SGP	Intra	Y	Y	Y	Y	Y	Y
4	E	UK	Inter	Y	Y	Y	Y	Y	Y
5	E	UK	Intra	N	N	Y	N	N	N
6	F	UK	Inter	Y	Y	Y	Y	Y	Y
7	F	UK	Intra	N	N	Y	N	N	N
8	G	SGP	Intra	Y	Y	Y	Y	Y	N
9	G	SGP	Intra	N	N	Y	N	N	N
10	H	SGP	Inter	Y	Y	Y	N-Y	N-Y	N
11	H	SGP	Inter	N	N	Y	N	N	N
12	I	UK	Intra	Y	Y	Y	Y	Y	N
13	I	UK	Intra	N	N	Y	N	N	N
14	J	UK	Intra	Y	Y	Y	Y	Y	N
15	J	UK	Intra	N	N	Y	N	N	N
16	K	UK	Intra	Y	Y	Y	Y	Y	N
17	K	UK	Intra	N	N	Y	N	N	N
18	L	UK	Intra	N-Y	N-Y	Y	N-Y	N-Y	N

Keys (for Table 1 entries):
Y: Yes
N: No
N-Y: Improved over time
GER: Germany
SGP: Singapore

Table 2 below details the common features and behaviors observed within the global virtual teams described in the interviews and identifies the inter-dependencies between the factors shown in Figure 1. These are explained in the following sections.

Table 2. Differences between High Trust and Low Trust Global Virtual Teams

Hight Trust Global Virtual Teams	***Low Trust Global Virtual Teams***
Factors related to Shared Goals:	**Factors related to Shared Goals:**
*Awareness of shared goals	*Lack of awareness of shared goals
*Take time to build shared goals	*Lack of shared goals
*Open debate for shared goals up front	*Opinions of others not considered
*Team-based goals have primacy	*Individual goals take primacy
Factors related to Power:	**Factors related to Power:**
*Availability of facilitators	*Power battles
*Facilitators focus on win-win	*Coercion
*Recognition of knowledge as power	*Misunderstandings and conflict of interests
*Recognition that power moves; power in many places	*Use of positional power
*Power differentials are minimised	*Perception of 'I have Power'
Communication:	**Communication:**
*Face-to-Face where possible	*Asynchronous CMC
*Regular synchronous CMC	* Time difference matters
*Social interaction	*Little or no social interest

4.1 Shared Goals and Trust

Shared goals were evident in all of the 'high trust' teams and these teams were also found by the interviewees to have worked well. High levels of trust are arguably key to effective communication (Dodgson, 1993) as trust facilitates challenge, debate, learning and innovation, and "improves the quality of dialogue and discussions … [which,] facilitates the sharing of … knowledge" (Ichijo et al, 2000: p.200), and committed relationships. Conversely in all of the scenarios where trust was described as low, shared goals were lacking. In the situations where team members were of the opinion that trust had been broken the level of emotion was high:

> *"At that point in time, because it was my neck on the line, I really felt that this had been done deliberately. You know, I was pretty mad. I was sure it was deliberate. I didn't have anything to back that up but when you are working in a team where things change very quickly, people do think that others hold on to information deliberately. You start to wonder whether their motivation is the same as yours. Whether their motivation has dropped back. When you are not sat next to somebody you can't see how hard they are working or whether they are working on the things that are [more] important to you" (Team 1, Interviewee B).*

A number of the scenarios included situations where facilitators had made use of shared goals to encourage collaboration. In all of these examples shared goals were used to create a higher level or an overriding goal or vision. These goals were focused on the success of the team as a whole, and in some cases attempted to combat the individual or non-complementary goals. An example of this is shown below:

> *"Initially the folks in the US perceived that they had the power and at the same time we felt that we had the power in the UK. So I was very sensitive to that, because if I had acted as though I had the power that would have caused problems. And I was not in a position to alter the power due to the political situation. So I looked for something that was important than our individual political needs, and used the customer as the central focus. This worked well in the majority of situations as everybody could easily relate to meeting the customers' needs" (Team 1, Interviewee A).*

Participants described situations in which facilitators were attempting to both rebuild and improve trust (Teams: 2, 4, 6, 14, 16, 18). In these examples the use of shared goals features prominently:

> *"We had a very definite vision of how we wanted the relationship to work. We were keen to engage and excite the other companies. We gave them an overview of our business and worked hard to try and give them the full picture to create a vision if you like" (Team 4, Interviewee E)*

> *"At the very start of the project the project managers from each company got together and put together a comprehensive contract...It was developed jointly and was very comprehensive. We went through a lot of*

iterative discussions to make sure that the document was extremely well thought out" (Team 6, Interviewee F).

The use of precisely defined, mutually beneficial projects had also been adopted by a number of interviewees in an attempt to begin building trust:

"...By focusing carefully on the things that are hurting both of us, I think we can start to rebuild trust. The new product introduction process reinvention is a good example" (Team 18, Interviewee L). In this scenario, ground rules were included in the shared goals definition sheet, such as "Think [Kappa] First, Front-end, Back-end Second" as well as explicitly listing the expected benefits.

The preciseness with which shared goals are defined is influenced by the levels of dependency and exposure that exist within a relationship. In team 3, shared goals were apparent but had not been defined. This was considered to have been appropriate given that the relationship was seen as a low risk one. Similarly, where the relationships had been devised specifically for the creation of knowledge (Teams 4 and 6) the shared goals were carefully developed.

For example, interviewee F refers to a team that was set up to specify and design a specific integrated circuit for a new product and involved designers in the UK and the US: *"We were detailing the specifications but even this tended to be a back and forth iterative process. At times the specification was modified based on feedback from the designers and at other times the specification would determine the design"* (Team 6, Interviewee F/UK R&D engineer)

These findings support the proposition depicted in Figure 1 that the use of shared goals has a positive influence on the establishment and building of trust. Moreover, they reveal that the *process* of constructing these shared goals, and not just their existence, contributes towards trust development within a virtual team. In what follows, we examine the relationship between power and shared goals.

4.2 Power and Shared Goals

Power differentials were acknowledged in all of the scenarios, even where the teams were considered to have worked well. It was apparent that where the 'more powerful' parties had identified the success of the team as their primary consideration, they consciously minimised the use of coercive power:

"We were very aware that we needed the technical expertise and we only had one company that could provide this. We were therefore very careful

and deliberately tried not to throw up our weight around. We did not want to wield our power (Team 4, Interviewee E).

"[..] because we had a good relationship we were not pushy, we did not force the supplier in any way. I guess some of this is down to our company culture, but not all. Where we have power and need to use it, we do. It was just not necessary in this relationship because we were all working towards the same things" (Team 10, Interviewee H).

Five interviewees described the power within their team as originating from knowledge and noted that at any given point in time the most powerful member was the individual with the most relevant information. In these situations coercive power was rarely used and significant emphasis was placed upon collaboration and the use of persuasive power:

"Power tended to move based on whatever activities were going on at that time. I guess it followed those that were most knowledgeable at any point in time. This is not surprising as the reason we selected the external design company was because of their knowledge" (Team 6, Interviewee F).

In those cases where the use of coercive power was adversely affecting the relationship, it was apparent that shared goals did not appear to exist. The exercising of coercive power was made visible through behaviors such as imposing one's own views on the team, ignoring the views of other team members and adopting the view that the other team members were 'wrong'. Within these situations the range of responses varied: from the 'weaker' team members adopting a more passive involvement or reluctant cooperation, through to more dramatic 'turf battle' examples where individuals appeared to be pursuing their own, often conflicting goals regardless of the negative impact this had on the effectiveness of the team as a whole (e.g. Team 8, Interviewee G, see 'Introduction').

These findings support the view that the use of shared goals can positively influence the use of persuasive power whilst it can discourage the use of coercive power within global virtual teams.

4.3 The Role of CMC in Global Virtual Teams

Many of the interviewees supported the view that face to face interaction was critical. However, it has also been recognised that the opportunities to meet face to face have been severely limited by economic pressures and more recently the terrorist attack on the world trade centre. As an interviewee put it: "*We have a travel freeze at the moment and I haven't met any of the global team for more than a year now*" (Team 1, Interviewee C).

Under these circumstances, it is found that those virtual teams that worked well tend to undertake regular communications via synchronous, 'live' communication technologies such as telephone and Microsoft NetMeeting. Participants confirm that synchronous media offered more feedback and therefore facilitated understanding more effectively than asynchronous technologies such as voicemail and email. The use of asynchronous technologies was, however, regularly used for documenting and recording agreements and providing brief, simple updates to work progress. The teams that worked well were also found to include a social and fun element in their interactions which appeared to have helped in creating a stronger shared social context.

Teams that did not work well or did not experience trust in the initial stages of the project were more likely to identify time zone differences as a drawback in their team's effectiveness. As such, they presented the asynchronous nature of the technology as a constraint to their virtual interactions:

> *"I was assigned to work on this team as the US order management representative....the team included supply chain, order management, business management with representatives form each region. ...The most difficult aspect was the time zone. I am in the UK and the customers are on a different coast to me. So if I get an urgent customer request late in the afternoon, I am unable to get a response from the UK until the next day. This could often appear to the customer as we are being very slow because I would have to wait a whole day before I could get an answer and then really it to the customer" (Team 1, Interviewee B/US order management specialist).*

5 DISCUSSION

Shared goals are and should be a key characteristic of virtual teams. They could provide a means to developing a common sense of identity for team members, which can be of particular benefit to those global virtual teams who meet infrequently or perhaps not at all. These benefits include the establishment of a foundation upon which to build trust and minimise the use of coercive power in pursuit of a collaborative and productive relationship.

However, the study finds that even though shared goals are important for the success of virtual teams, these should not be taken for granted. Indeed, goals may not be shared either because they do not exist at all, team members have not become aware of them, have their own priorities or share different interpretations of the team's role. Furthermore, the study has also shown that the construction of shared goals is often not a one-off activity, but rather it is a process that requires the participation of all parties involved. Though this could be a time consuming, iterative and difficult process, the findings of our study allow us to argue that it is far better to invest in it and as up front in the project as possible than deal with the vicious, destructive downward spirals that result from team members with conflicting goals and poor levels of trust.

In considering power within virtual teams there is an increasing recognition in the literature that knowledge is indeed power and that teams are often formed to create knowledge through combination and exchange. Within these teams, the team member with power at any given time is the one with the most relevant knowledge at that time. Our study found that in high trust teams power differentials do not disappear; rather, power shifts from one member to another throughout the life cycle of a project depending on the stage and requirement of each stage.

Moreover, facilitators are found to have an enabling role in constructing shared goals and minimizing destructive power differentials. The role of a facilitator is to help in team building techniques at the early stage of the virtual work project (Duncan and Panteli, 2001). For example, the facilitator may be valuable in both designing and conducting team sessions but also structuring team discussions by using group graphics/process templates and other forms of collaborative technology. The competence and skills of a facilitator in bringing individuals and organisations together and in encouraging the use of collaborative technologies will foster an atmosphere of collaboration and trust building for the duration of a team project. This becomes more challenging when managing teams in highly-knowledge intensive environments that operate in the global market and we suggest that future research examines this topic more systematically.

This leads us to our final insight on the dynamics of trust development, which highlight the importance of continuous inter-relation and interaction between virtual team members. Within the context of a global virtual team environment, it is within such computer-mediated interactions that members jointly construct their trust development. This joint construction is influenced by the presence of each other and the role of the facilitator in the project. Within a business environment where conflict and power differentials prevail, building trust is not always a swift process, but a focus on shared goals may provide the initial impetus to teamwork and provide the 'glue' to hold team members together long enough to begin the virtuous spiral of building a collaborative relationship.

6 CONCLUSIONS

The study presented in this paper reinforces arguments in the existing literature on the significance and complexity of trust dynamics in building effective virtual teams. It goes further than the existing research, however, to identify and illustrate the significance of shared goals and power in influencing trust development. A model, depicted in figure 1, has been developed to represent the interactions between these three factors. The interrelationships identified within the model have been supported by empirical research data gathered from the experiences of eighteen global virtual teams. In addition to providing support for the model, it has also become apparent that whilst the agreement of shared goals provide a mobilising force for the members of global virtual teams, the process of developing these goals holds significant value in terms of the exchange of information, learning, improving understanding and an opportunity to demonstrate trustworthiness. For this reason, we have argued that trust within a global virtual team context is not always swift. Based on our findings, we challenge the prevailing assumption that global virtual team members experience swift trust.

It is readily acknowledged that what has been attempted here is only an exploration of contingencies to provide a better understanding of trust development within global virtual environment. Further research is required to merit the conceptual and empirical work that is lagging in the existing literature.

REFERENCES

Allen, D., Colligan, D., Finnie, A., and Kern, T. "Trust, Power and Inter-Organisational Information Systems: The Case of the Electronic Trading Community TransLease", *Information Systems Journal*, (10), 2000, pp. 21-40.

Cook, J. and Wall, T. "New Work Attitudes Measures of Trust: Organizational Commitment and Personal Need Fulfilment", *Journal of Occupational Psychology*, (53:1), 1980, pp. 39-52.

Currall, S. and Judge, T. "Measuring Trust Between Organization Boundary Role Persons", *Organization Behaviour and Human Decision Processes*, (64:2), 1995, pp. 151-170.

Davidow, W. H. and Malone, M. S. *The Virtual Organization: Structuring and Revitilizing the Corporation for the 21st Century*, NY.: Harper Business, 1993.

Dodgson, M. "Learning, Trust and Technological Collaboration", *Human Relations*, (46:1), 1993, pp. 77-95.

Duncan, E and Panteli N. "Virtual Team Working: A Design Perspective", in *People in Control - An International Conference on Human Interfaces in Control Rooms, Cockpits and Command Centres*, UMIST, Manchester, June 18-21, 2001.

Handy, C. "Trust and the Virtual Organisation", *Harvard Business Review*, May-June 1995, pp. 40-50.

Hart, P. and Saunders, C. "Power and Trust: Critical Factors in the Adoption and Use of Electronic Data Interchange", *Organization Science*, (8:1), Jan-Feb 1997, pp. 23-42.

Jarvenpaa, S. L. and Leidner, D.E. "Communication and Trust in Global Virtual Teams", *Organization Science*, (10), 1999, pp. 791-815.

Ichijo, K., von Krogh, G. and Nonaka, I. "Knowledge Enablers", in *Knowing In Firms: Understanding, Managing And Measuring Knowledge*, von Krogh, G., Roos, J. & Kleine, D. (eds), London: Sage Publications Ltd, 2000, pp. 173-203.

Kanter, R. "Collaborative Advantage: Successful Partnerships Manage the Relationship, Not Just the Deal", *Harvard Business Review*, July-August 1994, pp. 98-108.

Kasper-Fuehrer, E. C. and Ashkanasy, N. M. "Communicating Trustworthiness and Building Trust in Inter-Organizational Virtual Organizations", *Journal of Management*, (27), 2001, pp. 235-254.

Lipnack, J. and Stamps, J. *Virtual Teams: Reaching Across Space, Time, and Organizations with Technology*, NY: John Wiley & Sons, 1997.

Lipnack, J. and Stamps, J. *Virtual Teams: People Working across Boundaries with Technology*, NY: John Wiley and Sons, 2000.

Meyerson, S., Weick, K. E. and Kramer, R.M. "Swift Trust and Temporary Groups", in *Trust in Organizations: Frontiers of Theory and Research*, R.M. Kramer and T.R. Tyler (Eds), CA, Thousand Oaks: Sage Publications, 1996.

Platt, L. (1999), "Virtual Teaming: Where is Everyone?", *The Journal for Quality and Participation*, (22:5), 1999, pp. 41-43.

Rassingham, P. "Risks in Low Trust Among Trading Partners in Electronic Commerce", *Internet Research: Electronic Networking Applications and Policy*, (10:1), 1999, pp. 56-62.

Rousseau, D. Sitkin S., Burt R and Camerer C. "Not so Different After All: A Cross Discipline View of Trust", *Academy of Management Review*, (23:3), 1998, pp. 393-404.

Shaw, E. "A Guide to the Qualitative Research Process: Evidence from a Small Firm Study", *Qualitative Market Research: An International Journal*, (2:2), 1999, pp. 59-70.

Tsai, W. and Ghoshal, S. "Social Capital and Value Creation: The Role of Intra-Firm Networks", *Academy of Management Journal*, (41:4), August 1998, pp. 464-476.

Van Der Smagt, T. "Enhancing Virtual Teams: Social Relations and Communications Technology", *Industrial Management and Data Systems*, (100:4), 2000, pp. 148-156.

About the Authors

Robert Tucker is a Supply Chain Strategy Development Manager for a global Information Technology corporation. He was born and lives in the UK and holds an Executive MBA from the University of Bath. Rob can be reached via his personal e-mail at robtuckerhome@hotmail.com.

Niki Panteli is a lecturer in Information Systems in the School of Management, University of Bath, UK. She was born in Cyprus and achieved her PhD in Information Systems at Warwick Business School, University of Warwick. Broadly defined, her research lies in the field of information and communication technologies and emergent organizational arrangements. She is currently involved in research on virtuality, virtual organizing and interactivity in computer-mediated communication. Niki can be reached by e-mail at mnsap@management.bath.ac.uk or through her web site at: http://www.bath.ac.uk/~mnsap.

Part 2:

Organizations, Corporations and Institutions

7

HAVE YOU GOT ANYTHING TO DECLARE?

Neo-Colonialism, Information Systems, and the Imposition of Customs and Duties in a Third World Country

Mariyam S. Adam[1] and Michael D. Myers[2]

[1]*Maldives College of Higher Education, Maldives;* [2]*University of Auckland, New Zealand*

Abstract: Many researchers agree that information and communications technologies are deeply implicated in the process of globalization. There is much less agreement, however, on what globalization means in particular situations and contexts. This paper places the development of a specific information system in a Third World country within the context of neo-colonialism and post-colonialism. It shows how the implementation of an integrated software package, developed by United Nations Conference on Trade and Development, imposed its own cultural logic on the Maldives Customs Service. This paper can be seen as one response to the call to study particular individuals, groups, organizations, or societies in detail, and in the context of globalization.

Keywords: Culture, customs, information systems, IS implementation, interpretive case study, colonialism, neo-colonialism, post-colonialism, Maldives, globalization.

1 INTRODUCTION

Over the past decade there has been increasing interest in the IS research literature in the subject of globalization (Gallupe and Tan, 1999; Ives and Jarvenpaa, 1991). Many researchers agree that information and communications technologies (ICT) are deeply implicated in the process of globalization. This process includes the increasing interconnectedness of different societies, the compression of time and space, and an intensification of consciousness of the world as a whole (Walsham, 2001).

There is much less agreement, however, on what globalization means in particular situations and contexts. Walsham (2002) says that, for some, globalization and the spread of ICT throughout the world means the increasing homogenization of culture. This is sometimes referred to as the McDonalds effect. For others, globalization is interpreted as a process whereby imported ideas, concepts and artefacts are "indigenized". This is sometimes referred to as the process of "globalization", where local cultures and societies adapt and modify imported ideas (Robertson, 1992). In the latter case, IS researchers have focused on the diversity and distinctiveness of particular cultures and how people adapt ICT for their own purposes (e.g. Korpela et al., 2000; Walsham, 2002).

While we do not disagree with the idea that local people in local contexts appropriate ICT to suit themselves, we suggest that ICT developments in the Third World need to be seen in a much broader context. The issues associated with colonialism, neo-colonialism and post-colonialism in the Third World have been discussed extensively in anthropology, sociology, history and political science (e.g. see Ashcroft et al., 1989; Ashcroft et al., 1995; Ashcroft et al., 1998), yet there has been little if any discussion of ICT in these terms in the IS field. Most IS research in "developing countries" has focused more narrowly on the impact and use of ICT on economic development or public administration (Bhatnagar, 2000; Fleming, 2002). One of the primary contributions of this paper, therefore, is to introduce this discourse into the IS research arena.

A secondary contribution of this paper is to provide one empirical example of IS development and use in a specific organization in a Third World country, in the context of globalization. This paper shows how the implementation of an integrated software package, developed by United Nations Conference on Trade and Development, imposed its own cultural logic on the Maldives Customs Service. The research method used was that of the interpretive in-depth case study following the principles suggested by Klein and Myers (1999).

The paper proceeds as follows. The following section discusses some of the issues associated with colonialism, neo-colonialism and post-colonialism in the Third World. Section 3 describes the interpretive case study research method that was used. In the fourth section, the empirical evidence from the case study in the Maldives is discussed. Section five presents an analysis of the case. The final section is the discussion and conclusions.

2 NEO-COLONIALISM AND POST-COLONIALISM IN THE THIRD WORLD

Myers and Tan (2002) note that much of the literature concerned with cultural and cross-cultural issues in the IS field has relied on Hofstede's proposed dimensions of national culture. They analyzed 36 studies from the cross-cultural IS literature, and found that 24 of these used some or all of Hofstede's dimensions. However, Myers and Tan (2002) suggest that the very concept of 'national culture' is problematic. They argue that there is no necessary alignment between culture and the nation-state, and propose a more dynamic view of the relationship between culture and information systems in a global context – one that sees culture as contested, temporal and emergent.

In a recent article, Walsham (2002) takes this argument further. He proposes a structurational analysis based on Giddens (Giddens, 1979; Giddens, 1984) that can accommodate elements such as the links between structural contradiction and conflict, cultural heterogeneity, an analysis of detailed work patterns, and the dynamic and emergent nature of culture. He says that his analysis "lies squarely within the literature which considers that globalization, facilitated by ICTs, is not leading to simple homogeneity of culture and approach". He argues that, whilst technologies such as GIS have features that reflect their cultural origins, "technology has a degree of 'interpretive flexibility' (Pinch and Bijker, 1987), and can be adapted and used in different ways" (Walsham, 2002: p.378).

Generally speaking, we agree with the arguments offered by Myers and Tan (2002) and Walsham (2002). We agree that culture is dynamic and emergent, and that information and communications technologies are appropriated by people in local contexts to suit themselves.

However, we suggest that it is instructive to see ICT developments in the Third World within a much broader context. This context is the long-standing historical relation of colonialism, neo-colonialism and post-colonialism in the Third World (Ashcroft et al., 1989). We will now define our usage of these terms (although we acknowledge that there is much debate in the literature about the precise meanings of these terms).

We use the term "Third World" as it is used in the classic sociological text by Worsley (1970). Worsley defines the Third World as including all those countries in Africa, Asia and Latin America that were colonised. In his words:

> *What the Third World originally was, then, is clear: it was the non-aligned world. It was also a world of poor countries. Their poverty was the outcome of a more fundamental identity: that they had all been colonised (Worsley, 1970).*

In this definition, we note that the term "Third World" does not imply any kind of economic pecking order (first, second, third). Rather, it is a political phrase describing a fundamental historical fact: that all the countries of the Third World were, at one time or another, colonised by the West (Hudfor, 1993). Although many prefer the words "development" and "underdevelopment" instead (e.g. see Bhatnagar, 2000), Hudfor (1993) notes that many people in these countries prefer to speak of themselves as part of the Third World. In our case, this describes exactly the preference of one of the authors of this paper, who is herself a citizen of the Maldives. The Maldives was a British Protectorate until 1965.

We use Loomba's definition of the word "colonialism" as follows: "The conquest and control of other people's land and goods", e.g. British colonialism (Loomba, 1998). It describes a system of political, economic, and social dominance over the colony by another country. Neo-colonialism is "A state whereby while formal political independence has been granted to a former colony, it remains economically dependent and subordinate to its original imperial power, or more commonly, to the world economy at large" (Loomba, 1998). The main distinguishing feature of neo-colonialism is formal political independence but continued economic dependence. Post-colonialism refers to the ways in which colonialism continues to affect the former colonies after political independence. It describes a period where the official colonial structure has ceased to exist yet colonial influences persist in many different ways.

We can see from these definitions that there is a continuum from colonialism through to post-colonialism, where the colonial influences become less overt over time. However, many scholars argue that the influences in the post-colonial era are still very powerful. These influences can be of a social and cultural nature rather than overtly political or economic. Globalization is often seen in this context as one of the many powerful post-colonial forces influencing Third World countries (Ashcroft et al., 1989).

The research project reported on here examined the development of a new information system for the Maldives Customs Service (MCS) within the broader context of neo-colonialism and post-colonialism. We examined how European social and cultural values were imported along with the importation of the software itself. Thus our study was designed to document the cultural changes that occurred when this new system was introduced. We suggest that these cultural changes are perhaps one of the most important aspects of the use of ICT in a global context. Our study adds to the growing body of literature on the development, use and impact of ICT in developing countries (Avgerou, 1990; Bhatnagar and Bjorn-Anderson, 1990; Cain, 1996; Mundy, 1996).

3 RESEARCH METHOD

As was stated earlier, the research method used was that of the interpretive in-depth case study (Myers, 1997a; Myers, 1997b; Walsham,

1995a; Walsham, 1995b). In using this method we followed the principles suggested by Klein and Myers (1999). We considered the interpretive case study to be the most appropriate method since it allowed the researchers to study the richness of the social and cultural changes that occurred when the new information system was introduced. The second principle proposed by Klein and Myers (1999) – the principle of contextualization – was particularly appropriate to this research project. This principle "requires critical reflection of the social and historical background of the research setting, so that the intended audience can see how the current situation under investigation emerged" (Klein and Myers, 1999: p.72). This describes exactly what we were seeking to do in this research project, to reflect on the neo-colonial and post-colonial influences in the research setting.

The case study research was conducted by one of the authors over a three-month period in 1999. Data were obtained from interviews and documentary sources. A total of 27 semi-structured interviews were carried out with all the key players in the project within the Maldives Customs Service. These interviews were carried out with personnel at different departments or sections including Information Technology and Statistics, Documentation, Cargo, Valuation, Tariff and Classification, Passenger Terminal, and Finance, and at different hierarchical levels. One interview was also conducted with an executive of the Ministry of Communication, Science and Technology. Additionally, 10 informal interviews were conducted with people who receive services from MCS. These people included exporters, large importers and small local traders.

All interviews within MCS lasted from 45 minutes to two hours. All interviews were tape-recorded. The informal interviews with the service receivers lasted from 15 to 30 minutes. Since much of the data is of a sensitive nature, real names have not been used for anyone who participated in the project.

The documents obtained included letters, proposals, organisational charts, reports, internal and external documents, and newspaper clippings. Additionally, there was substantial informal contact with the project director and other senior members of the technical team and system user group during the research project.

4 CASE STUDY: THE IMPLEMENTATION OF SYCUDA AT THE MALDIVES CUSTOMS SERVICE

This case study concerns the implementation of an information system at the Maldives Customs Service (MCS) in Male', Maldives. MCS is responsible for the administration of all the activities related to the collection of import and export revenues. In addition, MCS is solely responsible for the control of restricted and banned imports, the checking of vessels, the compilation of trade data, and dissemination of trade statistics to various authorities and the trading community.

The IS project in question involved the implementation of a generic Customs software package in the 1990s, namely, Automated System for Customs Data (ASYCUDA).

4.1 The Maldives

The Republic of Maldives is a group of 1190 small coral islands, 200 of which are inhabited. It is located in the Indian Ocean not far from India and Sri Lanka. The population of Maldives was estimated in 1997 at 258,678, more than a quarter of whom live in the capital, Male' – the largest city of Maldives.

The Maldives was for over 800 years ruled by a series of dynasties that continued until 1965 (Ellis and Amarasinghe, 1997). From 1887 to 1965 the Maldives was a British Protectorate. The Maldives became a fully independent nation on July 26, 1965, and a Republic on November 11, 1968. As a Third World country, the Maldives pursues a policy of non-alignment in international affairs (Maloney, 1980).

4.2 Culture and Processes in the 1980s

In the late 1980s (before ASYCUDA was introduced) the process of verifying and processing documents was a paper-based manual process.

The valuation process, the first step in the cargo clearance, determines the value of imported goods subject to tariffs. This was the most tedious and the main cause of delays in processing declaration documents. First, traders had to submit various trade declaration documents, such as a declaration form, commercial invoice, certificate of value and origin, import permit, insurance certificate, and bill of lading to the respective entry/exit point. However, before submitting the trade declaration documents to the MCS, all traders had to obtain import/export licences from the Ministry of Trade and Industries (MTI), and other licences from the appropriate controlling ministries. For instance, a trader could not import alcoholic beverages without a liquor permit from the MTI.

Once the declaration was submitted, three different officers would take responsibility for ensuring that all the necessary documents including licences were in order, and that the importer had enough quota value remaining in the import licence. Failure to meet the criteria meant a halt in the documentation processing until the importer had renewed the quota. The valuation process began only after all the declaration documents had been accepted.

Unlike some countries that accept invoice prices as the basis for valuation, MCS used a so-called 'market value' based on previous records or the intuition of the officers working at the Valuation Section. If these officers were unable to determine the value of certain goods, it was then decided by a Valuation Committee. This committee was composed of MCS staff and members from other related government ministries and departments. The need for the Valuation Committee arose because traders often submitted invoices, which had prices far below the actual cost (in an attempt to deceive customs staff and reduce the applicable tariff).

As can be seen, the valuation process was rather subjective. The subjectivity of the matter often led to misconduct among traders, corrupt behaviour among valuation officers, and lengthy arguments between importers and MCS officers. One small trader described the situation in the early 1980s as follows:

> *Customs was a mess - A real headache and a place where graft was the norm of the day. It was impossible to clear anything without going through unwritten regulations all applicable to the whim and fancies of the individual customs officers (Interview with small trader).*

After the valuation process, the goods were classified in terms of the tariff, and duty was affixed. However, classifying the goods was not easy. The nature of the problem is described by one of our interviewees as follows:

> *There are certain items that would fit in two or more different tariff categories. For example, let's say musical instruments. They are rated at 50% in the coding system specified in the import/export law, and all those products related to recording are rated at 5%. These products would be hard to classify because people use mikes [microphone] for recording purposes as well as for singing.*

The problems were compounded when individual traders declared 'hundreds of items' which could be classified into several different categories. A trader might import, in the same consignment, a variety of products such as chemicals, food, and hardware items.

Once the documents were processed and the duty amount had been calculated and affixed, the forms were sent to officers at the Examination Section. The officers scrutinised the documents to verify that the goods did not come within the provision of the restricted or prohibited import law. Since these items were subject to different interpretations, vigorous examination of each item was necessary to ensure that none had import restrictions placed on them. This thorough examination was very time-consuming and caused major congestion at points of entry and exit.

4.3 Project Initiation

In the late 1980s MCS was invited to participate as an observer at a meeting of the World Customs Organization held in Washington DC, USA. During the meeting a Customs software system, ASYCUDA, was introduced to all participants. It was an integrated software system developed by the United Nations Conference on Trade and Development (UNCTAD) to help Customs automate clearance procedures. (As an aside, by the end of 1999 more than 76 countries had adopted this software for trade administration purposes). UNCTAD said that ASYCUDA's functional design was based on the Kyoto Convention, and was modular, adaptable and configurable.

This system impressed the Maldives delegation and on return to the country the idea of adopting it was recommended to senior management of MCS. Approval from higher authorities was sought for its acquisition.

4.4 The Implementation of ASYCUDA

In early 1992 the ASYCUDA project was initiated with the provision of more than half of the project funding by the UNDP, and the rest by the Maldives government. A project national team was formed, an ASYCUDA regional expert was brought in to MCS, and the ASYCUDA software was purchased with technical support to be provided by UNCTAD.

The project was scheduled to be completed in late 1993, however there were endless problems and difficulties. For example, the ASYCUDA coding scheme did not match the coding system identified in the country's import/export law. An amended coding scheme was proposed, but progress was slow. When the regional ASYCUDA expert left in early 1993, much of the system work was incomplete and the amended coding system was yet to be approved. Another problem that emerged was the requirement of UNCTAD for ASYCUDA adopting countries to have a Single Administrative Document (SAD). Following complaints from various traders, a change in the design of the SAD was made.

Despite the problems and difficulties, in April 1994 MCS was able to implement a working system. The system went live at both the head office in Male' and the Cargo Section at the International Airport.

4.5 The Consequences of ASYCUDA

By late 1994 the effects of ASYCUDA's performance were beginning to be noticed. The processing time for trade documents had been reduced from a period of 45 minutes to 7 days to as little as 15 minutes. The collection of data and the generation of corresponding statistics also improved significantly. Although there were several remaining problems, its greatest benefit was seen in the rational and structured method it afforded users to perform daily work activities.

In 1995 MCS became a fully-fledged member of the World Customs Organisation (WCO). WCO requires its member countries to have certain standards in a number of areas. For example, it is obligatory for all member countries to adopt WCO's Harmonised Commodity Description and Coding System (HS) – a standard product classification and nomenclature system.

4.6 Migration to A++

In 1996 some severe problems were experienced with the ASYCUDA system. These problems were related to the outdated operating system software, poor response time, outdated hardware, and limited system support (UNCTAD was the sole supporter of advanced training for the operating system). As of 1996, only one person amongst the IT staff had a tertiary level educational qualification. Almost all the IT staff lacked higher education in computer usage and maintenance.

In 1997 the UN interregional advisor made a visit to MCS and concluded that the MCS ASYCUDA system was out of date. He recommended that MCS should migrate to the latest version of the software, called ASYCUDA++, or A++ for short. Subsequently in that same year a three-year project document was signed between the Government of Maldives and UNDP. MCS embarked on an ambitious reform program, entitled "Institutional Strengthening of Customs Data Processing System". The overall objective of the project was to provide an efficient service to the public, the trading community and the relevant government authorities.

However, there were numerous delays in getting the project started. One of the main problems was UNCTAD's unwillingness to provide the software without sufficient funding for technical support and software development. It was only in November 1998 that UNCTAD signed an agreement with MCS for the supply of the new software, after funding was obtained by the Maldives Government from the UN's LDC (Least Developed Countries) trust funds.

4.7 The Implementation of A++

In April 1999 a project team comprising 10 members was established with responsibility for the implementation of the A++ system. In the same month a project Steering Committee of 15 people from related public and private sector organisations was formed to advise and assist MCS in the project implementation, and to ensure that the project met the expectations of all stakeholder groups.

During the project two processes took place almost concurrently. First, since A++ required a single administrative document or SAD, the project team redesigned the declaration form in compliance with A++ SAD. This was then proposed for approval from various parties such as the Steering Committee, the concerned government authorities, UNCTAD and the trading community. Second, a new Customs Act was formulated which emphasised full utilisation of the WCO's Harmonised Commodity Description and Coding System (HS). That is, the previously modified HS in use was revised, and necessary changes were made to comply with the A++ HS and included in the Customs Act. The system, with the declaration and manifest modules in operation, went live on 9 September 1999.

5 CASE ANALYSIS

We have seen that the adoption and implementation of the ASYCUDA and A++ system for the Maldives Customs Service was but one part of a global effort to automate and standardize customs clearance procedures. The A++ software was developed by an arm of the United Nations (UNCTAD) and by the end of 1999 more than 76 countries had adopted this software. In adopting this software, however, MCS had to change its work practices and procedures, and even had to change the law of the country. This was done in order to "harmonize" the procedures followed in the Maldives with those recommended by the UN. In addition, many social and cultural changes were mandated as a consequence of adopting this system.

Before the introduction of ASYCUDA, the valuation process was very subjective and individual customs officers exerted considerable personal power over the traders and the valuation process. With low pay relative to the private sector, most employees in MCS were relatively uneducated (by Western standards).

The culture of MCS was for MCS staff to do everything in their power to please their superiors and in particular those from the President's Office or other government "super-powers" such as the MTI and Ministry of Finance and Treasury. Senior officials in these "super-power" organisations would often act as if they were above the law and demand special treatment from those organisations such as MCS that they considered to be inferior to them. As one MCS employee explained

> *Nobody wants to follow the procedure. . . Even the government authorities [don't want to follow the rules and regulations]. . . A government company requested us to release their cargo on the basis*

> *that this cargo is particularly for transport . . . (Interview with MCS customs officer).*

All government organisations are under the direct control of the President's Office, where the President has the constitutional power to overrule the regulations. For example, the president has the authority to reduce or to exempt goods imported from duty, as long as these are for personal use.

All of these issues go some way to explain why the project team had such a hard time in configuring the ASYCUDA system to comply with the tariff coding system identified in the country's import/export law. The system could not be used to its full potential until the import/export law was changed to fit ASYCUDA's requirements. However, "harmonizing" the Commodity Description and Coding System (HS) was as much of a cultural change within MCS as it was a change in the law. The new system required that the personal and subjective way in which MCS had operated before had to give way to more impersonal and "objective" formalized rules of procedure.

For example, in the past there was considerable difficulty in classifying a product according to an appropriate tariff category. This involved a certain amount of subjectivity and guess work. However, the new system did not allow guess work. The appropriate tariff category was now determined by the system.

As another example, the ASYCUDA system did not allow MCS staff to please their superiors from the President's Office or other government "super-powers". The introduction of the newly designed Maldives Single Administrative Document and changes to the import/export law meant that individual favors were ruled out.

Just like the GIS technology described by Walsham and Sahay (1999), the ASYCUDA software can be described as having the societal interests and attitudes of its Western developers inscribed within it. This is because ASYCUDA was designed and developed in terms of the WCO's standards for Customs procedures. Implementing ASYCUDA required the developers, users and management of MCS to take on the attitudes and culture embedded within the system.

For example, some of the dominant features of Maldivian culture include placing more importance on family and kinship ties, and respecting and obeying one's superiors. Maldivians respect the chief of an atoll, the head of an organization, and especially the President of the country. However, in order to implement the ASYCUDA system, MCS had to place more value on written rules and regulations instead of the 'word' of the leader. Work procedures were documented and then standardized so as to make them conform to the software. All MCS staff, traders, and other interested parties were forced to accept and follow this new set of standardized procedures.

6 DISCUSSION AND CONCLUSIONS

There were significant benefits resulting from the use of the new system. One benefit is that the new system helped MCS to analyze the data and produce statistics in a consistent manner. Another is that the system actually gave more power and recognition to MCS within the Maldives Government. A third benefit is that corruption was effectively stamped out within MCS as government officials were no longer given discretionary power within ASYCUDA.

These positive benefits do not alter the fact, however, that there were powerful economic, social and cultural influences affecting the development of this system within the Maldives Customs Service. These neo-colonial and post-colonial influences included the continued dependence of the Maldives on external sources of funding, and the software supplier (an agency of the United Nations) mandating how the second version of the software should be installed. Further, MCS had to change its work practices and procedures, and even had to change the law of the country. These changes were made in order to "harmonize" the procedures followed in the Maldives with those recommended by the UN and the World Customs Organization.

Returning to our earlier discussion of "globalization", we can say that local people in the Maldives appropriated ICT to suit themselves. The first version of the software in particular was modified to suit the requirements of the Maldives Customs Service. However, the successful deployment of this software meant that MCS was now "locked into" and dependent upon this software. When serious problems emerged with the ASYCUDA system in 1996, the most logical solution was to upgrade and migrate to A++. Migrating to this new system, however, required that many of the modifications made to the first version of the software be discarded. The end result is that the processes and procedures used by the Maldives Customs Service were "harmonized" with those recommended by the United Nations.

We can see that many social and cultural changes occurred as a consequence of adopting this system. In retrospect not only were customs and duties imposed by MCS on importers and exporters, but many social and cultural changes were imposed by UNCTAD and the WCO on MCS. This was globalization facilitated by software.

We freely acknowledge one limitation of this study i.e. only one Third World country was studied. However, we believe our findings may be relevant to the study of information systems in other Third World countries. We suggest there is a need for more in-depth research into the social and cultural implications of information systems in these countries. We especially recommend further research of a historical or longitudinal nature looking at the neo-colonial and post-colonial influences on globalization in other Third World countries.

ACKNOWLEDGEMENTS

We would like to express our sincere appreciation to the staff at the Maldives Customs Service who participated in the data collection phase of this study. We are especially grateful to the Executive Director of the Maldives Customs Service and the National Project Director of the ASYCUDA project. We also wish to thank the New Zealand Ministry of Foreign Affairs and Trade for providing financial assistance.

REFERENCES

Ashcroft, W. D., Griffith, G., and Tiffin, H. (eds.). *The Empire Writes Back: Theory and Practice in Post-Colonial Literatures*, London: Routledge, 1989.

Ashcroft, W. D., Griffith, G., and Tiffin, H. (eds.). *The Post-Colonial Studies Reader*, London: Routledge, 1995.

Ashcroft, W. D., Griffith, G., and Tiffin, H. (eds.). *Key Concepts in Post-Colonial Studies*, London: Routledge, 1998.

Avgerou, C. "Computer-Based Information Systems and Modernisation of Public Administration in Developing Countries", in *Information Technology in Developing Countries*, N. Bjorn-Anderson (ed.), Amsterdam: Elsevier Science Publishers, 1990, pp. 243-250.

Bhatnagar, S. "Social Implications of Information and Communication Technology in Developing Countries: Lessons from Asian Success Stories", *Electronic Journal on Information Systems in Developing Countries*, (1:4), 2000, pp. 1-10.

Bhatnagar, S. C., and Bjorn-Anderson, N. *Information Technology in Developing Countries*, Amsterdam: Elsevier Science Publishers, 1990.

Cain, P. "Making the Transition to the Electronic Age: Managing Electronic and Paper Records as a Strategic Resource for Good Government in Developing Countries", *Information Technology for Development*, (7:4), 1996, pp. 159-167.

Ellis, R., and Amarasinghe, G. *A Maldives Celebration*, Singapore. Times, Singapore, 1997.

Fleming, S. "Information and Communication Technologies (ICTs) and Democracy Development in the South: Potential and Current Reality", *Electronic Journal on Information Systems in Developing Countries*, (10:3), 2002, pp. 1-10.

Gallupe, R. B., and Tan, F. B. "A Research Manifesto for Global Information Management", *Journal of Global Information Management*, (7:3), 1999, pp. 5-18.

Giddens, A. *Central Problems in Social Theory*, Basingstoke, UK.: Macmillan, 1979.

Giddens, A. *The Constitution of Society: Outline of a Theory of Structure*, Berkeley. University of California Press, 1984.

Hudfor, K. B. *Dictionary of the Third World Terms*, London: Penguin, 1993.

Ives, B., and Jarvenpaa, S. L. "Applications of Global Information Technology: Key Issues for Management", *MIS Quarterly*, (15:1), 1991, pp. 32-49.

Klein, H.K., and Myers, M.D. "A Set of Principles for Conducting and Evaluating Interpretive Field Studies in Information Systems", *MIS Quarterly*, (23:1), 1999, pp. 67-93.

Korpela, M., Soriyan, H. A., Olufokunbi, K. C., and Mursu, A. "Made-In-Nigeria Systems Development Methodologies: An Action Research Project in the Health Sector", in *Information Technology in Context: Implementing Systems in the Developing World*, G. Walsham (ed.), Aldershot: Ashgate Publishing, 2000, pp. 134-152.

Loomba, A. *Colonialism/Postcolonialism*, London: Routledge, 1998.

Maloney, C. *People of the Maldive Islands*, Bombay: Orient Longman, 1980.

Mundy, D. "IT in Developing Countries: A Loss of Independence", in *Information Technology, Development and Policy*, M. J. Blaire (ed.), Avebury: Ashgate Publishing, 1996, pp. 27-45.

Myers, M. D. "Interpretive Research Methods in Information Systems" in *Information Systems: An Emerging Discipline*, F. Stowell (ed.), London: McGraw Hill, 1997a, pp. 239-266.

Myers, M. D. "Qualitative Research in Information Systems", *MIS Quarterly*, (21:2), 1997b, pp. 241-242.

Myers, M. D., and Tan, F. "Beyond Models of National Culture in Information Systems Research", *Journal of Global Information Management*, (10:1), 2002, pp. 24-32.

Robertson, R. *Globalization: Social Theory and Global Culture*, London: Sage, 1992.

Walsham, G. "The Emergence of Interpretivism in IS Research", *Information Systems Research*, (6:4), 1995a, pp. 376-394.

Walsham, G. "Interpretive Case Studies in IS Research: Nature and Method", *European Journal of Information Systems*, (4:2), 1995b, pp. 74-81.

Walsham, G. *Making a World of Difference: IT in a Global Context*, Chichester: Wiley, 2001.

Walsham, G. "Cross-Cultural Software Production and Use: A Structurational Analysis", *MIS Quarterly*, (26:3), 2002, pp. 359-380.

Walsham, G., and Sahay, S. "GIS for District-Level Administration in India: Problems and Opportunities", *MIS Quarterly*, (23:1), 1999, pp. 39-65.

Worsley, P. *The Third World*, Chicago: The University of Chicago Press, 1970.

About the Authors

Mariyam S. Adam is the Acting Dean of the Faculty Management and Computing of the Maldives College of Higher Education, Male', Maldives. She was born in Male', Maldives. She did her Bachelor of Information Science degree at the University of Newcastle, Australia and achieved her Master of Commerce in Management Science and Information Systems (Hons) at the University of Auckland. Her research interests include information systems in developing countries and information technology in education. She can be reached by e-mail at mariyamadam@yahoo.co.uk.

Michael D. Myers is Professor of Information Systems in the Department of Management Science and Information Systems at the University of Auckland, New Zealand. He currently serves as Senior Editor of *MIS Quarterly,* Associate Editor of *Information Systems* Research, Editor of the *ISWorld Section on Qualitative Research,* and Editor of the *University of Auckland Business Review*. He also serves as Vice-Chair of IFIP Working

Group 8.2. His research articles have been published in many journals and books. He won the Best Paper award (with Heinz Klein) for the most outstanding paper published in *MIS Quarterly* in 1999. He also won the Best Paper Award (with Lynda Harvey) for the best paper published in *Information Technology & People* in 1997. He can be reached by e-mail at m.myers@auckland.ac.nz. .

8

INNOVATION THEORIES IN RETROSPECT

The Case of Electronic Commerce Adoption in Small Business in New Zealand

Nabeel A. Y. Al-Qirim

Auckland University of Technology (AUT), New Zealand; Deakin University, Australia

Abstract: A review of the technological innovation adoption literature on small to medium-sized enterprises (SMEs) provides useful insights into factors influencing innovation adoption but points to the need to introduce more determinants of innovation adoption to SMEs research. This research is interested in identifying these factors and hence, introducing more potential determinants to electronic commerce (EC) adoption research in SMEs. Therefore, this research attempts to extend the technological innovation theories to EC adoption research in SMEs by identifying potential constructs and factors from these theories and then checking their face validity using three case studies in New Zealand. This research endeavours to shortlist and discuss the most important determinants of EC adoption and to eliminate the least relevant ones.

Keywords: Innovation theories, electronic commerce adoption, SMEs, determinants of adoption, case studies.

1 INTRODUCTION AND JUSTIFICATION

The adoption of information technologies by individuals and organisations is part of the process of information systems implementation, which is a prominent area in IS research (Kwon and Zmud, 1987; Moore and Benbasat, 1991; 1996). Measuring potential adopters' perceptions of information technology innovation has been termed a "classical issue" in the innovation diffusion literature and a "potential key" for integrating different findings within diffusion research (Kwon and Zmud, 1987).

In view of the innovation adoption research in small business, the available research provides valuable insights into the adoption criteria of IS in small business (Premkumar and Roberts, 1999; Thong 1999; Thong and Yap, 1995; 1996). Therefore, extending the innovation adoption factors implemented earlier in countries such as the U.S and Singapore to new country-settings such as New Zealand would highlight significant features that are unique to the adoption culture of that country (Thong, 1999). The uniqueness of the New Zealand perspective however, stems from the fact that 84 percent of the New Zealand sector is dominated by micro enterprises employing up to five employees only (MOED, 2000), from the geographical isolation and from the time differences which separates New Zealand from the rest of the northern hemisphere, and from the small population in New Zealand (3.82 million) (NZStat, 2001). On the other hand, past research found that facilitation factors vary according to the innovation type (Swanson, 1994). Therefore, extending the prior innovation theories in small to medium sized enterprises (SMEs) to eCommerce (EC) adoption is unwarranted as EC introduces features that are unique to its perspective, which may not necessarily relate to IS as such. Features such as the external focus of EC with suppliers and buyers, security and legal concerns are some of the features that are unique to EC only.

2 RESEARCH PROBLEM AND OBJECTIVES

In supplementing factors developed in earlier IS adoption research in small business different researchers have highlighted the importance of introducing and testing the significance of new variables on innovation adoption (Fichman, 1992; Thong, 1999). Accordingly, this research is confronted with the following question: *How can factors extracted from the innovation literature influence EC adoption in SMEs?* That is, what are the most probable determinants of innovation adoption and then, how can those determinants explain EC adoption in SMEs? Therefore, the research objectives are to introduce additional determinants of innovation adoption and then to discuss their significance on EC adoption in SMEs in New Zealand. Adopting such an approach in this research could contribute significantly to the technological innovation literature, the development of appropriate factors that relate to EC adoption in SMEs and hence, safeguard against any failure in missing out the effect of an important determinant/s on EC adoption.

3 POTENTIAL DETERMINANTS OF INNOVATION ADOPTION IN SMEs

In view of the innovation theories, Rogers' (1995) model appeared to be the most widely accepted model by researchers in identifying "perceived" critical characteristics for innovations in IS research in general, and on SMEs research in particular (Iacovou et al., 1995; Kaplan, 1999; Karahanna et al., 1999; McGowan and Madey, 1998; Moore and Benbasat, 1991; 1996; Premkumar and Roberts, 1999; Thong 1999). Rogers (1995) identified five significant characteristics of the innovation that influences its adoption: relative advantage, compatibility, complexity, trialability, and observability. However, the same researchers who endorsed Rogers' (1995) model (above) contend that Rogers' model should be blended with other contexts/factors in order to provide a more holistic adoption model (For full details about criticisms for Rogers theory refer to Attewell 1992; Chau and Tam, 1997; Larsen and McGuire, 1998; Moore and Benbasat, 1991; 1996). However, Van de Ven (1991), Fichman and Kemerer (1993), and Kwon and Zmud (1987) argue that the innovation attributes not only play an important role on its adoption in organisations, but also support its post- adoption stages as well. They even extend their argument to include the adoption of complex technologies. Understandably, the strength of Rogers' model could be complemented with other contexts in order to gain a more holistic understanding about IS innovations in organisations (Fichman and Kemerer, 1993). This is necessary to minimise the risk of adoption failure (Larsen and McGuire, 1998).

The innovation literature also emphasises the importance of environmental, organisational, and technological characteristics on IS adoption (Chau and Tam, 1997; Rai and Bajwa, 1999; Tornatzky and Fleischer, 1990). Kwon and Zmud (1987) in their review of the technological innovation literature identified the importance of Rogers' (1995) characteristics and introduced five influencing contexts: innovation, organisational, environmental, individual, and task (task structure, autonomy, uncertainty) factors. In review of the IS adoption literature in small business, the available adoption models provide essential influencing factors on IS/IT adoption in SMEs (Table 1). As there are few adoption studies tackling innovation adoption in SMEs, it is necessary to revisit the different contexts in light of the innovation theories in order to propose potential determinants of EC adoption in this research.

Table 1. Research on IS/IT adoption by SMEs

Contextual Effects						
Author	The manager	Innovation characteristics	Organisational characteristics	Environmental characteristics	Essential influencing factors	Most influential factors
Thong and Yap (1995)	The CEO: CEO's innovativeness CEO's IT knowledge CEO's attitude towards adoption of IT		Size. Information intensity.	Competition (Rivals).	CEO's IT knowledge. CEO's attitude towards adoption of IT. Size. CEO's innovativeness.	Size.
Thong and Yap (1996)	The CEO: CEO's innovativeness. CEO's Attitude towards IT.		Employee's IT knowledge. Information intensity.	Competition (Rivals).	Employee's IT knowledge. CEO's Attitude towards IT. CEO's innovativeness.	Extent of adoption: Employee's IT knowledge. Information intensity.
Thong (1999)	The CEO: CEO's innovativeness. CEO's IS knowledge.	Relative advantage. Compatibility. Complexity.	Size Employee's IS knowledge. Information intensity.	Competition (Rivals).	Size. Relative advantage +Compatibility. CEO's IS knowledge. Employee's IS knowledge. CEO's innovativeness.	Extent of adoption: Organisational characteristics in general (Size, employees' IS knowledge, and information intensity) and size in specific.
Premkumar and Roberts (1999)		Relative advantage. Compatibility. Complexity. Cost.	Size. Employee's IS knowledge. Top management support.	Competitive pressure from rivals. External pressure from suppliers/buyers. External support from IT vendors. Vertical linkages (tight integration with another firm)	Relative advantage (+Cost). Top management support. Size. Competitive pressure.	Relative advantage.

3.1 Technological (Innovation) Context

As noted in Table 1, some of Rogers' (1995) factors were considered by IS adoption research in SMEs and hence, including the remaining factors of Rogers' (1995) model is deemed appropriate in this research. Tornatzky and Klein (1982) examined the relationship between innovation characteristics and innovation adoption and found cost as a significant factor on adoption as well. Although Rogers (1995) suggested that "image" could be explored within the relative advantage characteristic, it was highlighted as an autonomous determinant in recent IT research (Moore and Benbasat, 1996).

3.2 Organisational Context

Research on IT adoption identifies many organisational factors that would influence IT adoption. Kwon and Zmud (1987) found the following as the main factors that influence IT adoption: top management support, size, quality of IS, user involvement, product champion, and resources. These findings are also endorsed by recent literature tackling SMEs (Premkumar and Roberts, 1999; Thong 1999; Thong and Yap, 1996). Damanpour (1991) found that organisational innovativeness (adopting innovations) correlated positively with business specialisation and external (publishing and media) and internal (from peers, employees, friends, etc.) communications. On the other hand, in small businesses the CEO (usually the owner) is the central authority and the decision-maker and the one who provides support and resources for the adoption and diffusion of IS (Blili and Raymond, 1993; Thong, 1999; Thong and Yap, 1995). Therefore, the top management support and the product champion characteristics are investigated under the individual (CEO's) context next. However, top management support would be addressed if any of the studied cases were found managed by a group of managers.

3.3 Individual Context

Individual characteristics of the CEO, such as education, age, experience, and psychological traits have been found to strongly influence innovation adoption (Rogers, 1995). Thong (1999) and Thong and Yap (1995; 1996) considered individual characteristics represented by the chief executive officer (CEO) as an essential part in IT adoption in SMEs. They found that the CEO's innovativeness (e.g., introducing new original ideas, always looking for something new rather than improving something existing, and risk taking) and IT knowledge (computer experience and computer awareness) has a positive effect on IT adoption. Therefore it was important to explore the impact of the CEO's innovativeness and the CEO's IT knowledge on EC adoption.

3.4 Environmental Context

Thong (1999) found competition influencing IT adoption in small businesses to be insignificant and interpreted that on the basis that Singapore SMEs existed in similar competitive environments and hence, possessed similar perceptions about competition on IS adoption. On the other hand, Premkumar and Roberts (1999) found the competitive pressure factor influencing IT adoption to be significant. They found vertical linkages were tightly correlated with son-parent type organizations and external support (from consultants, vendors) to be insignificant in IT adoption. This research is interested in exploring the government's role on EC adoption in SMEs in New Zealand because research in SMEs showed that government incentives lower the barriers to computerisation and make it more attractive to SMEs (Yap et al., 1994).

4 RESEARCH METHODOLOGY

This research is exploratory in nature in the sense that there is no prior research in New Zealand to guide this research endeavour. Accordingly, this research attempts to use the case study approach to investigate the effect of the developed determinants on EC adoption in SMEs in New Zealand and to explore the EC adoption criteria using three case studies.

Historically, researchers tended to categorise methods hierarchically and argued that case studies were appropriate for the exploratory phase of an investigation only (Yin, 1994). Vidgen and Braa (1997) introduced an IS research framework where they classified the different methodologies according to their intended research outcomes: positivist, interpretivist, and interventionist. An overlap among these purified research disciplines (paradigms) would depict hybrid methodologies: quasi experiments, hard case, and action case. According to the preceding taxonomy, Yin's (1994) case study approach matched the one depicted by the hard case methodology. In an introduction to Yin's (1994) book, Donald Campbell (in Yin 1994: pp. ix-xi) indicated that near the positivist stance, Yin (1994) "epitomizes a research method for attempting valid inferences from events outside the laboratory while at the same time retaining the goals of knowledge shared with laboratory science". He commented that near the interpretivist school, Yin (1994) adopted an implicit positivist stance in describing case study research and hence, immersed himself in the classical social science cases studies. Yin's (1994) views are that case studies are the preferred research strategy to answer how and why type questions and using interviews would also is acceptable by the interpretivist school (Walsham, 1995). Thus, attempting to provide a balance between understanding and prediction, of subjectivity and objectivity. Accordingly, this research adopts Yin's (1994) multiple-case (comparative) design in studying three single units of analysis (holistic). Semi structured and structured interviewees were conducted with the managers of three SMEs in New Zealand between July 2000 and December 2000. Table 2 provides different organisational information about the three cases. Interviews were recorded on audiocassettes and notes were taken. The Web sites of the different cases have been analysed as well. The interviewed SMEs reviewed a draft of the research (during the composition phase) to validate interpretations and conclusions made.

Table 2. Organisational information about the cases

No.	SMEs	SelfStor	ShipBrok Limited	AerialMap
1	Base	Auckland	Auckland	Auckland
2	Branches	5 branches – Auckland based	One in Sydney	One in the South Island (Flying base)
4	Business description	Self storage services	Shipping brokerage (ShipBrok) Commodity trading	Aerial photography and mapping
5	Annual turnover (NZ$Million)	4.6	4	3
6	Organisation size based on number of employees (FTEs)	23	Auckland: 4 Sydney: 3	42
7	Age of Web site (months)	12	6	4
8	Customers	General/retailer	Specific	Specific

Table 3 shows the different adopted EC technologies across the three cases. Most of the cases were found to be adopting email for internal and external communications and Web sites. Technologies such as FTP, Telnet, email lists, Bulletin Boards were not quite clear to all of the cases in terms of their definitions and functionalities, and the researcher had to explain those to the interviewees in order to get their responses.

Table 3. Adopted Internet technologies across the cases

I	Internet technologies	SelfStore	ShipBrok	AerialMap
1	Communication technologies	Adopted (duration)	Adopted (duration)	Adopted (duration)
	Internal email	X (1.5 years)	X (1.5 years)	X (3 years)
	External email	X (1.5 years)	X (1.5 years)	X (3 years)
	Email lists (List servers)	-	-	-
	Bulletin boards (Usenet)	-	-	-
	Others	-	-	-
2	Searching/retrieving tools			
	FTP	-	-	X (2 years)
	Telnet	-	-	X (2 years)
	WWW browsing (through Microsoft explorer or Netscape)	X (1.5 years)	X (1.5 years)	X (3 years)
	Others	-	-	-
3	Communication infrastructure and applications			
	Intranet	-	-	X (1 year)
	Extranet/VPN	-	VPN (1 year)	-
	Internet based EDI	-	-	-
	Web site	X (1 year)	X (6 months)	X (4 months)
	Others	-	-	-
4	Internet enabled technologies for commerce			
	Mobile data systems Teleconferencing Video conferencing	None.	None.	None.

5 RESEARCH ANALYSIS

In view of the depicted determinants above, it is worth noting here that it is quite impossible to investigate the effect of all of the depicted determinants on EC adoption in great detail in this research. Upon completing the interviews and data analysis, the researcher was able to discard the insignificant factors (Table 4) earlier on. Such an approach gave the researcher more confidence in not missing out a potential determinant of adoption and more space to concentrate and expand on the most important factors on the adoption decision of eCommerce (Table 5, detailed next).

Table 4. Insignificant factors on EC adoption

Technological context

- Observability: although, the degree to which, results of using EC are observable to SMEs through the media, that did not mean anything to the adoption context of each of the three SMEs. The interviewees highlighted that by reviewing relevant magazines to their businesses, there were some advertisements for EC but there was nothing specific about EC to their business.
- Trialability: this factor was seen by the cases as irrelevant, not applicable, and simply they would not do it. They do not have the time to trial EC and suspected that the EC vendor/supplier would accept that. None of the cases were found to be adopting such an approach. ShipBrok commented, "as we were already in an environment using electronic mailing, it was not necessary to experiment or trial with the internet before switching over".

Organisational context

- User involvement: the different SMEs indicated that computerisation in their organisations was not sophisticated to be of any concern to employees and the fact that only few employees are running the computer system.
- External/Internal communications: external/internal communications from peers or/and media, internal networking, etc. were not perceived by any of the cases as influencing their adoption decision and therefore, were not represented in any of the three cases. Maybe due to the busy nature, centrality of decision-making (CEO), and to the small number of employees in SMEs (Blili and Raymond, 1993), this factor did not appear as significant.
- Quality of IT systems and capability: none of the cases retained complete or integrated IS system/s in house. The cases pointed to the existing gap between EC and IS and the fact that both were related but quite differently. Firstly, despite the simpler IS systems (and not integrated) in place, none of the cases reported that this could have prevented them from exploring and adopting EC. Secondly, the managers of the different cases indicated that in order to have full functional EC, integrating EC with the different IS systems in place (e.g., accounts, HR, inventory, etc.) is essential in order to deliver full EC functionality and benefits. However, the cases envisioned achieving such integration in the long run and hence, did not stop them from exploring and adopting EC in the first place. However, it was not clear when the SMEs would achieve such integration (or whether they would do it!)
- Specialisation: the different cases did not perceive EC assisting in furthering their specialisation. It seems that this factor relating to IS more than to EC and this in turn found to be related to industry/product specifics like in the case of AerialMap as opposed to ShipBrok as explained above. For instance, AerialMap's is a specialised business and relied heavily on information processing. This specialisation is built on state of the art technologies such as networking and GIS system and not on EC.
- Top management support: from the analysis made, it was suggested that the presence of EC in the cases correlated positively with the presence of an enthusiastic manager (usually an owner as well) in order to guarantee EC adoption. Thus, making the adoption decision for EC not related to top management support as such.

Environmental context

- Vertical linkages: none of the cases were engaged in son-parent type business relationship. Future research targeting a larger SMEs sample could test for this factor.
- The role of the government on EC adoption: there was no apparent role played by the government on the adoption decision of the three cases. The interviewees highlighted that they were not aware of any government initiative or training programme aiming at introducing EC to businesses in New Zealand.

Table 5. A theoretical framework made of potential determinants of EC adoption

Innovation characteristics:	**The Environment:**
Relative advantage Cost Complexity Compatibility Image	Competition from other companies in the business (Rivalry) External pressure (from Suppliers/buyers) External Support (Technology vendors)
Individual characteristics:	**Organisational characteristics:**
CEO's innovativeness CEO's IS/IT/EC knowledge	Size Information intensity

5.1 Technological Innovation Context

5.1.1 Complexity

All cases indicated that the different software tools and packages within EC (i.e., Microsoft outlook email software, VPN, browsers, Intranet, Web site) overwhelm SMEs with a huge knowledge stream that they cannot cope with. The cases indicated that they were not used to dealing with such a range of technologies before. Above all, such technologies kept changing all the time and coping with these changes was quite challenging and would raise complexities and incompatibilities issues pertaining to EC within their organisations. For instance, AerialMap had to change ISPs several times because they were not able to configure the ISDN connection properly and effectively. AerialMap described the preceding experiments as complex and as very expensive as they had to pay dearly for an open ISDN connection (24X7), which should not be the case and they should have been able to use the dial-up option. AerialMap emphasised that registering their domain name with different search engines was a very complex task as their trading name and business details were very difficult to search for in the different search engines. ShipBrok commented, ".... Mastering EC vocabularies and terminologies and understanding and knowing EC products is a very difficult task".

5.1.2 Compatibility

SelfStor indicated that most of their resident employees at the five branches in Auckland were between the ages of 50-55. Those employees were not used to working with computers in the first place, thus resistance was expected, and this highlights incompatibility in the case of SelfStore. However, the manager of SelfStor emphasised that the adoption decision is an organisational issue and employees had to comply with management decisions concerning the adoption of new technology. AerialMap employees are mostly computer programmers and operators and hence, viewed EC as compatible with them. ShipBrok had few employees who felt any incompatibilities with EC. However, they raised the lack of receipt confirmation over the Internet for legal purposes as a concern when adopting email. Other than the preceding concern, details concerning legal, lack of standards and security concerns from hackers and viruses were not quite clear to the cases, as they were not extensively involved enough with EC to feel the effect of these factors on their online businesses (e.g., selling, buying, collecting money online, etc.). Thus, showing little knowledge about some of EC's major implications. During the interviews, the cases highlighted that such issues did not interfere with their adoption decision for EC. However, the cases revealed that this would represent a hurdle in the long run when their Web sites start selling and generating revenues. The cases referred to viruses and spam as concerns but would not hinder EC adoption. Over all the interviewees looked at this factor as a deterrent in the long run.

5.1.3 Relative Advantage

"EC is expected to generate more enquiries about our products and services from potential customers. There are no immediate positive results but EC allows for greater exposure of our company to the wider world" (ShipBrok). All cases emphasised the cost-benefit analysis as being important in making the adoption decision for EC. They perceived investing on a Web site as a long-term investment and immediate return on investment (ROI) was not anticipated in the near future. AerialMap perceived that they would run their Web site with losses for two years. ShipBrok even questioned the investments made on EC infrastructure and upgrades as being too expensive to justify financially. However, ShipBrok highlighted one advantage by commenting, "...EC is more effective than traditional methods of advertising". SelfStore expected the number of potential sales through their Web site to be less that 3 percent, which cast doubt about the effectiveness of such EC initiatives in generating sufficient profits to SMEs. On the other hand, the interviewees suspected that the preceding depressing perceptions about the effectiveness of EC could prevent them from making investments on their current EC initiatives. They all emphasised the importance of email as an efficient external communication tool with their business partners. The cases emphasised that the Web sites could provide new opportunities by advertising and publishing relevant information about the company's products and services on the Web.

ShipBrok stressed that, "our business is quite different, we cannot sell our products over the Web as our business is based on personal contacts and established relationships.......yes, the Web might be suitable to sell our commodity range of products but only time can tell if we can expand on this business or not through the Web". It was suggested that the industry/product specific perspective is quite apparent in adopting certain EC technology more than the others? For ShipBrok it was the communication perspective (email) and remote and secure login (VPN) to the company's internal network. For instance, when ShipBrok's interviewee was asked about the reason for not adopting Extranet and/or EDI he commented, "....because the nature of the business does not require or encourage too close a bonding...". For AerialMap, the Web site could introduce different business opportunities and revenues and therefore, AerialMap was very keen to develop a professional Web site. As they retained larger number of employees, adopting the Intranet was quite logical to AerialMap in order to streamline lots of the internal manual and paper-based transactions.

EC was perceived to enhance organisational image in the different cases and therefore, EC would influence adoption positively. The interviewees highlighted the importance of the image factor on their adoption decision for EC. Having an email address on business cards and letters and most importantly a Web site (URL) would project high organisational image and would position the cases at a higher position than their competitors. It indicates a lead in technology, which could highlight professionalism as well. ".....It is the buzzword that attracts the attention of media, customers and competitors....", as highlighted by AerialMap.

5.2 Organisational Context

5.2.1 Size

The researcher in this research crossed the organisational size of the different cases with the adopted EC technologies (Tables 2 and 3) and found that organisational size correlated positively with EC adoption and with adopting more EC technologies in the case of AerialMap. AerialMap had 42 employees and their environment relied heavily on information processing of aerial photos and satellite imageries. SelfStor like AerialMap had a large number of employees and therefore, it was expected that SelfStor would adopt EC more than ShipBrok, which maintained a few employees only. However, ShipBrok's EC initiative was more serious and extensive than SelfStor. SelfStor emphasised that the main objective of having the Web site is to lure customers to their physical stores and therefore maintained simple Web design in terms of the number of pages and the information provided. On the other hand, the business needs (their customers) of the charter broker (ShipBrok) encouraged the introduction and the integration of email into their business, which points again to the above industry/product specifics and to the dominant significance of certain factors on EC adoption "customers" and "individual context" (discussed below). Therefore, size was not found to be conclusive in the case of EC adoption in this research.

5.2.2 Information Intensity

The analysis suggested the triviality of the information intensity factor on the adoption decision of EC. SelfStore's information processing environment was not intensive and they perceived this factor as irrelevant to EC adoption. AerialMap's information processing environment was intensive and relied heavily on IT in processing imagery data (Geographical Information Systems (GIS)) but that did not relate to EC adoption. Even in the case of adopting the Intranet, AerielMap stressed that the Intranet (once fully developed!) would be used in different types of applications other than processing and sharing imagery data. ShipBrok's processing environment was less intensive but relied heavily on communications with their international partners using email and VPN which, points to the product-specific perspective. The last two cases pointed to the information part pertaining to their products and services. For instance, AerialMap enthusiastically emphasised the possibility of selling their library of New Zealand photos through their Web site very easily due to the possibility of digitising (scanning) these photos where customers could easily pay for these photos and download them at the same time. Thus, making this factor not relating to EC as such and hence would not influence EC adoption. However, looking alongside the information content of products and services (the product, the process and the delivery agent (Choi et al., 1997)) could yield more useful results.

5.3 Individual Context

5.3.1 CEO's Innovativeness

Despite the lack of tangible benefits in the short term, the managers of the different cases showed keen interest in adopting EC and in embracing the different automation and EC initiatives in their organisations from the initiation phase till adoption, which further endorses the innovative perspective across the cases in adopting EC. It was suggested from the interviews, the manager of AerialMap showed the highest enthusiasm for EC followed by the manager of ShipBrok and SelfStor, respectively. Driven by his keen enthusiasm for EC, the manager of AerialMap received training courses on developing Web sites and involved himself in developing the organisation's Intranet along with another technical person in the organisation and he even shared in designing and developing the pages of their Web site with their contracted Web designer. He was still considering establishing a VPN link with the remote branch and enhancing and integrating the Intranet and the Web site with their existing legacy systems. However, the manager stressed that he was not rushing things concerning the different EC initiatives and he is taking things slowly.

5.3.2 CEO's IT Knowledge

It is suggested that adopting EC in the case of AerialMap is driven by the manager's interest and enthusiasm about the technology otherwise, he would not have allocated all that time to train himself and to trial with EC. The manager confirmed earlier that he expected to incur losses from the Web site for the next two years before it breaks even and even then, he was not sure about the future. This may justify his relaxed mode in experimenting with EC rather than allocating professional resources and expertise behind EC. The other cases pointed to their lack of knowledge about EC and hence, would rely on technology vendors to grasp some of the perspectives and to provide the necessary information to make the adoption/rejection decision. Therefore, making their adoption decisions for EC was not related to their knowledge of EC as such. There are two issues emerging alongside this factor. Firstly, for the less knowledgeable managers in EC, this will not impede them from adopting EC, as they would resort to experts in the field to complement this lack of knowledge about EC. Secondly, for the knowledgeable manager this would accelerate the decision making process pertaining to adoption, whether to adopt or not. Being a knowledgeable person with technology, AerialMap's manager opted not to reject EC but however, pursued an experimental path in having EC. Therefore, it is suggested that knowing about EC was not a main driver for effective adoption of EC, simply because the managers would not have that great a knowledge about EC and business models in the first place, as confirmed by recent research in New Zealand (Chapple, 2002; Deloitte, 2000; PWHC 1999), or about its returns to the business.

5.4 Environmental Context

5.4.1 Supplier/Buyer Pressure

All the cases indicated that their buyers drove their automation processes including the adoption of EC and hence, perceived it significant. However, their suppliers did not influence their adoption decision of EC, as none of the cases were found to be dealing with major suppliers. The cases indicated that it is quite logical to take whatever steps necessary to adopt technologies that could attract more business from their existing customers and maintain their loyalty by making their shopping experience more convenient or introduce new business opportunities and online customers. On the other hand, if the different customers were not EC ready or interested in conducting business over the Internet, SMEs would not be motivated to go into EC or at least would impede the progress of their EC initiatives. SelfStor and AerialMap believed that their Web sites could attract more customers to their business in the long run however, were not in a position to confirm that. ShipBrok adopted EC as their buyers dumped Comtext and started using Internet email. However, their Web site proves to be a failure and it did not generate any business for them but ShipBrok kept it running due to the low hosting costs. The content of their Web site is not updated as such.

5.4.2 Competition

The cases were major players in the marketplace and hence, maintained that leadership in the marketplace required a closer look at their competitors but however, the cases emphasised that competition is not significant on their adoption decision for EC as such. SelfStore indicated that once they adopted a local software solution to manage their storage systems the rest of their competitors imitated them and adopted the same software from the same supplier. SelfStor perceived that adopting EC could endorse their leadership in the marketplace but however, not as a direct response to their competitors. EC was perceived by AerialMap as a necessary step in order to keep their leadership in the industry. ShipBrok on the other hand, did not perceive this factor significant on their EC adoption decision and related that to the international nature of their business and the fact that adopting email was a strategic necessity in order to stay in business.

5.4.3 Technology Vendors

The cases maintained negative perceptions about the performance of technology vendors. They confirmed that they dealt with different vendors for the ongoing support of their equipment but the deliverables were very disappointing. ShipBrok indicated that they were forced to live with that shortcoming for a long time. Their internal applications stayed dormant for many years and reached a level where they could not invest any more in supporting such systems. The VPN server kept failing all the time without any obvious reasons or justifications from the vendor's side. AerialMap delayed the live launch of their Web site for a couple of months due to technical problems with the ISDN providers and changed ISPs three times until they settled eventually with a local wireless radio communications provider. It is suggested that the preceding argument and comments could be further aggravated by the lack of detailed knowledge about EC at both the technology vendors and the cases.

6 DISCUSSION AND FUTURE WORK

In addressing the professional implications in this research, EC is recent innovation and hence, represents a complex phenomenon to SMEs. There is not detailed knowledge about its different perspectives and models (Abell and Black, 1997; Abell and Lim, 1996; Deloitte, 2000; PWHC, 1999). The interviewees' raised different "compatibility" issues facing the wide success of their EC initiatives in the long term, e.g., the age of their employees, legal and security concerns. The SMEs did not anticipate many advantages out of their EC initiatives in the short term. However, they perceived EC to be an efficient communication tool (e-mail) with their customers, a secure access (VPN) to their remote internal databases, a Web presence to promote the company's physical location and products, and as image enhancement tool. It was suggested that larger organisations are more capable of adopting EC and different EC technologies than smaller ones. The SMEs' viewed their products as suitable to EC generally. The role of the CEO's innovativeness in adopting EC has been demonstrated in the different cases and in the case of AerialMap specifically. Suppliers and buyers could influence adoption significantly, but the cases did not have any major suppliers or many online buyers to further or to speedup their EC initiatives. They would not adopt EC as a direct pressure from their competitors and the cases perceived technology vendors as incapable of providing adequate services.

The interviewees did not report these impediments as hindering their adoption decision of EC. Their EC initiatives were not complex or incompatible or costly. This could be attributed to the simple EC initiatives in the cases. The suggested significance of the CEO's innovativeness makes the adoption decision in the hand of the CEO of the SME. The other suggested significant influencers such the relative advantage, size, product suitability to the Internet, and supplier/buyer pressure did not appear as conclusive to the adoption decision. This again could be attributed to the simple and experimental EC initiatives and hence, leading to the conclusion that EC in the cases was not viewed as critical or as a strategic tool. Due to the reported lack of tangible benefits in the short term by the different cases, it is suggested that at this limited level of EC adoption across the different cases, EC could not achieve any competitive edge to any of the cases. On the other hand, the reported enthusiasm about adopting EC and being in the EC playing field by the different cases points to the strategic necessity of EC to the cases and not to its strategic advantage. The cases enthusiastically stressed the need to explore EC and to have an on online presence in order not to be left alone when eCommerce diffuse in the future. Therefore, they were interested in maintaining investing on their simple EC initiatives, as long as such initiatives did not exhaust their resources. This points to the CEO's innovativeness in identifying this reality and in taking the risk to proceed with the EC investment. It was suggested that complexity, compatibility, and cost could play significant role on EC adoption in the long run when the Web sites are successful and generating revenues. Size and the product specifics could play a motivating role on adoption. It was suggested that the supplier/buyer factor could play the main driver for adoption. Although it could be argued here that certain industries (suitable products to the Internet) could adopt EC more than others and that the lack of online buyer/supplier and support from technology could further aggravate the adoption decision, however, more work is needed to further the CEO's innovativeness to take the next step towards the advanced EC initiatives.

These issues are of significant importance to SMEs, policymakers, and professional (consultants, education institutions) in New Zealand and elsewhere. The rich insight provided by the different cases, is of valuable resource to those interested parties and to SMEs specifically who are interested in adopting EC. This could be of interest to large organisations as well. The impediments pertaining to the EC technology (e.g., Rogers' (1995) factors, e.g., complexity, compatibility) are not limited to SMEs only and larger organisation could face similar circumstance alongside these factors. However, the difference between large and small businesses is fundamental as highlighted alongside the other contexts in this research. Initially, larger enterprises have better resources (finances, employees, experts) than smaller ones, which could lessen the impact of many of these impediments. For instance, larger enterprises are more capable to invest extensively, allocate time and experts, trial with EC technologies, and to sustain these experimental and risky investments much longer than smaller enterprises.

In addressing the theoretical implications in this research, technological factors such as the observability and trialability play no significant role on EC adoption in SMEs and hence, only the complexity, the compatibility, and the cost factors play a significant role on EC adoption in the long term. The insignificance of Rogers' (1995) observability and trialability on adoption is found to be consistent with the findings of prior research (Tornatzky and Klein, 1982). Recent IS adoption research in small business reported the same as well (Premkumar and Roberts, 1999; Thong and Yap, 1995; 1996; Thong, 1999). Image was suggested to play a supplementary role on EC adoption in this research. At the organisational level, factors such as user involvement, external/internal communications, quality of IT systems and capabilities, and specialisation were found to influence adoption insignificantly. The information intensity and support from top management were found to point to information content of products (point to industry-specific as well) and to the CEO's role on adoption, respectively. The size factor was not conclusive but pointed to the importance of the organisation's size in adopting EC and in adopting different EC technologies. Future empirical research that addresses the information content of products and the organisation's size and development of more accurate measures could yield more insights. Controlling for the industry-specifics could yield more accurate results to the different SMEs' sectors. At the individual level, CEO's innovativeness played an important role on EC adoption in SMEs. CEO's knowledge with EC was viewed as influencing adoption positively but SelfStor and ShipBrok pointed to their lack of detailed knowledge about EC. Thus, suggesting that the CEO's innovativeness was more significant to adoption than the CEO's EC knowledge. Having an enthusiastic CEO with a vested interest in EC and a willingness to involve himself/herself in EC from initiation until finalisation will increase the chances of EC adoption and success in SMEs. At the environmental level, factors such as the government's role in promoting EC adoption in SMEs and vertical linkages did not play any significant role? The cases emphasised the importance of the pressure from supplier/buyer on their adoption decision of EC. However, the cases did not report having any major suppliers or many online customers to justify the adoption decision or in having huge investments on EC. Their competitors did not drive EC adoption and technology vendors would influence adoption negatively.

This research attempted to revisit the technological innovation literature and to introduce relevant significant factors to EC adoption research in SMEs. Researchers in different countries are encouraged to revisit these factors and to assess their importance from the context of their countries. This research suggested the importance of certain factors and the irrelevance of other factors to the adoption decision of EC in SMEs in New Zealand. Such an approach could create the foundation for future research aiming at investigating EC adoption in SMEs utilising the suggested determinants and hence, build on this research's results. The case study approach showed its strength in exploring the impact of innovation factors on EC adoption in SMEs and in generating rich insights around the different proposed determinants of adoption. This is of great importance to parties interested in SMEs and to researchers and policymakers specifically interested in identifying potential determinants of EC adoption in SMEs. However, concluding the significant factors or the most significant ones on EC adoption was not possible from limited cases in this research. It is worth investigating these suggestions further by undertaking large empirical work targeting large SME's sample (e.g., survey, more case studies) either in New Zealand or elsewhere.

REFERENCES

Abell, W. and Lim, L. "Business Use of the Internet in New Zealand: An Exploratory Study", 1996. Available at: http://www.scu.edu.au/ausweb96/business/abell/paper.htm (Accessed at: August 8, 2000).

Abell, W. and Black, S. "Business Use of the Internet in New Zealand: A Follow-Up Study", 1997. Available at: http://www.scu.edu.au/ausweb96/business/abell/paper.htm (Accessed at: August 8, 2000).

Attwell, P. "Technology Diffusion and Organisational Learning: The Case of Business Computing", *Organisational Science*, (3:1), 1992, pp. 1-19.

Blili, S. and Raymond, L. "Information Technology: Threats and Opportunities for Small and Medium-Sized Enterprises", *International Journal of Information Management*, (13), 1993, pp. 439-448.

Chapple, I. "Small Firms Drag Feet with Internet", The New Zealand Herald, Wednesday August 14, 2002. Available at: www.nzherald.co.nz/storyprint.cfm?stotyID=2347435. (Accessed at: August 14, 2002).

Choi, S., Stahl, D. and Whinston, A. *The economic of electronic commerce*, Indiana.: Macmillan Technical Publishing, 1997.

Chau, P. and Tam, K. "Factors Affecting the Adoption of Open Systems: An Exploratory Study", *MIS Quarterly*, March 1997.

Damanpour, F. "Organisational Innovation: A Meta-Analysis of Effects of Determinants and Moderators', *Academy of Management Journal*, (34), 1991, pp. 555-590

Deloitte Touche Tohmatsu "Deloitte E-Business Survey: Insights and Issues Facing New Zealand Business, 2000. Available at: http://www.deloitte.co.nz/images/acrobat/survey.pdf (Accessed at: August 8, 2000)

Fichman, R. G. "Information Technology Diffusion: A Review of Empirical Research", *Proceedings of the Thirteenth International Conference on Information Systems*, Dallas, 1992, pp. 195-206.

Fichman, R.G. and Kemerer, C. "Adoption of software engineering innovations: The case of object orientation", *Sloan Management Review*, Winter 1993, pp. 7-22.

Iacovou, C., Benbasat, I. and Dexter, A. "Electronic Data Interchange and Small Organisations: Adoption and Impact of Technology", *MIS Quarterly*, December 1995.

Kaplan, A. "From Passive to Active about Solar Electricity: Innovation Decision Process and Photovoltaic Interest Generation", *Technovation*, (19), 1999, pp. 467-481.

Karahnna, E., Straub, D. and Chervany, N. "Information Technology Adoption Across Time: A Cross-Sectional Comparison of Pre-Adoption and Post-Adoption Beliefs", *MIS Quarterly*, (23:2), June 1999, pp. 183-213.

Kwon, T., and Zmud, R. "Unifying the Fragmented Models of Information Systems Implementation" in *Critical Issues in Information System Research*, Borland, R. and Hirschheim R. (Eds), New York. John Wiley, 1987, pp. 252-257.

Larsen, T. and McGuire, E. (Eds.) *Information Systems Innovation and Diffusion: Issues and Directions. Hershey*, London: Idea Group Publishing, 1998.

McGowan, M. and Madey, G. "Adoption and Implementation of Electronic Data Interchange", in *Information Systems Innovation and Diffusion: Issues and Directions*, Larsen, T. and McGuire, E. (Eds.), Hershey, London: Idea Group Publishing, 1998, pp. 345-372.

Moore, G., Benbasat, I. "Development of an Instrument to Measure the Perceptions of Adopting an Information Technology Innovation", *Information Systems Research*, (2:3), 1991, pp. 192-221.

Moore, G., and Benbasat, I. "Integrating Diffusion of Innovations and Theory of Reasoned Action Models to Predict Utilisation of Information Technology by End-Users", in *Diffusion and Adoption of Information Technology*, Kautz, K., and Pries-Heje, J. (Eds.), London: Chapman and Hall, 1996, pp. 132-146.

(MOED) Ministry of Economic Development "SMEs in New Zealand: Structure and Dynamics", *Firm Capability Team,* update Report, January 2000. Available at: http://www.MOED.govt.nz/gbl/bus_dev/smes2/index.html#TopOfPage (Accessed at: May 5, 2000).

(NZStat) Statitics New Zealand "A report on the Post-enumeration survey 200". Available at: www.stats.govt.nz/domino/external/pasfull/ (Accessed at: September 11,2002).

Premkumar, G., and Roberts, M. "Adoption of New Information Technologies in Rural Small Businesses", *The International Journal of Management Science* (OMEGA), (27), 1999, pp. 467-484.

PWHC (PriceWaterHouseCoopers) "SME Electronic Commerce Study" (TEL05/97T), September 24, 1999. Available at: http://apec.pwcglobal.com/sme.html (Accessed at: April 10, 2000).

Rai, A. and Bajwa, D. "An Empirical Investigation into Factors Relating to the Adoption of Executive Information Systems: An Analysis of EIS for Collaboration and Decision Support", *Decision Sciences*, (28:4), Fall 1997, in *Emerging Information Technologies: Improving Decisions, Cooperation, and Infrastructure*, Kendall, K. (Eds.), Thousand Oaks, California: Sage Publications, Inc, 1999, pp. 205-239.

Rogers, E. *Diffusion of Innovation*, New York: The Free Press, 1995.

Swanson, E. B. "Information Systems Innovation among Organisations", *Management Science*, (40:9), 1994, pp. 1069-1092.

Thong, J. "An Integrated Model of Information Systems Adoption in Small Business", *Journal of Management Information Systems*, (15:4), 1999, pp. 187-214.

Thong, J., and Yap, C. "CEO Characteristics, Organisational, Characteristics and Information Technology Adoption in Small Business", *Omega, International Journal of Management Sciences*, (23:4), 1995, pp. 429-442.

Thong, J., and Yap, C. "Information Technology Adoption by Small Business: An Empirical Study", in *Diffusion and Adoption of Information Technology*, Kautz, K., and Pries-Heje, J. (Eds.), London. Chapman and Hall, 1996, pp. 160-175.

Tornatzky, L., and Klein, K. "Innovation Characteristics and Innovation Adoption implementation: A Meta-Analysis of Findings", *IEEE Transactions on Engineering Management*, (29:11), 1982, pp. 28-45.

Tornatzky, L.G. and Fleischer, M. *The processes of technological innovation*, Lexington, Mass.:Lexington Books, 1990.

Van de Ven, A. "Managing the Process of Organisational Innovations", in *Changing and Redesigning Organisations*, Huber, G. (Ed.), New York: Oxford University Press, 1991.

Vidgen, R & Braa, K. "Balancing interpretation and intervention in information system research: the "action case" approach", in Proceedings of IFIP 8.2: Information Systems and Qualitative Resaerch, Philadelfia, 31 May - 3 june, 1997, pp 524--542.

Walsham, G. "Interpretive Case Studies in IS Research: Nature and Method", *European Journal of Information Systems*, (4), 1995, pp. 74-81.

Yap, C., Thong, J. and Raman, K. "Effect of Government Incentives on Computerisation in Small Business", *European Journal of Information Systems*, (3:3), 1994, pp. 191-206.

Yin, R. *Case Study Research Design and Methods*, California: Sage Publications, 1994.

About the Author

Nabeel A. Al-Qirim is a lecturer in the School of Information Technology, Faculty of Business, Auckland University of Technology, Auckland, New Zealand. He has a Bachelor degree in Electrical Engineering, MBA (UK), InfoSys. (Hons. with distinction), Cert. (Education), and PhD (Current). His research interests and publications are in IT and eCommerce in small business, SCM, mobile commerce, and telemedicine. He is currently editing one book in the area of eCommerce in SMEs and co-editing another book addressing policy implications pertaining to eCommerce, Government, and SMEs. He worked in the IT industry for 12 years as a consultant and in managing total IT solutions with international companies: IBM, Compaq, Data General, Group Bull, and Siemens Nixdorf. He can be reached by e-mail at nabeel.alqirim@aut.ac.nz.

9

ORGANIZATIONAL AND NATIONAL ISSUES OF AN ERP IMPLEMENTATION IN A PORTUGUESE COMPANY

Jose Esteves, Joan Pastor, João Carvalho

Universidad Politecnica de Catalunya, Universidad Internacional de Catalunya, Universidad do Minho, Spain

Abstract: This paper describes a case of an Enterprise Resource Planning (ERP) implementation in a Portuguese SME. We focused on the identification of organizational factors that affected the ERP implementation project. We also analyze the ERP implementation project from a national cultural perspective using Geert Hofstede's dimensions. These dimensions are used to explain some of the attitudes and behaviours during the ERP implementation project. Our findings enforce that some of the problems in ERP implementation projects are not of technological nature but may be attributed to organizational factors while some issues are related to national culture.

Keywords: Enterprise Resource Planning, ERP implementation, grounded theory, organizational culture.

1 INTRODUCTION

In recent years, it has been demonstrated that Small and Medium sized Enterprises (SMEs) strongly contribute to national economies. SMEs constitute around 95 percent of enterprises and account for up to 70 percent of employment in most countries around the world. The adoption of ERP systems is now reaching SMEs, bringing up problems in ERP implementation projects which are specific to this type of companies. This paper describes the results of a case study carried out in a Portuguese SME that implemented the SAP R/3 system in 1998 by following a big-bang implementation approach. The big-bang approach is characterized by the installation and go live of all implemented ERP modules at the same time. The interpretive perspective adopted in our research reflects our aim for understanding the phenomenon of ERP implementation in a SME within the organizational and national contexts where it occurs. The paper is structured as follows. First we present the research methodology; in this study we combined the case study method with the grounded theory method. Next, we describe the case study background followed by a presentation of its findings. Hofstede's (1991) dimensions of national culture are then related to the case study. Finally, the conclusions and the implications for further research are outlined.

2 RESEARCH METHODOLOGY

As we believe that the understanding of ERP implementation cannot be achieved without considering the organizational context where it occurs, the chosen research method was the in-depth case study method (Yin, 1994). In order to identify the organizational factors that affect an ERP implementation in SMEs we opted for an interpretive research approach. Interpretive research does not predefine dependent or independent variables and it attempts to explain phenomena through the meanings that people assign to them (Orlikowski and Baroudi, 1991).

We started the case study by defining a plan to collect data. We first analyzed documentation created during the SAP implementation project that was provided to us by the project manager. The documentation helped to understand the project background and to prepare the questions for the interviews. The main technique chosen for data collection was semi-structured interviews. Following the contact with key informants in the company, interview schedules were agreed upon. Interviews were tape recorded and transcribed to ensure accuracy of written data, and to minimize researcher's bias. Initially, three interviews were made with the project manager, another member of the project team and a key-user. Then, we interviewed the remainder members of the project team. Data from interviews was triangulated with the documentation so far accumulated. In order to build theory from this case study, we adopted the Grounded Theory (GT) method.

GT is a general method developed by Glaser and Strauss (1967) for building theories that are grounded in data systematically gathered and analyzed (Glaser and Strauss, 1967). Strauss and Corbin (1990: p.23) explain that by using GT "a theory is inductively derived from the study of the phenomenon it represents. That is, it is discovered, developed, and provisionally verified through systematic data collection, analysis, and theory stand in reciprocal relationship with each other. One does not begin with a theory, and then prove it. Rather, one begins with an area of study and what is relevant to that area is allowed to emerge". The coding process of all interviews and documentation allowed major themes/categories to emerge. Then, we used the paradigm model proposed by Strauss and Corbin (1990) to relate these categories. Briefly, the paradigm model encompasses the following elements: causal conditions, the phenomenon, the context, the intervening conditions, strategies and actions, and finally the consequences. Finally, in order to analyze the national culture perspective, we used Hofstede's (1991) dimensions. The paradigm model context was thus extended with an additional category: national context. The national perspective is explained in Section 5.

3 CASE STUDY BACKGROUND

PhotoPics S.A. (fictitious name) is the Portuguese subsidiary of the PhotoPics multinational company. The PhotoPics Group was created in Germany around 1849 to develop lenses and microscopes. The company started in Portugal in 1973 and the Portuguese unit is now the main one of the PhotoPics Group worldwide. It has a total of 660 employees, most of them wage earners with approximately 100 salaried employers. Table 1 describes the different actions taken to improve information technology infrastructure in PhotoPics.

Table 1. Information technology evolution in PhotoPics

Period	Actions
Until 1997	- Only dumb terminals connected through a dedicated infrastructure. - Comet Top software package to support logistics, production planning, payroll and accounting. - Only the administrator and the financial director used PCs. - No internet connection.
Beginning 1997	- Analysis of Comet Top upgrade or adoption of an ERP system due to Y2K problem and Euro conversion.
First quarter 1997	- Decision to adopt SAP system.
1997	- Implementation of a computer network. - Upgrade of Comet Top to a Unix version until SAP go live. - Training of users on using PCs (average age of users was 40 years old).

3.1 ERP Implementation Phases

PhotoPics followed a typical ERP implementation lifecycle with a big-bang approach. The project goal was the implementation of SAP R/3 system, version 3.1H with the main SAP modules. The number of expected end-users was 30 to 40. At the beginning of the project, it was estimated that no be-spoke development would be made and that only enhancements to forms and reports would be tailored. Table 2 describes the implementation phases.

Table 2. ERP implementation phases in PhotoPics

Phase	Description
Planning	This was the basis for the entire project. The goal of this phase was to detail the project definition and its functional needs. The project structure was defined. This phase was arduous due to three main aspects: the definition of all processes that attempted to be implemented in the new system, contact with all the process stakeholders, and the difficulty to obtain information.
Design	The goal of this phase was to produce the technical specification of how to implement the chosen solution and the beginning of the parameterization and the preparation of a prototype that allowed the demonstration of the system working for each planned situation. This phase was felt as fundamental for the system comprehension since the internal project team took its first contact with the SAP system.
Realization	The goal of this phase was to obtain the configuration of the SAP system according to the design, the development of some complementary programs that served as interfaces to SAP, and the creation of training manuals and final tests.
Go live & Support	The goal of this phase was to put the new system at work. The go-live phase was started a month behind schedule given some changes in the scope of the project. The expressions "the company will stop" or "it will not work" were in the mind of everyone, but everything worked perfectly. At the end of this phase an analysis of the general difficulties of the SAP implementation project was made.

4 ERP IMPLEMENTATION PROJECTS IN SMES: ORGANIZATIONAL PERSPECTIVE

In this section we describe the paradigm model (see Figure 1) that we developed. The paradigm model includes both, the technological and organizational perspectives.

4.1 Phenomenon

Strauss and Corbin (1990: p.101) stated that phenomena are "the central ideas in data represented as concepts". According to their account, the purpose behind naming phenomena is to enable researchers to group similar events, happenings, and objects under a common heading or classification. The phenomenon in the paradigm model is represented by the central category (sometimes called core category), which represents the main theme of the research. The phenomenon addressed in this study is the implementation of an ERP system in a SME. ERP implementation at PhotoPics was considered to be a successful one.

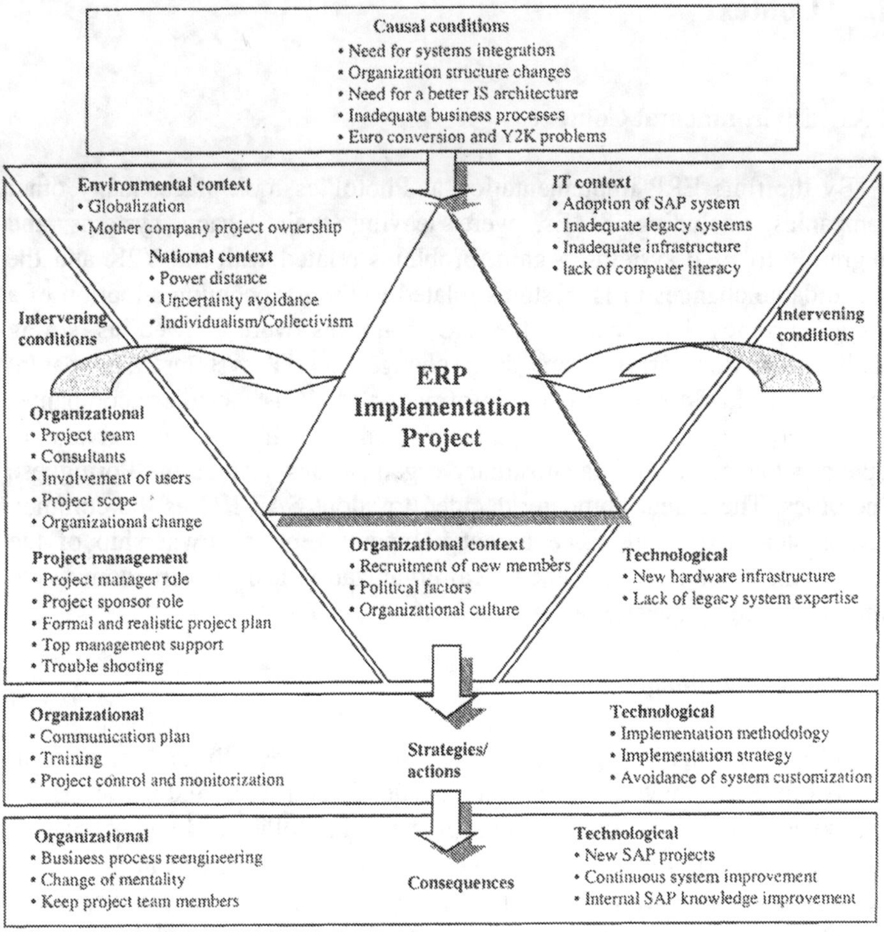

Figure 1. A model for ERP implementation projects in SME

4.2 Causal Conditions

The circumstances that led to the ERP implementation at PhotoPics were: the need for systems integration worldwide at PhotoPics Group; the need for a better IS architecture at PhotoPics SA; most of the business processes were inadequate and they needed a strong reengineering effort; and the mentioned Euro conversion and Y2K problems. In the first quarter of 1997 the company selected the SAP R/3 system. The main reason was that this software seemed to be the best answer to the company needs. The fact that SAP is a German software certainly transmitted security to a German company and also contributed to the decision.

4.3 Context

4.3.1 Environmental Context

By the time ERP implementation at PhotoPics took place, many other companies, including SMEs, were leaving their legacy systems and migrating to ERP systems. Again, problems related with the Y2K and the unavoidable changes in IT systems related to the forthcoming adoption of a single currency by most of the EU countries were viewed as strong justifications for carrying out deep changes on IT systems. The parent company decided to integrate the information of all its subsidiary companies worldwide. Therefore, the company decided to unify information and to have a better control of its subsidiary organizations such as the Portuguese PhotoPics. The parent company decided to adopt SAP R/3 as the common ERP system worldwide. The parent company kept the ownership of the project implementation, but each local organization had total freedom in the implementation process.

4.3.2 Organizational Context

Before the SAP implementation project started, the administration decided to recruit new members, with academic education and proved skills and experience in their field. At that moment, PhotoPics had a conservative organizational culture with power and influences concentrated in just a few members with long careers within the organization. Almost all the senior staff had no academic education and their whole career had been made within the organization. The channels of power and communication were quite complex. Most of the business processes were complex and inefficient. Technology and innovation were associated with costs and troubles. Political factors had a strong impact on this project. The interests of stakeholders within the organization were challenged as senior and middle level managers lost power and influence. Many employees feared losing their jobs or that such an expensive project could originate an organization rupture. They also feared the expected increase on their expected efficiency with performance control measures. Some managers intentionally concealed information about their intentions with the SAP system to team members, some other managers disseminated false information in attempts to discredit the SAP project, while others avoided participating in the project. Top executives faced an internal conflict as they were being forced to carry out changes that they felt could be against their interest. Conflicts among senior managers arose as they tried to dominate and impose their view in the SAP project, since most of them were aware that the results of the SAP project could influence their future career within the organization.

4.4 Intervening Conditions

4.4.1 Organizational

Internal project team – The implementation project structure is represented in Figure 2. Most of the internal team members were new to the organization. The project team included young university graduates that recently completed their studies. Although these people had no organizational nor SAP knowledge, they had very good skills and they were very motivated to learn. The purpose of hiring these people was to prepare them for performing to work in the organization and to develop internal knowledge in SAP. This also minimized the risk of expertise turnover.

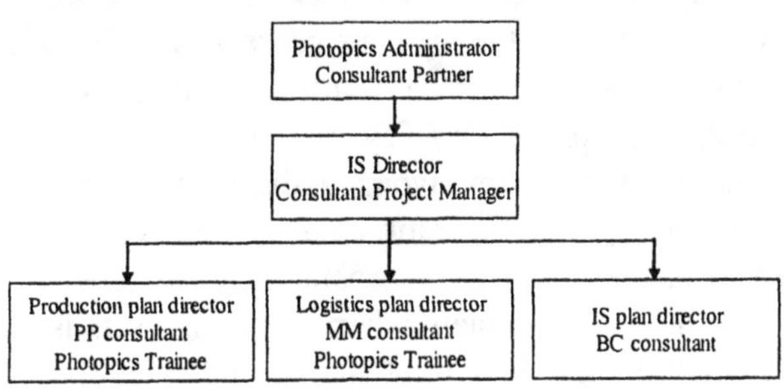

Figure 2. The ERP implementation project structure

Since the team members, which were hired specially for the SAP implementation, did not know the company well, they could carry out the analysis process from a more neutral position and work on the improvement of business processes without bias. The project team included key-users and chief department directors. These directors were assigned the responsibility of the SAP modules corresponding to their own departments. According to the project manager, the responsibility for implementing each module was given to each current domain manager, but logistics and IS directors were new within the organization. Currently, they admit as weaknesses their lack of knowledge about the business and organizational culture and politics.

External consultants – Consultants played an important role in the whole project. There was a careful selection of consultants according to their skills. The consultant selection made by PhotoPics was due to: SAP recommendation, the geographic location, and because the PhotoPics financial auditors knew the consulting company. However, at the beginning of the project, some of these consultants abandoned the project as they moved to other consulting companies. Due to the consulting company lack of experts at that moment, most of the newly allocated consultants were junior consultants with lack of knowledge in SAP projects and the PhotoPics business. The positive side of these junior consultants was their motivation and dedication to the project. As soon as the team members started getting SAP knowledge, they decided which consultants should remain and which new consultants should be hired. There was a lot of turnover in relation to consultants with some of them remaining only part-time in the project. Since the project manager felt quite insecure with the consulting company, he decided to hire other consultants from a second consulting company to cooperate in specific activities of the project. The role of these 'impartial' consultants was crucial to improve the quality of the project and to help the project team to make decisions in aspects about which it lacked knowledge.

User involvement – Users were involved in the project even though some claimed the contrary. Due to their lack of business knowledge, team members interviewed users in the analysis process, and managers approved the processes procedures. In most cases users themselves did the process analysis for their specific processes. Key-users were particularly relevant in this task and during tests. Their involvement also helped them to adapt to the new system. They brought in operational knowledge and they served as enablers of change in their part of the organization. Members of the project team reported that the major problem when involving users is that each user thinks his/her opinion is more important and valuable than that of others (see political factors in organizational context section).

Organizational change – Probably, due to the lack of experience in change management from the project manager and the consultants, there was not a formal organizational change plan. The organizational change affected all levels and aspects such as organizational structure and culture. Probably the most important change was the creation of the logistics department that took over logistics functions carried out before by the financial department. Most interviewees mentioned that organizational change procedures should exist, since several job descriptions were changed and/or eliminated and in some cases training needs could have been predicted. In most cases that aspect was analyzed a *posteriori*. Top management started, facilitated and supported the organizational changes and implied decision-making tasks during the project. Another aspect was that the changes were visible only within a one year span after the system go-live. The organization needed one year to adapt, as well as to understand the new organizational model.

Project scope – The project scope definition was quite limited. It focused mainly on the modules and the functions to be implemented. The lack of project strategic vision was a weakness. Since the SAP system was imposed, the project manager and top management did not carry out a strategic analysis of the implementation, especially in what concerns long-term impact. Goals were not properly defined at the beginning. The single goal to be expressed was that SAP implementation would be carried out within a determined period.

4.4.2 Project Management

Project manager – The project manager was recruited specifically for the SAP implementation project. He made the connection between the implementers (internal team members and external consultants) and top management. He now admits that someone with experience in SAP would have performed better the job, especially at the early stages of the project. The project manager stresses that skills to manage conflicts and people are crucial in a project of this kind. He says that a SAP project is a task of managing willingness, expectations and conflicts rather than an IT implementation project. As PhotoPics is owned by a German company, mastering the German language was an important competence for the project manager although that was not considered at selection time. This was fundamental to communicate and share knowledge with the parent company and it contributed to solve many problems and to improve collaboration.

Top management support – Top management support and commitment were critical to achieve success in the project. Top managers were always available to discuss doubts and trouble-shooting and to make prompt decision-making in order to avoid bottlenecks. This commitment was vital for disseminating the project to the whole organization and especially for dealing with SAP opponents. It was the organization CEO who played the role of project sponsor. According to most interviewees he was the person that drove forward the SAP project.

Trouble shooting – Several problems appeared along the SAP project mainly due to project team inexperience and to the political factors described above. Two main issues affected the SAP project: lack of knowledge about the legacy systems and the difficulties with data conversion. In what concerns the legacy system, there was lack of documentation and there were no experts in the system. Most of the information was obtained via the parent company. The data dealt with by the legacy system was not enough for the new system, or it was not structured according to the new business model. Therefore, there was the need to create several conversion programs and for some manual data insertion. With the SAP knowledge acquired meanwhile, the project team admits there were better data conversion solutions that those adapted in the project.

4.5 Action/ Interaction Strategies

Communication plan – The communication plan was divided in two types: communication inwards (between team members and with top management) and communication outwards (with the rest of the employees). Inwards communication worked very well. A special room was allocated to team members that facilitated knowledge sharing and cooperative work. In what concerns outwards communication,. the project team regularly presented newsletters, invitations to participate in promotion events, and an intranet was implemented. However, the interest of employees was minimal. As the project manager and some team members mentioned, employees were more worried on how much money was being spent.

Training – The consultants proposed a training plan that was accomplished. However, team members complained about their training: training was made at the end of the design phase instead of the end of the planning or at least the beginning of analysis phase; training was very basic and focused in the potential of SAP, and trainers had little training experience. Therefore, team members developed their own skills based in self-study during the project. The first step in end-users training was to define a training plan that included basic training in computers use because almost all the users were computer illiterate. The training started early for end-users. This helped to keep them in contact with the system and to start the process of adaptation to SAP. The project manager noted that starting training too early might be a problem because some users tend to forget what they learn, which implies continuous refreshment training courses. On the other hand, late training will bring problems to the go-live phase and extra effort for team members. He also recommends to use a SAP parameterization as close as possible to the final one since this provides a better and quick adaptation to what users will have to deal with after go-live phase. One thing that failed in the training plan was new or changed business domain training. The lack of such training affected organizational change. The project manager admits that an ERP training plan must be a mix of technical and organizational training.

Project monitoring and control – Project monitoring and control mainly consisted on verifying the accomplishment of the project plan and schedule. With no goals defined at the beginning and with the lack of experience in project management, there were no project metrics being used. However, project meetings were used to continuously monitor and control the project. There were weekly meetings with project team and monthly meetings with top managers. At the end of each phase there was a meeting with all managers. The estimated duration of the project plan was not achieved but all the scheduled tasks were performed.

4.6 Consequences

Business process reengineering (BPR) – Although BPR started in the gap analysis, its effective change occurred in the go-live phase. Parallel to the implementation of the SAP system, managers started changing the business processes while explaining it to the organization. The project team admits that this should have been done before the implementation since some processes were totally obsolete or inadequate. BPR is not finished yet. Nowadays, during the post-implementation period and using the knowledge that the project team has at the present, they are always improving processes by extending processes functionality through the current discovery of functionality in the SAP system. The SAP system brought in new business processes and helped in the reengineering of the old ones. As the project manager mentioned, PhotoPics is a manufacturing organization with no unique processes. Therefore, they would not loose competitiveness by adopting SAP best practices. On the contrary, the adoption of practices that SAP provides helped to reengineer the existent business processes and simplified implementation and future maintenance efforts.

Change of mentality – One of the most interesting consequences of this SAP implementation was the induced change of mentality. From the beginning, most of the users disagreed with the implementation of the system but their attitude reversed progressively after the go-live phase. Nowadays, they recognize that SAP is useful in having helped to improve the business and they think that more business analysis should still be done, with managers demanding more people to improve the work. The main reasons for the changeover were: top management commitment, the continuous training and the support of team members. Some users were moved from their old functions while others had intensive training. Now, only a few employees are still not using the system. Some of them were middle managers that, due to their age and education, had difficulties to adapt to the new system. The solution was to train some subordinates of these middle managers that would then help them in the tasks that demand the use of SAP. Nowadays, users are no longer afraid of changes. They are aware that the system is something about continuous improvement process.

Keep team members – In relation to the project team, the young graduates were incorporated into the organization with substantial increase of salaries. This helped to keep and improve the SAP knowledge in the organization and avoid dependency on consultants. A continuous learning process was developed with continuous training in SAP in order to improve SAP internal knowledge.

5 ANALYSIS FROM A NATIONAL PERSPECTIVE

After we created the ERP implementation model, we decided to try understanding better the organizational context of PhotoPics. Based on Hofstede's (1991) study of national cultures and organizations, we analyzed the findings of PhotoPics case study from three dimensions: power distance, uncertainty avoidance and individualism-collectivism. Table 3 describes the dimensions and the values for Portugal.

Table 3. Cultural dimensions suggested by Hofstede (1991) with score and mean for Portugal

Dimension	Description	Score	Mean
Power distance	Power distance is the degree to which members of a society accept as legitimate that power in institutions and organizations is unequally distributed (Hofstede 1991). Higher power distance scores indicate an order of inequalities, special privileges for those of higher status, superiors consider subordinates as a different kind of person.	63	51
Uncertainty avoidance	Uncertainty avoidance is the degree to which members of a society feel uncomfortable with uncertainty and ambiguity, which leads them to support beliefs that promise certainty and to maintain institutions that protect conformity (Hofstede 1991).	104	64
Individualism / Collectivism	Hofstede (1991) proposes a single dimensional structure called individualism-collectivism, where those cultures that emphasize the autonomy of the person are grouped under individualism, while those cultures whose most important values place emphasis on the dependency of the individual with respect to groups are clustered under collectivism.	27	51

Next, we relate and discuss each dimension with the PhotoPics SAP implementation project.

5.1 Power Distance

The score of Portugal is 63 (mean 51), which means that Portugal has a high power distance score. The political and culture issues emerged as important factors in the organizational environment in our case study of PhotoPics (see context section above). In this case, we emphasize the issue of losing organizational power and influence. Most of these managers had a large career within the organization and they controlled their departments with authority like 'feuds' as one of the interviewees mentioned. Each department director filtered the information for the organization owner and the one shared among managers. Conflicts among senior managers arose as they tried to dominate and impose their view in the ERP project, since most of them were aware that the results of the ERP project could influence their future career. Since managers were not happy with the project, they passed the negative image to their subordinates, which caused more general disagreement with the project.

5.2 Uncertainty Avoidance

According to Hofstede (1991), the Portuguese society has a high level of uncertainty avoidance, scored 104 (mean 64). In the case of PhotoPics the uncertainty was very high during the whole project. Uncertainty was mainly due to the people being afraid of losing power (top and middle management) and lower levels, where people were afraid of losing their jobs because of organization bankruptcy or the decision to reduce the number of employees. This uncertainty made some managers express their disagreement during the project and some of them difficulting the process by omitting some information, lack of participation in some implementation tasks, no appearance to some important meetings, and decision-making delays in order to delay the whole ERP project.

5.3 Individualism-Collectivism

According to Hofstede (1991), the Portuguese society is a collectivist society. PhotoPics workers had been doing their tasks in the same way during many years and they were not happy in changing the way of working. One of the reasons for this disappointment was their age, most of the employees were over 50 with more than 20 years working at PhotoPics. Some of them were also expecting retirement (the legal age is 65 years). The work relationships were in some sense like being part of a family. Hofstede (1991) states for this case that the relationship between owner and employee is viewed in a moral perspective, and is like a familiar relationship were both have obligations: protection in exchange of loyalty. Most of the employees of PhotoPics doubted about the real intentions from the new top management with the introduction of this new expensive SAP system. There was a notorious lost of trust on their managers what made that most employees started showing indifference to the SAP project.

6 CONCLUSIONS AND IMPLICATIONS FOR FURTHER RESEARCH

We think that the main conclusion of this study is the evidence that organizational, cultural and national issues have a strong impact in ERP implementation projects. In the particular case addressed in this study if those organizational and cultural issues had been taken into account during the planning phase, it is very likely that some problems would have been avoided or mitigated during the ERP implementation project.

Skok and Doringer (2000) pointed out the need to study the relevance of macro and micro-level cultural issues in the successful operation of ERP systems focusing on the function versus process view of ERPs. Our findings suggest that the relevance of these studies goes beyond the use of ERPs. The analysis of Hofstede's dimensions in our case study suggests that these dimensions could be helpful in addressing some ERP project issues and, consequently, to define a strategy for ERP implementation. For instance, one of the main issues in ERP implementation projects is the project duration which is often used as one of the ERP project success evaluation indicators. Kale (2000) evidences the impact of cultural readiness on ERP project duration, which supports our preliminary findings. These findings can also help in defining the determinants of Critical Success Factors (CSFs) for ERP projects (Esteves and Pastor, 2000). Research on CSFs identification does not provide the reason why some CSFs are more relevant in some organizations than others. In this study we analyzed Hofstede's dimensions one by one. Future research should also consider the influence of associations of dimensions like power distance/uncertainty avoidance or uncertainty avoidance/individualism effects. The literature on IS evidences a lack of research on corporate and national culture. The influence of culture has mainly been focused in the IS solution and how it fits with the organizational culture, or the cultural changes. Another aspect to consider is whether ERP implementations significantly differ from the implementation of other types of systems in the past or currently (e.g., datawarehousing,

CRM, SCM). This research path can bring up conclusions regarding the specificity of ERP implementation projects or the identification of general trends in IS implementation.

REFERENCES

Esteves, J., and Pastor, J. "Towards the Unification of Critical Success Factors for ERP implementations", *10th Annual BIT conference*, Manchester, UK, 2000.

Glaser, B., Strauss, A. *The Discovery of Grounded Theory: Strategies for Qualitative Research*, New York: Aldine de Gruyter, 1967.

Hofstede, G. *Cultures and Organizations: Software of the Mind*, New York: McGraw-Hill, 1991.

Kale, V. *Implementing SAP R/3: The Guide for Business and Technology Managers*, Indiana, USA: SAMS publications, 2000.

Orlikowski, W., and Baroudi, J. "Studying Information Technology in Organizations: Research Approaches and Assumptions", *Information Systems Research*, (2:1), 1991, pp. 1-28.

Skok, W., and Doringer, H. "Potential Impact of Cultural Differences on Enterprise Resource Planning (ERP) Projects", *Electronic Journal on Information Systems in Developing Countries*, (7:5), 2000, pp. 1-8.

Strauss, A, and Corbin, J. *Basics of Qualitative Research: Grounded Theory Procedures and Techniques*, Newbury Park, CA: Sage Publications, 1990.

Yin, R. *Case Study Research: Design and Methods*, Thousands Oaks, CA: Sage Publications, 1994.

About the Authors

José Esteves is a doctoral student in the IS doctoral program of the Technical University of Catalonia, Barcelona, Spain. He is computer engineer, has a master in IS, and a diploma in business administration. His research interests include ERP systems, knowledge management and qualitative research. He can be reached by e-mail at jesteves@lsi.upc.es.

Joan Pastor is an associate professor and Dean of the IT school of Universidad Internacional de Catalunya, Barcelona, Spain. He holds a degree in computer science and a Phd in software. He is also leading the "Twist" group on IS qualitative research, addressing topics like ERP procurement and implementation. He can be reached by e-mail at jap@unica.edu.

João Carvalho is an associate professor and Head of the Information Systems Department at University of Minho, Portugal. His research interests include the fundamentals of information systems, knowledge management and research methodology. He can be reached by e-mail at jac@dsi.uminho.pt.

10

ICTS AND ORGANIZATIONAL CONTROL ACROSS CULTURES

The Case of a UK Multinational Operating in China

Wei Liu, Chris Westrup

School of Accounting and Finance, University of Manchester, UK

Abstract: This chapter examines issues on ICT-related coordination and control between the head offices of a UK manufacturing multinational in England and Hong Kong, and one of its joint ventures in Mainland China. Giddens' theory of modernity and an anthropological view of culture are used to analyze and interpret the case. As an expert system, in Giddens' terms, partly leading to globalization (Giddens, 1990; 1991), ICTs facilitate the disembedding of information and communication transfer in different time-space domains without the limitation of place and therefore enable the head office to control its worldwide operations at a distance. However, this is not simply the 'death of distance' (Cairncross, 2001). ICT-enabled coordination is only effective when linked with other mechanisms such as expatriates and face-to-face contacts. Furthermore, the chapter shows that national culture characteristics appear to be more dynamic and less homogenous than the existing literature suggests. ICTs act as an arena where organizational culture change is often expressed.

Keywords: ICTs, MNEs, Disembed, Reembed, Expatriates, Face-to-face contacts, Culture.

1 INTRODUCTION

We live in a globalizing world where companies, consumers and governments interact in chains of association that encompass the world. One of the key components enabling these arrangements is seen as the advent of advanced information and communication technologies (ICTs) (see Castells, 2000; Gore, 2000) though clearly many other issues are of prime importance – for example world trade agreements such as GATT and now the World

Trade Organization. An investigation of ICTs within multinational enterprises (MNEs) can enable insights into one of the mechanisms of globalization.

How do these ICTs work and what needs to be in place for organizing using ICTs to take place over large distances, in different time zones and cultural milieus? This chapter seeks to give some tentative answers to these questions based on an in depth case study of a UK based MNE and one of its joint ventures (JVs) in China. This example is perhaps a limiting case in terms of using ICTs for two reasons – China, in general, remains a region where market disciplines are not well known and, for this JV, the MNE has refrained from implementing its sophisticated world wide SAP and in-house financial reporting systems preferring instead to manage through a regional office in Hong Kong. Thus, the debates around the use of ICTs in this case can illuminate issues on the conditions that promote, reduce, and are needed for ICTs to *work* in these circumstances. The chapter uses Giddens' work on modernity as a way to theorize this situation that as one expert system or disembedding mechanism ICTs disembed social relations in one place and reembed them in another place. Another candidate theory would be the work of Latour (see Latour, 1999) however Giddens' theory is chosen because it is used elsewhere in the IS literature (see Barrett et al., 2001), because it particularly addresses the dynamics of modernity leading to globalization, and because we wish to identify limits of Giddens' theorization.

A limited range of work on ICTs and MNEs exists (Roche, 1996; Torre and Moxon, 2001). King and Sethi (1999; 2001) argue that the design of information systems in MNEs is determined by their transnational strategies. However, they fail to show much more complex processes (rather than cause and effect relationships) and the social nature of MNEs' employing different mechanisms and strategies.

The literature of ICTs and culture (see Myers and Tan, 2002) tends to follow Hofstede's arguments (1991) that culture differences vary greatly between nations, and organizational culture is determined by national culture. But cross-cultural changes as well as conflicts have frequently increased during the rapid process of globalization. Nation states are facing more difficulties in restraining culture within their borders. Culture itself is not a fixed entity to explain what is happening, but the condition produced and used by people in and out of their activities. An anthropological view of culture is employed here to explain the dynamics and complexity of ICTs usage in this MNE.

The chapter begins by discussing theoretical concepts on disembedding, reembedding and culture; followed by descriptions of the research method and the case. The next step is one of analysis and finally conclusions and some implications are given.

2 MODERNITY, CULTURE AND ICTS

Giddens (1990; 1991) point out that three main dynamics of modernity are sources of 'globalizing' or 'world-embracing' modern institutions: the separation of time and space, disembedding mechanisms, and reflexivity. The separation of time and space is the primary condition of disembedding by breaking up the links between social activity and its embedding in the particularities of place. Furthermore, it enables modern organizations to control and coordinate social relations of many people physically absent from one another instead of using face-to-face interactions within one place.

Disembedding "means the 'lifting out' of social relations from local contexts of interaction and their restructuring across indefinite spans of time-space" (Giddens, 1990: p.21). Two types of disembedding mechanisms (also referred to as abstract systems) are 'symbolic tokens' and 'expert systems'. Expert systems are defined by Giddens (1990: p.27) as "systems of technical accomplishment or professional expertise that organize large areas of the material and social environments in which we live today"; he further explains that they "bracket time and space through deploying modes of technical knowledge which have validity independent of the practitioners and clients who make use of them" (1991: p.18). Expert systems therefore remove social relations from localized contexts and penetrate many aspects of social life in a continuous way. A typical example of a network of expert systems is hospitals. Doctors acquire professional expertise trained in universities. They can be allocated in various local hospitals and apply their systematized and codified knowledge to every patient. Other associated expert systems involve hospital departments and accountants.

Both types of disembedding mechanisms depend vitally on trust. Giddens asserts, "[t]he nature of modern institutions is deeply bound up with the mechanisms of trust in abstract systems" (1990: p.83, italic deleted). Trust is what derives from 'faith' in the reliability of a person or system. Trust in abstract systems needs faceless commitments, in which faith is developed and maintained in the working of knowledge, which, for laypersons, are largely unknown. Trust in persons involves facework commitments, which refer to "trust relations which are sustained by or expressed in social connections established in circumstances of copresence" (1990: p.80). "*Reembedding* refers to processes by means of which faceless commitments are sustained or transformed by facework" (1990: p.88), that is, "the reappropriation or recasting of disembedded social relations so as to pin them down (however partially or transitorily) to local conditions of time and space" (1990: pp.79-80).

In the ICT literature, Barrett and Walsham (1999), Barrett et al. (2001) are a couple of examples where Giddens' theory has been used. Hayes

(2001) and Hanseth et al. (2001) have also applied this form of theorization in a more local situation. This paper develops this literature by specifically exploring ICTs as expert systems removing social relations out of local contexts, transferring them into other time-space configurations, and facilitating organization control without presence. Meanwhile, we look at how cultural characteristics manifest themselves in processes of disembedding and reembedding in MNEs.

Culture is normally conceptualized as shared beliefs, values and concepts for individuals and groups to act and live. Hofstede's work is particularly well known as an exemplar of different types of national culture (Hofstede, 1980; cf. Myers and Tan, 2002). Our starting point is different: we look for evidence of cultural characteristics in action consciously and unconsciously made by members. Thus, we do not use culture as a resource to explain what is happening ... they are doing this because they are Chinese (or English) ... rather we look for the way culture is handled by people to justify actions, to argue against proposals and so on. In exploring culture this way, we find that cultural characteristics are both being reinforced (for example, national schooling develops national characteristics) and being changed (see Kuper, 1999). The implication for ICTs is that organizational members constantly (re) interpret, and (re) create culture while interacting with ICTs (Avison and Myers, 1995: p.52).

3 RESEARCH METHOD

Over one year period, one of the authors[1] conducted seventeen formal semi-structured interviews at four research sites (see Table 1): the UK head office of a UK manufacturing multinational called MAC (pseudonym), one UK factory, the regional head office in Hong Kong (HK), one Chinese joint venture named as SUB. Interviewees involved financial controllers, expatriates, general managers (GMs), IS manager, accountants, and administration/production/marketing managers. Most of interviews lasted about one hour and a half. A majority of the interviews in the UK were taped-recorded. Except for one, all the interviews in China were documented by extensive notes. Participant observation included visiting two factories in the UK and China, looking at how a financial reporting system works through an intranet, sitting at a temporary desk in the Chinese factory. Some internal documents were collected.

[1] The other author attended the initial meeting, two formal interviews and a factory visit.

Table 1. Formal interview list in MAC

Research Sites	Managers (e.g. financial controllers, expatriates, GMs)	IS manager & technician	Others (administration officer, accountant)	Total
The UK head office	3	1		4
The UK factory	2			2
The HK regional head office	1		1	2
SUB	7	1	1	9
Total	13	2	2	17

4 THE CASE

MAC is a large UK based multinational manufacturing automotive parts that began its investment in China in the early 1980s. At the moment it holds three JVs together with two associated undertakings in different parts of China (see Figure 1). The Chinese business only occupies two percent of MAC's worldwide business. SUB was set up in 1995. MAC owns 60% of SUB plus 30% from an associated undertaking, and 10% owned by two local partners.

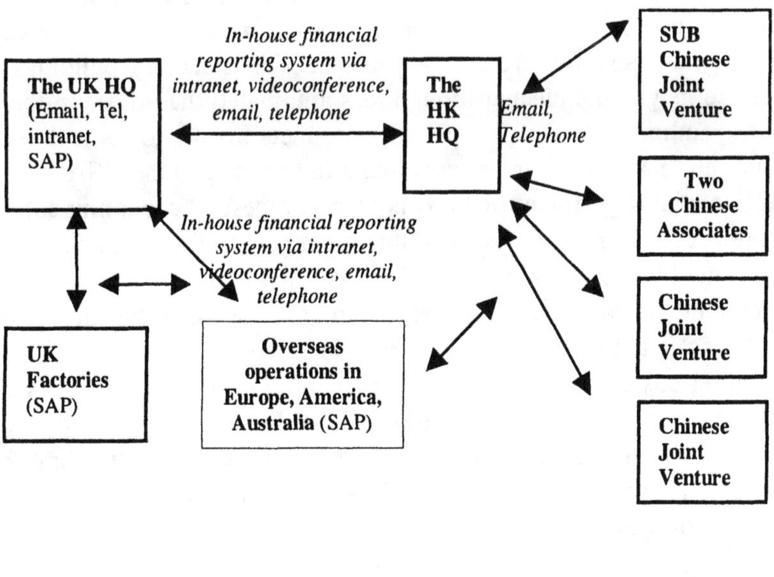

Figure 1: The organizational structure and ICTs of MAC and its relations to SUB in China

MAC's group headquarters in the UK delegates the control of the JVs to a regional head office in HK responsible for the business in China. An in-house accounting system standardizes and transfers all financial information via an intranet between the two sites. In addition, a country report is submitted by the General Manager in HK every month. Videoconferencing enables the GM to present the report in further detail. Occasionally, a few UK senior managers visited the Chinese factories. Chinese managers have not visited the group head office although they sometimes go to the UK market. The British GM and one accountant[2] based in HK, and two expatriates in Shanghai are controlling the three JVs and the associated undertakings. They are not involved in their detailed operations that are managed by local Chinese. Email, fax and telephone are major ICTs to contact and control the JVs in conjunction with the visits to each JV usually once a month.

The Chinese employees of SUB are employed directly by MAC and the local partners' influence is very limited. A strict work environment is enforced in SUB: e.g. signing one-year contracts and no factory accommodation is provided. The Chinese are eager to learn technology and management skills from expatriates and other sources, and are efficient in adapting them to the local situation. In the other two JVs, the local partners appoint the local management teams and, in these, expatriates have problems in implementing some MAC policies and transferring management skills.

SUB is still a recent undertaking. Its Western production technology is planning to continue for another six years without major renewal. Although SUB is the smallest among the three JVs, its management, production indicators, and cash flow are in good shape. SUB's factory buildings and facilities are newer and cleaner though less automated than those in the UK factory the authors visited. It supplies its products mainly to some domestic companies and partly to overseas customers in Europe and the USA.

The financial department of SUB is using domestic accounting software to prepare accounts. Parts of the accounts are manual. Final financial reports are reformatted in Excel and then emailed or faxed to the HK head office for the accountant to check them under the requirements of the group head office. The software only deals with financial accounts and has no function of calculating costs and executing management control in such areas as the material flow, purchase planning, sales and production. There is no costing information system connected with other departments notwithstanding there is an intranet in which some selective data can be shared between different departmental offices.

---

[2] A Hong Kong Chinese, her predecessor was a British accountant.

5 ANALYSIS AND DISCUSSION

We will focus our discussion on three areas: how ICTs act as an expert system (in Giddens' terms) between the two head offices and SUB; secondly on other conditions that need to be in place; thirdly, the desire of Chinese managers for an ERP indicating unexpectedly dynamic cultural characteristics.

5.1 ICT-Related Disembedding and Reembedding

Facilitated by ICTs, senior managers and accountants in MAC's head offices do not have to travel to their worldwide subsidiaries that might be thousands of miles away to obtain financial and non-financial information. Social relations between MAC's headquarters and subsidiaries are strongly mediated by ICTs which are the means of standardized information transfer such as financial reports, scorecards. Via an intranet, a standardized financial reporting system is used between the UK and HK offices. The system can be regarded as an expert system because it represents 'best practice' of management and financial accounting from the perspective of the head office (Jones and Dugdale, 2001: p.43). Its format includes classical profit/loss accounts, balance sheets and cash flow statements. An accounting manual is produced to guide how to use the system. At the HK end the accountant collects financial data from the JVs, revises and enters it into the blank areas of the system under fixed headings such as sales, operating profits, tangible assets. The whole procedure takes her about four or five days. After the data is inputted in the system, the people at the UK end can read the data a few minutes later through the intranet. They can recognize and consolidate the data in the same format as seen by the accountant; compare the performance of subsidiaries in different places, and the financial results with previous ones. Through the shared financial system, the activities in the JVs are accounted for the head offices over time and space, and communication between the UK and HK offices can have a common 'language'.

The intranet increases the disembedding of the financial language by further breaking down time and space links through place and can (re) combine the time-space configurations of information transfer. Email also has the same function of replacing one-to-one communication at the same time as well as at the same place, and increasing or delaying communication without the constraints of place and time. The financial controller in the UK commented:

Email, intranet systems certainly help us because we can turn around information we request over night. I can leave a request in the evening, and have an answer when I come back to work in the morning.

The intranet, the reporting system and the ERPs (SAP) used in MAC's European, American and Australian factories are not extended to the joint ventures in Mainland China. Five hindrances are seen as: the lack of Chinese employees' skills, worries about confidentiality with Chinese partners, the small size of the Chinese operation, the cost and complexity of the systems, and IT infrastructure. The small human resource in HK becomes a 'transfer center' to implement corporate strategies; monitor the performance of the JVs; and collect their information via email and fax monthly without staying there. Otherwise, for instance, it takes more than five-hour flight to one factory in Northern China from HK. On the other hand, visits are important for the GM. He explained:

Out of our five general managers [of the JVs and associates], only two speak English. So again email is not an easy way to communicate with them directly.

The disembedding and reembedding process via ICTs as expert systems cannot be impersonal and systematic totally out of local contexts. The power of their universalization and mobilization (Giddens, 1994: p.85) across time and space cannot get rid of language barriers and other influences such as human resources, local employees' abilities, confidentiality, size, and long travel journeys. Expatriates, face-to-face contacts, different interpretation and cultural values discussed later further illustrate the social and local importance of the process.

5.2 ICTs, Expatriates and Face-To-Face Contacts

In addition to ICTs and financial control, expatriates are another essential part of expert systems for the corporate head office to exert control. In the last few years, expatriates have brought their expertise developed in the UK or other Western countries to SUB employees. In particular two expatriates have lived in SUB for two years at different times since the setup of SUB. They have been an important conduit to transfer Western management and technology knowledge to those in the JV in a way that is hardly achieved through ICTs particularly when tacit knowledge is considered (Polanyi, 1967). In the JV, expatriates' working style or behavior communicates more vivid knowledge than manuals or technology itself. For instance, being easy going with workers without giving themselves airs and encouraging Chinese employees to challenge them have impressed local employees a lot and expressed different cultural values. One manager recalled that one expatriate organized managers to play games to learn coordination and time management. Another Chinese manager was surprised that the same expatriate encouraged managers to be a coordinator rather than a professional technician who must know everything.

As mentioned before, visits are an alternative way to ICTs for the people in the HK office to control the JVs. The previous British accountant explained:

> *If I fax a question to one factory, I would generally get an answer. But unless I went there, sat down, discussed it and explained why I thought it was strange, dug into, for example, some measurements, what was included there and what was omitted.*

Business processes and work practices embedded in and through ICTs and financial control require extensive facework commitments from expatriates when they are reembedded in another different locale. It seems difficult to introduce an expert system and manual to Chinese employees without extensive coaching and socialization. The people in HK try to 'educate', as the previous British accountant said, local Chinese employees to observe the same 'game rules' or speak a similar 'management language' to ensure the trustworthiness of information that is understandable to the managers in the UK and themselves. On the other hand, it is essential for managers particularly in HK to question and investigate the source of abstract information disembedded from local contexts. Visiting the plants frequently is one major way to achieve that aim. If the people in HK failed to explain what differences lie behind common patterns, the managers in the UK would make wrong decisions while simply comparing numbers.

British senior managers, expatriates and Chinese managers have different understandings of the operation of SUB. The controller and the IT manager in the UK have never visited the JVs and their understanding is mediated by ICTs, financial reports and expatriates. The abstract information disembedded from SUB loses local meanings and contexts as discussed above. For instance, in a monthly report from the GM in HK, the British managers got two short sentences about the financial performance of SUB:

> *Return on Sales at 12%, and strong cash flow. A quality problem will have an adverse impact on profits later this year, but the result is still forecast on budget.*

The financial controller of SUB said that she and her colleagues spent a lot of time on, for example, collecting money. The three words of 'strong cash flow' do not disclose the process of that effort. In China, some customers do not want to pay money on time even if there are formal contracts. Strong personal relationships with customers could reduce the problem. The British controller and even some expatriates who have stayed in China for several years could not understand this and think that contracts (legal relationship) should be honored based on their experience in Western countries. Thus, we find that ICTs are an important part of expert systems linking the UK and HK head offices and the JV, but that the local contexts they link are themselves mediated by ICTs especially at the UK headquarters. We cannot find any pure 'social' relations that Giddens' theorization might suggest. For the ICT-enabled systems to be trusted to work they require both the presence of expatriate managers and the continuing shuttling back and forth of accountants seeking to standardize and control the ways in which data is entered into corporate systems. To explain more about the human side of this process, we move to the area of the interactions between cultural characteristics, and ICTs and other expert systems.

5.3 Chinese Cultural Characteristics and ICTs

As noted above, the collection of financial information is still made manually in SUB. The Chinese managers are very keen to have a small ERP type system for production and other functions. The major reasons to install an ERP are to control costs, improve production planning, facilitate information transfer, increase the materials and money flow, and formalize management. However, the GM in HK is not keen on the change because the system is costly and too advanced for the factory. The IT senior manager in the UK has a little knowledge of the Chinese JVs and is afraid that putting a system together with the global intranet in China would disclose MAC's technology know how to local Chinese partners although SUB should be regarded as a wholly-owned subsidiary rather than a JV. Moreover, the financial controller is worried about the abilities of local employees stating that there is no reason to put a complex system in the JVs since parts of accounting and production process are still manual, and there were a lot of difficulties in implementing SAP in the European business.

In the literature, Chinese managers are seen as preferring to low context information and hold information as a major instrument of personal power and avoiding ICT systems (see Martinsons and Westwood, 1997: p.224). In contrast, the Chinese managers in SUB are very keen on an ERP despite doubts from Western managers. The previous introduction of expensive production technology by the Chinese side is beyond the expectation of the British side and the technology proves to be efficient in the factory. One aim of the call for an ERP is to reduce personal intervention such as information corruption by people. At the same time, the Chinese managers do not downplay the significance of relationships for business since they live in a Chinese society that does still promote relationships. They actively seek and maintain relationships with customers and government officials. But they do not want relationships to intrude on data collection and analysis within the factory. Thus, we find evidence of changing cultural characteristics that do not apply homogeneously across Chinese society, where change is in part connected with ICTs and the promise of more advanced ICTs. Equally, in the literature, there is an equating of western management with a preference for more formal and systems based approaches (*ibid.* p.217). We found evidence that some expatriates were able to recognize the importance of relationships in transacting Chinese business and were active in helping local government officials use the factory as a showcase. They were prepared to use a variety of approaches to influence Chinese workers rather than relying on standard business processes.

6 CONCLUSIONS

This chapter examined issues in cross-cultural coordination and control related to ICTs in MAC, a UK manufacturing multinational operating in China, using Giddens' theory of modernity and an anthropological view of culture. As an expert system defined by Giddens, ICTs facilitate the disembedding of communication and information transfer in different time-space domains without the limitation of place and therefore enable the head office to expand its business around the world and control its worldwide operations at a distance. However this is not simply the 'death of distance' (Cairncross, 2001). The ICT enabled co-ordinating mechanisms only 'work' when other features such as expatriates, face-to-face contacts and interpretations are present. For instance, the British accountant from Hong Kong spent considerable time going to these factories seeking to ascertain trust by making sure the figures entered into the systems follow established patterns.

Interestingly, ICTs in the form of ERPs and the in-house financial reporting system are seen not to work in China by MAC executives and so all systems between China and other parts of MAC are mediated through the Hong Kong office. Their reluctance to use ERP systems in China is based on a lack of trust of Chinese employees' ability to use these systems properly; in the costs of installing and maintaining the systems; in the difficulty in training staff to 'understand' the ERPs. In short, the benefits of ICTs tend to flow from a trusted, well-trained workforce who can use these systems 'properly'.

On the other hand, Chinese managers are very keen to use an ERP system as they see it as an advanced Western technology that they would like to understand and which they believe would bring benefits to their company. This is in contradiction to many of the cultural characteristics presumed of Chinese in other analyses (cf. Martinsons and Westwood, 1997). It is interesting to compare the case of ICTs in SUB with that of the production technology being used at this site which is comparable with that in Europe, and which appears to be working without major difficulty. Therefore, ICTs act as an arena where organizational culture change is often expressed. Of course, we do not mean that cultural differences would disappear since political and local environments still require different actions.

Returning to the issue of ICTs as a mechanism of globalization set out at the start of this chapter, we would like to emphasize four features. Firstly, ways are needed to embed locally centered issues and then to disembed them: these are central to make ICTs work. This is an insight that Giddens' analysis illuminates. Secondly, that ICTs are also material artifacts continually rubbing up to issues such as upgrades, cost, infrastructure and so on. Only when these seemly mundane issues work does even the possibility of ICT-mediated interactions emerge. Thirdly, that ICTs, as expert systems in Giddens' terms, are increasingly penetrating social relations in different locations. However, the seemingly 'closeness' through ICTs 'filters out' or misrepresents much of local culture and knowledge that occurs in different local contexts. Finally, though Giddens' argument centers on modernity originated from Western countries, we suspect that the appropriation of both technologies and management practices in domains such as China are likely to lead to reinterpretations that are far reaching and unknowable. Rather than homogeneity we will find the reassertion of locality surrounding the use of the mechanisms of globalization such as ICTs. This remains an interesting agenda for the future.

ACKNOWLEDGEMENTS

We would like to acknowledge the participants in the anonymous UK manufacturing multinational who have been extremely helpful in giving their time and assistance. The two reviewers' useful comments are appreciated.

REFERENCES

Avison, D. E., and Myers, M. D. "Information Systems and Anthropology: an Anthropological Perspective on IT and Organizational Culture", *Information Technology and People*, (8:3), 1995, pp. 43-56.

Barrett, M., and Walsham, G. "Electronic Trading and Work Transformation in the London Insurance Market", *Information Systems Research*, (10:1), March 1999, pp. 1-22.

Barrett, M., Sahay, S., and Walsham, G. "Information Technology and Social Transformation: GIS for Forestry Management in India", *The Information Society*, (17:1), January 2001, pp. 5-20.

Cairncross, F. *The Death of Distance 2.0*, Harvard, Mass.: Harvard Business School Press, 2001.

Castells, M. *The Rise of the Network Society*, Oxford: Blackwell Publishers Ltd., 2000.

Giddens, A. *The Consequences of Modernity*, Cambridge: Polity Press, 1990.

Giddens, A. *Modernity and Self-Identity: Self and Society in the Late Modern Age*, Cambridge: Polity Press, 1991.

Giddens, A. "Living in A Post-Traditional Society", in *Reflexive Modernisation*, U. Beck, A. Giddens, and S. Lash (eds), Cambridge: Polity Press, 1994, pp. 56-109.

Gore, C. "The Rise and Fall of the Washington Consensus as a Paradigm for Developing Countries", *World Development*, (28:5), 2000, pp.789-804.

Hanseth, O., Ciborra, C. U., and Braa, K. "The Control Devolution: ERP and the Side Effects of Globalization", *The DATA BASE for Advances in Information Systems*, (32:4), Fall 2001, pp. 34-46.

Hayes, N. "Boundless and Bounded Interactions in The Knowledge Work Process: The Role of Groupware Technologies", *Information and Organization*, (11:2), April 2001, pp. 79-101.

Hofstede, G. *Culture's Consequences: International Differences in Work-Related Values*, London: Sage, 1980.

Hofstede, G. *Cultures and Organizations*, London: HarperCollinsPublishers, 1991.

Jones, T. C., and Dugdale, D. "The Concept of An Accounting Regime", *Critical Perspectives on Accounting*, (12:1), February 2001, pp. 35-63.

King, W. R., and Sethi, V. "An Empirical Assessment of the Organization of Transnational Information Systems", *Journal of Management Information Systems*, (15:4), Spring 1999, pp. 7-28.

King, W. R., and Sethi, V. "Patterns in The Organization of Transnational Information Systems", *Information & Management*, (38:4), February 2001, pp. 201-215.

Kuper, A. *Culture: The Anthropologist's Account*, London: Harvard University Press, 1999.

Latour, B. *Pandora's Hope*, Cambridge, Mass: Harvard University Press, 1999.

Martinsons, M. G. and Westwood, R. I. "Management Information Systems in the Chinese Business Culture", *Information and Management*, (32:5), October 1997, pp. 215-228.

Martinsons, M. G., and Hempel, P. S. "Chinese Business Process Re-engineering", *International Journal of Information Management*, (18:6), December 1998, pp. 393-407.

Myers, M. D., and Tan, F. B. "Beyond Models of National Culture in Information Systems Research", *Journal of Global Information Management*, (10:1), Jan-March 2002, pp. 24-32.

Polanyi, M. *The Tacit Dimension*, London: Routledge & Kegan Paul, 1967.

Roche, E. M. "Multinational Corporations – The Emerging Research Agenda", *Journal of Strategic Information Systems*, (5:2), June 1996, pp. 129-147.

Torre, J. D. L., and Moxon, R. W. "Introduction to the Symposium E-Commerce and Global Business: The Impact of the Information and Communication Technology Revolution on the Conduct of International Business", *Journal of International Business Studies*, (32:4), Winter 2001, pp. 617- 639.

About the Authors

Wei Liu is a third-year PhD student at the School of Accounting and Finance of the University of Manchester. He was born in China. His PhD research project looks at how the head offices of three UK multinational corporations employ ICTs and other mechanisms to control and coordinate their subsidiaries and joint ventures in China. Giddens' theory of modernity in particular is used to explore the process of globalization in the three cases. Wei has finished his extensive fieldwork at thirteen research sites in the UK and China and is at the stage of writing up his PhD thesis. He can be reached by e-mail in msryjwl2@stud.man.ac.uk

Chris Westrup is a senior lecturer at the School of Accounting and Finance. He is interested in the various aspects of development and ICTs; in the use of large scale systems ICT systems and their development. He can be reached by e-mail in msryscw@fs1.acc.man.ac.uk

11

ON THE IMPLEMENTATION OF AN INFORMATION SYSTEM IN THE MOZAMBICAN CONTEXT

The EDM case

Esselina Macome

Faculty of Science, Department of Mathematics and Informatics, Eduardo Mondlane University, Maputo, Mozambique

Abstract: Information systems are widely acknowledged to be of central importance in contemporary organizations. The transfer of information systems designed for other places that are different from the implementation context, has been part of the global-local debate. In this paper it is argued that the local context should be considered in the implementation of such IS and that local stakeholders should be involved in the entire process. This paper attempts to explore the issue of global-local interaction, and we will use an example of an invoice information system, called Galatee, designed by an international team and implemented in the Mozambican Electricity Company (EDM). Some implications will be drawn for the conference theme on IS perspectives and challenges in the context of globalisation.

Keywords: Information System, technology transfer, adoption and use of ICT, actor network theory (ANT), translation, *Galatee*.

1 INTRODUCTION

The implementation process of ICT-based initiatives and the adoption of new management techniques provide organizations in developing countries with the prospect of actively participating in globalization processes which could contribute to national development efforts (Bada, 2000; Avgerou, 2001). In Mozambique many organisations are in the process of implementing different ICT-related projects in order to improve their efficiency. A systematic study on how a specific ICT-related initiative was implemented might help to guide future efforts in the country. It is from this perspective that a research study was conducted to address the question: How does a specific ICT-related initiative become institutionalised? To respond to this question a case study was undertaken describing the analysis of an IS adoption in the Electricity Company in Mozambique (EDM). The information system described here relates to the invoicing of electricity in Mozambique, and is called *Galatee* meaning *the* 'God of Water'. *Galatee* is being implemented in five EDM operational areas, namely Maputo City (south), Beira, Chimoio, (centre) Nampula and Nacala in the northern part of Mozambique. A French company called *SAUR* was contracted by EDM to design and implement this system.

This research is concerned with the process of introducing an innovation that is typically designed in the developed world and introduced to developing countries like Mozambique through a 'technology transfer' process. This process involves not only the transfer of technology but also various organizational and management practices that are often found to contribute to the tensions within the socio-organizational contexts of developing countries (Avgerou, 1996; 2001; Bada, 2000). The potential value and effectiveness of the transfer process is dependent on the local socio-organizational conditions and the ability of the organization to adapt the ICTs to address local priorities (Odedra, 1991; Avgerou and Walsham, 2000).

An approach to studying the process of technology transfer is provided by Rogers' (1995) theory of diffusion of innovations. Diffusion theory researchers have tried to understand the factors that influence ICT/IS innovation adoption decisions (Attewell, 1992; Fuller and Swanson, 1992; Cool et al., 1997), typically at the organizational and intra-organizational level, rather than with an inter-organizational focus (Prescott and Conger, 1995). Another assumption of the traditional diffusion theory perspective has been the homogeneity of the institutional environment across the adopting units, which tends to obscure variations across the units arising from geographic distance from the source of the innovation, local norms and regulations, and the availability or unavailability of factors to support the diffusion system. It has now been well established by IS researchers that contextual factors play an important role in influencing the ability of an organisation to adopt and effectively apply technological innovations (Williams and Edge, 1996; Du Plooy, 1998; Allen, 2000; Walsham, 2001). This view of diffusion theory is especially problematic in the current period of globalisation that is characterised by processes of rapid technological change.

In this paper, we attempt to address these limitations, drawing upon an alternative theoretical approach that is provided by Actor-Network Theory (ANT) (Vidgen and McMaster, 1996; McMaster et al., 1997; Büscher and Mogensen, 1997; Knights and Noble, 1997; McMaster et al., 1998). The topic of this paper might be considered as contributing to the debate on IS perspectives and challenges in the context of globalisation.

This paper is divided into six main sections. It starts with an introduction in which the research topic is outlined and the importance of the theme of the research is highlighted. The second section describes the research problem and the main objectives of the study. The next section describes the process of how the study was conducted and also presents the setting for the case study. The fourth section starts by describing the case study of the *Galatee* implementation and continues with the analysis of the case by using the translation phases from the ANT perspective. The findings of the case study based on the ANT perspective are discussed in Section 5. The final section sums up the paper by presenting the conclusion.

2 RESEARCH PROBLEM AND OBJECTIVES

Recent advances in ICTs are becoming central to the process of socio-economic change and development in different contexts, especially that of developing countries (Avgerou, 2001; Avgerou and Walsham, 2000; Walsham, 2001). While different organizations in Mozambique are in the process of implementing different ICT-related projects in order to improve their efficiency, little is known about their impact and the successes that are achieved on the ground. There is currently a dearth of studies that analyze such organizational experiences with new ICTs and associated management techniques, or deal with how these lead to practical improvements in organizational functioning. A systematic analysis of these experiences can help to guide future efforts in the country, and this research seeks to contribute to this body of knowledge.

The main aim of this research study is to analyze how an information system is implemented within a Mozambican organization. To this end an ANT perspective will guide the analysis.

3 RESEARCH SETTING, DESIGN AND METHODS

The main objective of this section is to describe how the research was conducted. The context within which this research took place will also be presented.

The research study was conducted in Mozambique, a developing country in Southern Africa. This country belongs to what are considered the poor countries of the world. Its Human Development Index was calculated at approximately 0,341 and its Technological Achievement Index as 0,066. The estimated population of this country is about 17.242.240 inhabitants.

In order to study the process of adoption and use of an ICT-related initiative, a case study of the implementation of an invoice information system at the national electricity company was conducted.

The historical context of EDM, the Mozambican electricity company, falls into two main periods, from its creation/inception in 1977 to 1995 as a parastatal company and subsequent to 1995, as a public enterprise. EDM is a public company and deals essentially with the production and distribution of electrical power to the whole country. It was formed as a parastatal company that was known as EDM_E.E., for the purpose of integrating all the electrical power generation centres that existed in the country, just after independence in 1975. This was expected to make a major contribution towards satisfying the basic needs of the country in terms of electricity, particularly in relation to the development of agriculture, industry, services and the public sector.

With the liberalisation of the economy, in 1995 EDM was transformed into a public company, its mission being to improve the quality of customer service, maximise revenues, introduce new management practices, and promote the autonomy and decentralisation of regional and operational offices. This company has about 2.860 employees and approximately 186.208 customers (EDM, 2000a).

The research approach selected was that of a case study based on the interpretive perspective in order to enable us to increase our understanding of the critical social and organizational issues relating to the adaptation and adoption of ICT/IS in organizations. According to Walsham (1993) the purpose of the interpretive approach in information systems (IS) is to produce an understanding of the context of IS and the process whereby IS influences and is influenced by the context. Interpretive approaches give the researcher greater scope to address issues of influence and impact, and to ask questions such as 'why' and 'how' particular technological trajectories are created (Orlikowski and Baroudi, 1991).

The fieldwork for this study took place in Mozambique where the EDM is situated. This fieldwork began in May 2000, six months after the *Galatee* implementation, and lasted for eight months. The historical reconstruction of the story of *Galatee*, its implementation and use was possible through interviews with key people involved in the project, system users and the analysis of secondary documents and observations, as well as formal and informal conversations. In general, the interviews provided explanations about the functioning of each operational area, challenges being experienced in the ICT initiation and how these are being addressed. This data-gathering process helped to reconstruct the history of each ICT initiative involved in *Galatee*.

Thus, the empirical component consisted of approximately 38 individual and group interviews, and one group meeting. Interviews typically lasted about 2 hours. The *Galatee* system demonstration took about 3 hours and much time was spent on informal discussions with many staff members as well as on making general observations.

4 RESULTS

4.1 The EDM case study - *Galatee*

4.1.1 The Technology Used

As a way to improve the invoicing system and to increase EDM's income, the previous Ingress invoicing system had been implemented in the early nineties by a private IT company, IC. However, it needed to be improved and extended and made Y2K-compliant. Hence a new information system, Galatee, was implemented at EDM. Galatee was based on Sybase in a Unisys environment.

At the head office, there are two powerful Solaris servers for use and backup systems. Data stored in the servers were also backed up on discs (albeit in the same building as the server). In each operational area there were at least two servers linked to the work stations located in the cashier's and client management's offices. The system was designed in such a way that the cashier can use an electronic pen to enter the invoice details at the moment of payment.

The Galatee system had previously been used in various countries such as the Ivory Coast, Brazil, France and Poland. There was a need to make certain changes in order to accommodate the local context, for instance the use of the Portuguese language. When the research commenced, the process of adapting changes was still in progress. The development team was based in Abidjan. For the implementation in Mozambique all adaptation occurred at the head office where supercomputer mainframes are centralized and control the network of sales work stations in all five operational areas. They are also connected to the EDM server, accessible via a wireless data communications network.

Galatee was developed as a technical tool that could also manage, control and invoice the customers in order to maximize profits. The Galatee system has the following modules:

- Production of the electricity invoice and handling of its payment
- Management of electricity cut-off for those who are not paying their invoices
- Control of partial payment of invoices
- Client management
- Management of off-line cashier.

4.1.2 Organizational and Political Conflicts

The new strategy is that *Galatee* belongs to the invoicing sector, but its operation and data communication form part of the IT Department's function. Based on this strategy, some technical personnel from the IT Department were transferred to the Commercial Department, so as to work closely with the *Galatee* system.

Some staff from the operational areas felt that there was not enough involvement of local technical people in some IT-based projects in the company. One of the interviewees expressed the following opinion:

> *I cannot understand why the Mozambican IT staff are not leading this (Galatee) project. This lack of involvement of local IT staff is creating some problems. For instance, all our problems must be reported to SAUR and we are not sure if the language translation of our concerns is well done or not. [Interview 30, pp. 29-30].*

Another difficulty relating to the lack of involvement of indigenous people in this project will have to be faced when the *SAUR* team is no longer at EDM. Since IT staff were not directly involved in the initial development phase of the system, and assuming that the IT Department will be responsible for maintenance of *Galatee,* there is clearly an urgent need to address the issue of IT Department involvement. This problem of lack of involvement of IT staff in the project had already been identified by EDM, but needs to be readdressed. It is important to bear in mind that the IT Department is an official part of the *Galatee* project team, but its involvement in the analysis and selection process of the software was minimal due to different organizational issues within EDM.

In anticipation of the problems caused by *SAUR* withdrawal, EDM has been requesting greater and more formal involvement of IT staff. To support this technology (transfer), a team from the IT Department visited the Ivory Coast in May 2001 in order to make a detailed study of *Galatee* with the development team.

The organizational climate, particularly in one of the operational areas, became tense and fraught with conflict, with rivalries between divisions, and so the staff became demotivated.

Two categories of training were organized, one for IT personnel at the operational area and the other for the billing staff (cashiers). In general, all participants considered that the training was too short and that it did not help them much. One of the interviewees commented 'For me it was not real training; it was a demonstration of the *Galatee'* [Interview 35, p.34].

The training sessions for the invoicing staff lasted 3 days, which were spent explaining the new commercial policies and the rationale for the new pricing structures. Staff judged it inadequate since it did not provide enough information on the user-computer interface aspects, and it was therefore seen as merely a system demonstration.

With the new computerized system, EDM also implemented a monitoring system to keep track of the number of transactions, the time taken for each transaction, and the types and amount of invoices dealt with by each cashier. However, the *Galatee* system did not differentiate between two cashiers working at the same work station at different times. This made it difficult to identify the number of invoices dealt with by each cashier. The first cashier who used the work station was simply considered to have handled all the transactions. The invoicing staff were therefore subjected to a change of computer system and its user interface together with changes in training, role and qualifications, working conditions, handling of performance monitoring and reporting.

There are different opinions regarding the *Galatee* invoicing system project. During the interviews, different users presented the practical problems that they were currently facing. These problems were similar in all operational areas and they were also reported in EDM (2000a). The problems presented can be grouped into different categories such as: (a) technical, (b) organizational and managerial, and (c) people-related.

4.2 The EDM Case as Viewed Through the ANT Lens: Translation Phases

This section analyses the case study from the ANT perspective by drawing upon the sociology of translation. The focus is on how the actor-networks grow, change and stabilise during the process of adoption and use of Galatee, examined in particular organizational and societal contexts.

The concept of the sociology of translation is based on Latour's (1991; 1997) alternative to technological determinism, in which things do not happen unless other actors make them happen. This implies that each actor who takes the project further may take it in a different direction than that intended by the previous actor. Latour (ibid.) uses the term 'translation' to describe this effect. Translation operates between actors: an actor gives a definition to another actor, and imputes to him/her/it/them interests, projects, desires, strategies, reflexes and afterthoughts (Callon, 1991).

ANT proposes the concept of 'translation' to replace 'diffusion' and sees innovation spreading as a result of how actors 'translate' the interest of others so that they become aligned with their own interests. ANT does not see technology as an independent artefact that 'diffuses' out from one central point, but views technology as part of a complex heterogeneous network of human actors and non-human artefacts. The effectiveness of the transfer process is then dependent on how these networks are created, stabilised and strengthened over time (Callon, 1986; Callon and Law, 1989; Latour, 1999; 2000). Callon (1986) characterizes four moments of translation:

- Problematisation or how to become an indispensable actor
- *'Interessement'* or how the allies are locked into place
- Enrolment
- Mobilisation

These are the translation phases that are going to be followed in the analysis of the EDM case in this paper.

4.2.1 Problematisation

Problematisation is the first 'moment' of translation where actors seek to identify and define the obligatory point of passage (OPP) and attempt to impose their definition of the problem on others. Problematisation is an indispensable moment as it implies that the problem resolution can only be negotiated through the obligatory point of passage.

Identification of actors/agencies is a way to find out who the actors are and what they are doing. Using Latour's language it was necessary to 'follow them' (Latour, 1987) in their daily interactions with *Galatee* and others actors in their attempts to enrol them. The initiators of the *Galatee* system at the EDM are the Commercial and Informatics Departments. The key actors in *Galatee* at the EDM consist of five main organisational groups: the initiators, the Invoice unit, Cashiers, the Training Unit at the Human Resource Department and the consultant company – SAUR. Other actors included the data items, the technological infrastructure, organizational and management routines and procedures, external influences like the Mozambican ICT policy, French Cashier Agency financing support, management policy and the expectation of the EDM Customers. The characterisations of some of these actors formed part of the description of the case study in the previous section.

Each of the actors has a role to play and an interest in being part of the actor network. The next part shows examples of the role and interests of some actors. The Commercial Department is responsible for all commercial issues involved in managing the sale of electricity; the basic product of EDM, to its customers throughout the country. One of the tasks of the Commercial Department is to collect data about electricity consumption, to produce invoices and collect payments. In addition, this department serves as the interface point between EDM and its clients. The IT Department is responsible for strategic and operational level IS/IT issues. The IT Department was marginally involved in the initial stage of the *Galatee* project, but subsequently became crucial at the point of conversion of data from the *Ingress* system to *Galatee*. The invoice personnel have the task of introducing the raw data (electricity consumption units) into the system for the production of invoices, which are then distributed to the customer's residences. The cashiers use the *Galatee* system to collect payments from the customers based on the invoice and receive all payments made to the company, for example, payments concerning electricity contracts, overdue payments and others.

The initiators of the *Galatee* system problematised the initiation of *Galatee* by pointing out that the *Ingress* system was not Y2K-compliant and that there was also a need to have a customer-oriented, integrated, computer-based information system. It was argued that *Galatee* would contribute to the efficiency and extension of control of EDM through improved payment collections. These arguments constituted the rationale behind EDM's decision to purchase the *Galatee* system.

Galatee became an OPP for the various departments, as it was only by using this system that electricity bills could be invoiced, and electricity sold to the public. For instance, the Commercial Department sees *Galatee* as a way to achieve efficiency in invoicing and client services and hence to achieve the EDM objectives of customer satisfaction and improving collection revenues. The commercial manager from Nampula described their aims:

> *Since transformation of the EDM into a public enterprise, we are becoming more and more committed to the improvement of quality service to our clients. So if we have an efficient information system, which supports us in this area, we think that we will change the EDM's present situation. This means that we will be able to collect more payments [Extract from the commercial report of the Operational Area Nampula, pp. 3-4].*

4.2.2 Interessement

At present the system has been partially implemented in all the EDM operational areas. However, there is a need to ensure that all its modules are properly used, and also to interest other user departments in making better use of *Galatee*.

Being part of the Commercial Department, the Invoice Unit naturally sees the use of *Galatee* as integral to its functioning, and is therefore interested in increasing *Galatee's* role.

There is a pervasive and accepted Mozambican cultural principle that dictates that subordinates must obey their managers, making it only a remote possibility that the cashiers (front-end users) might reject this system. The managers will try to ensure that the cashiers accept the system, and the latter will feel obliged to obey the decisions made by their managers. Therefore, there are few grounds for expecting any type of resistance from the personnel when it comes to using the new system. Administrative disciplinary techniques, such as the evaluation, monitoring system and control of transactions made by each cashier, were also introduced to ensure the staff's compliance. This institutionalised discipline of cashiers ensures the Commercial Department's successful *interessement*.

The process of monitoring and controlling transactions led to the involvement of the Finance and Audit Departments in using *Galatee* as a tool to control the collection of payments from the customers. For example, in Nampula we found the contracts of three cashiers with EDM being cancelled due to irregularities in the payment of electricity invoices (EDM, 2000b).

The IT Department used the opportunity presented by *Galatee* implementation to convince the EDM top management to upgrade the enterprise-wide ICT infrastructure and to define a new strategic role for ICT in rendering organizational change possible and improving efficiencies.

The managers presented the system to the EDM employees as an opportunity to overcome the Y2K problem and as a means of increasing efficiencies.

4.2.3 Enrolment

Enrolment requires the initiators to persuade and convince the other actors to join them. The motivation of all actors is of central importance to enrolment, since enrolment has to do with bonding elements together.

The initiators of *Galatee* decided to acquire an integrated customer-oriented, computer-based invoicing information system from the international market based on the argument of lack of adequate in-house IT skills to develop the system and to comply with the conditions of the funding source market. As a result, the *SAUR* was contracted to undertake the development and implementation of the required system for EDM.

The *SAUR* experienced many problems during the implementation of the *Galatee* system, including some that inhibited communication, as well as the fact that not all EDM conditions were included the system. For example, the cashiers could not differentiate between currencies (MZM - *Metical* or US $) in the invoices' payments. Presently the customer, depending on his/her contract, can pay the electricity invoice in the national currency or in American dollars. In addition, the *Galatee* system did not permit the cashiering of the overdue amount, forcing the cashiers to make these transactions manually. This made it difficult to handle all customers using the same procedures.

The IT Department's involvement during the procurement of the system was not a smooth process. Despite this situation, the project manager achieved a high enrolment of IT Department members during the implementation by emphasising their role in the success of implementation of this system after the *SAUR* finished its work and left.

All initiators continue to be part of the network implementing *Galatee*, and new actors also joined this network.

4.2.4 Mobilization

The mobilization of actors, according to Callon (1986: p. 216), results in rendering agencies mobile that were static beforehand, in order to traverse the defined OPP. This helps to strengthen the network of alliances and provides stable information systems, and renders possible the linking of agencies that were not linked before. For example, the cashiers who were not linked to the client management personnel earlier on (opening contract, etc.), now became linked. This implies that *Galatee* could only be stabilized when relations between these actors were established. Since not all modules of *Galatee* were implemented as yet, further and new relations between actors would need to be built up during the process of adopting and using other *Galatee* modules.

In summary, the initiators – the Commercial and IT Departments – proposed *Galatee* as the only solution to the Y2K problem and to improve invoicing efficiency. The implementation of *Galatee* helped to build a strong network of alliances between the Invoice Unit personnel, and the Finance and Auditing Departments. It seems that *Galatee* has become an indispensable resource in the commercial area of EDM, although the network for its implementation is not yet closed. The designers of this system are still in the process of adapting it to the EDM regulations and procedures. The personnel who are directly affected by *Galatee* are attempting to learn the required skills to use it. However, some personnel have showed dissatisfaction with project management practices, as well as with the level of *Galatee* adaptation and as regards the responsiveness of the development team in addressing their problems. This was due to the low level of consultation with the users in the adoption and use of *Galatee*. As a result, the new EDM network was not clearly defined and recognized by all users, thus influencing the level of institutionalization of *Galatee* within EDM.

5 DISCUSSION

The experience of the implementation of *Galatee* at the EDM suggests a number of insights for both theory and practice related to ICT/IS implementation in organizations. This case illustrates an aspect that has been emphasized by other researchers – the importance of involving users in the process of introducing new ICT initiatives. This brings to the debate the issue of team building for the ICT/Is implementation. In this case, it was clear that the level of user involvement was low and this could have contributed to the creation of some of the aforementioned problems.

The analysis of this case study has implications for the transfer of ICT and management techniques in the context of the debate about global-local interaction. The Information system analyzed in this paper was developed in other countries and applied in Mozambique as part of the globalization process. One of the implications is that global-local interaction does not happen in a homogenous context. This implies that the local context must be able to mediate global influences in order to derive benefit from this interaction process. Evidence from the case study also suggests that ICT-related initiatives implemented in the case as features of the global institutional environment, were adapted in an organizational context. From this point of view, in the implementation of a new ICT-related initiative the social context, and the history and tradition of doing things within the local organizational context must be taken into consideration. Consequently this implies that the organizations need to have skilled human resources with the ability to creatively adapt new technologies and global practices to the local context and also manage the whole process of implementation. This has significant implications for human resource development as well as for education.

From this case study, it is apparent that one of the major challenges of the process of adoption and use of ICT in organizations is related to the transformation of the organization in order to meet customer and competitive demands. This transformation has to be seen as facilitating the capacity to learn to manage ICT-enabled change.

New ways of working necessarily bring about shifts in organizational power, culture, process and structure. This also implies a redistribution and diffusion of learning in the organization caused by a new way of working. This study shows, as other studies have, how difficult it is to manage change. ICT-enabled change adds a new dimension of difficulty; new technologies stimulate unprecedented processes of change throughout the organization, including shifting the location of knowledge levels within the organization.

Another important role is to educate the organization's business units to be directly involved in the process and understand the implications of change before initiatives are implemented. This indicates the need to learn what change management is, and how to control new processes in the context of organizational transformation.

Overall, the above findings reinforce the value of adopting an interpretivist perspective and the use of ANT to focus on meaning and flow with the active actors in the building and stabilizing of a network.

6 CONCLUSION

The story described in the EDM case study in this paper is a story about active builders of socio-technical actor-networks, and how the actor-network grows, changes and stabilizes within the global and local interaction perspective.

The four steps in the translation of the adoption and use of ICT-related initiatives point to different skills that human actors (such as managers, ICT professionals, and users) may develop to enhance their interaction with the non-human actors. Problematisation and *interessement* demonstrate how important it is for the managers and ICT professionals to develop communication skills, which permit them to convince and persuade other agencies to traverse the OPP. Enrolment depends on the capacity for negotiation of the initiators to convince other actors to enrol in the initiative. This means that the negotiating skills of initiators (managers, ICT professionals) might increase the likelihood of enrolling other agencies. The mobilization phase is concerned with the identification of the spokesmen or representatives integrated in the network.

From this analysis three major issues emerge: (a) the role of actors' involvement and their efforts to introduce and institutionalize a new system such as *Galatee*; (b) the communication and negotiation skills of the actors to create, translate and stabilize the network of alliances; and (c) the management of the implementation process itself as an organizational change.

Through the description and analysis of this case study, we are attempting to contribute to the global-local debate by arguing that the implementation of the ICT-related initiatives in organizations should take into account the local context as a way to contribute towards the likelihood of the ICT initiative becoming institutionalized.

ACKNOWLEDGEMENTS

The author would like to thank the representative of the EDM who gave her the opportunity to research a real-world situation. I also thank the IFIP8.2&9.4 referees for their valuable and constructive comments. My grateful thanks also go to my PhD supervisor, Professor JD Roode, for his constructive comments during this research work.

REFERENCES

Allen, J.P. "Information Systems as Technological Innovation", *Information Technology & People*, (13:3), 2000, pp. 210-221.

Attewell, P. "Technology Diffusion and Organisational Learning: The Case of Business Computing", *Organisation Science*, (3:1), 1992, pp. 1-19.

Avgerou, C. "Transferability of Information Technology and Organisational Practices", in *Global Information Technology Socio-Economic Development*, edited by M. Odedra-Straub, Ivy League, Nashua, New Hampshire, 1996, pp. 106-115.

Avgerou, C. *Information Systems the Diversity of Globalisation*, UK: Oxford University Press, 2001.

Avgerou, C. and Walsham, G. Editors of *Information Technology in Context, Perspectives from Developing Countries*, UK: Ashgate, 2000.

Bada, A. B. "Global Practices and Local Interests: Implementing Technological-Based Changes in a Developing Country Context", *PhD thesis*, Department of Information Systems, LSE, University of London, 2000.

Büscher, M. and Mogensen, P.H. "Mediating Change: Translation and Mediation in the Context of Bricolage", in *Facilitating Technology Transfer Through Partnership Learning from Practice and Research* edited by T. McMaster, E. Mumford, E; B. Swanson, and D. Wastell, London: Chapman & Hall, 1997.

Callon, M. "Some Elements of a Sociology of Translation: Domestication of the Scallops and the Fishermen of St Brieuc Bay", in *Power, Action and Belief*: edited by J. Law, Routledge and Kegan Paul, London and New York, 1986, pp. 196-233.

Callon, M. "Techno-Economic Networks and Irreversibility", in *A Sociology of Monsters: Essays on Power, Technology and Domination*, edited by John Law, Sociological Review Monograph 38, London, England, 1991, pp. 132-161.

Callon, M. and Law, J. "On the Construction of Socio-technical Networks: Content and Context revisited", *Knowledge and Society*, (9), 1989, pp. 57-83.

Cool, K.O., Dierickx, I. and Szulanksi, G. "Diffusion of Innovations Within Organisations. Electronic Switching in the Bell System 1971-1982", *Organisation Science*, (8:5), 1997.

Du Plooy, N. F. "An Analysis of the Human Environment for the Adoption and Use of Information Technology", *PhD thesis*, University of Pretoria, South Africa, 1998.

Electricidade de Moçambique (EDM). *Relatório Interno: Problemas do Galatee*, EDM, Maputo, Mocambique, 2000a.

Electricidade de Moçambique (EDM). *Despacho de Cancelamento de Contratos de dois Funcionarios - Caixas*, Área Operacional de Nampula, Nampula, Mocambique, 2000b.

Fuller, M. K. and Swanson, E. B. "Information Centers as Organisational Innovation: Exploring the Correlates of Implementation Success", *Journal of Management Information Systems*, (9:1), 1992, pp. 47-67.

Knights, D. and Noble, F. "Networks and Partnerships in the Evolution of Home Banking", in *Facilitating Technology Transfer Through Partnership Learning from Practice and Research*, edited by T. McMaster, E. Mumford, E.B. Swanson and D. Wastell, London: Chapman & Hall, 1997.

Latour, B. *Science in Action: How to Follow Scientists and Engineers through Society*, Cambridge, Massachusetts: Harvard University Press, 1987.

Latour, B. "Technology is Society Made Durable" in *A Sociology of Monsters. Essays on Power, Technology and Domination*: edited by J. Law, Routledge, London and New York, 1991, pp. 103-131.

Latour, B. "On Actor-Network Theory: Few Clarifications", *Centre for Social Theory and Technology*, Keele University, UK, 1997.

Latour, B. "On Recall ANT", in *Actor Network Theory: and After*, edited by B. Latour, J. Law, and J. Hassard, Oxford: Blackwell, 1999, pp. 15-25.

Latour, B. "When Things Strike Back – A Possible Contribution of Science Studies", *British Journal of Sociology*, Special Millenium issue, 2000, edited by J. Urry. Available at: http://www.ensmp.fr/~latour/Articles/sciencesstudies.html, (Accessed at 25 June, 2001).

McMaster, T., Vidgen, R. T. and Wastell, D. G. "Technology Transfer: Diffusion or Translation?", in *Facilitating Technology Transfer Through Partnership Learning from Practice and Research*, edited by T. McMaster, E. Mumford, E.B. Swanson and D. Wastell, London: Chapman & Hall, 1997.

McMaster, T., Vidgen, R. T. and Wastell, D. G. "Networks of Association and Due Process in IS Development", *IFIP WG 8.2 &8.6 Joint Working Conference on IS: Current and future changes*, Helsinki, Finland, December, 10-13/1998.

Odedra, M. "IT Transfer in Developing Countries: The Three Case Studies from Kenya, Zambia and Zimbabwe", *Ph.D thesis*, LSE, London, 1991.

Orlikowski W. J. and Baroudi, J. J. "Studying Information Technology in Organisations: Research Approaches and Assumptions", *Accounting, Management and Information Technologies*, (1:1), 1991, pp. 9-42.

Prescott, M. B. and Conger, S. A. "Information Technology Innovations: A Classification by IT Locus of Impact and Research Approach", *Data Base Advances*, (26:2&3), 1995, pp. 20-41.

Rogers, E.M. *Diffusion of Innovations*, Free Press, (4th edition), 1995.

Vidgen, R. T. and McMaster, T. "Black Boxes, Non-human Stakeholders and the Translation of IT through Mediation", in *Information Technology and Changes in Organizational Work*: 239-249, edited by W.J. Orlikowksi, G. Walsham, M.R. Jones and J.I. DeGross, *Proceedings of the IFIP WG8.2 Working Conference*, Cambridge University, 7-9 December 1995, London: Chapman and Hall, 1996.

Walsham, G. *Interpreting Information Systems in Organisations*, New York, USA: John Wiley & Sons, 1993.

Walsham, G. *Making a World of Difference: IT in a Global Context*, West Sussex, England: John Willey & Sons, 2001.

Williams, R. and Edge, D. "The Social Shaping of Technology", in *Information and Communication Technologies*, edited by H. William and Dutton, England: Oxford University Press, 1996.

About the Author

Esselina Macome is a lecturer in Information Systems at the Eduardo Mondlane University, Maputo City, Mozambique. She was born in Maputo – Mozambique, and achieved her MSc in Information Systems at the University of London, LSE. She is presently in the process of completing her PhD at the University of Pretoria, South Africa. Her research interests include organizational and social aspects of ICT/IS. She can be reached by e-mail at macome@nambu.uem.uem or uffo@zebra.uem.mz, or through the Department of Mathematics and Informatics, Faculty of Science, UEM, P.O. Box 257, Maputo, Mozambique, Tel/fax 258 1 493377.

12

IMPLEMENTING GLOBAL INFORMATION TECHNOLOGY PRODUCT IN A LOCALIZED CONTEXT

An Exploratory Study of ERP System Implementation in China

Gary S C Pan[1], Shan L Pan[2], Xin-Xin Chen[2] and Donal Flynn[1]

[1]*Department of Computation, University of Manchester Institute of Science and Technology;*
[2]*Department of Information Systems, School of Computing National University of Singapore;*

Abstract: This multiple-case study examines the phenomenon of implementing a global information technology product in a localized context. The study attempts to explore issues related to the localization experience encountered by ERP adopters in China. Specifically, the paper explores ERP system implementation in five organizations in China. Despite its potential, ERP is considered one of the most difficult systems to implement to date. The risk of implementing an ERP package in China is even higher due to the relatively backward, nonstandard management mechanisms and business processes operating in this environment. Taking into account these weaknesses, this paper argues that BPR could play a significant role in ERP deployment. As the concept of ERP is still new to Chinese enterprises, it is natural that relatively few organizations were found to have carried out BPR as a preparatory step for ERP implementation. By developing an understanding of the importance of BPR, we derive five important findings based on the case studies. These comprise what literature suggests as key issues for an ERP implementation and integrate what our research revealed as essential issues for organizations implementing a global IT product in a localized context.

Keywords: Enterprise resource planning systems, business process reengineering, developing countries, multiple case study and global product.

1 INTRODUCTION

Enterprise Resource Planning (ERP), is an enterprise-wide package that integrates all necessary business functions into a single system with a shared database (Lee and Lee, 2000; Newell et al., 2003). Evolving from Material Requirement Planning (MRP) systems, ERP systems have flourished in recent years, and sales of ERP systems are expected to remain high at US$16 billion through 2004 (AMR, 2002). While there is wide acceptance of ERP in developed countries, developing countries such as China still lag far behind as many of its organizations have just started automating business processes using computer information systems (Reimers, 2002). Currently, China is undergoing a technological change with huge information technology (IT) investments in both public and private sectors. Its rapid economic growth has turned it into a major target for large foreign ERP vendors. These foreign vendors hold approximately 70% of the ERP market share in China (CCID, 2000). The ERP industry can be seen as an example of China's increased emphasis on IT. This paper argues that it is interesting to explore the implementation issues of ERP systems of both foreign- and locally-owned organizations in China.

While many Western enterprises have achieved successes in their ERP implementations (Scheer and Habermann, 2000; Somers and Nelson, 2001), ERP implementation should not be seen as a universal solution (Soh et al., 2000). This is because several complex issues need to be resolved for successful implementation, such as smoothing out irregularities in business processes and introducing management standards (Krumbholz et al., 2000), and overcoming massive resistance to change (Jin, 2001). In addition, research suggests that there are difficulties in localizing foreign ERP packages, which often results in ERP failures (Holland and Light, 1999; Krumbholz et al., 2000). Acknowledging the difficulties in implementing a global IT product in a localized environment, the paper proposes a bundling strategy for ERP implementation. Specifically, it is recommended that a company conduct a series of business process reengineering (BPR) activities before actual ERP implementations.

The research on which this paper is based involved a number of exploratory studies in the context of a developing country that has not been widely researched previously, i.e., China. Despite Chinese organizations' fast growing demand for ERP systems in recent years, there are challenges in developing countries that are related to economic, cultural, and basic infrastructure issues, which have been relatively neglected in the mainstream IS literature (Huang and Palvia, 2001). In particular, based on data collected from five Chinese organizations, we examine the possibility of bundling BPR with ERP implementation to help improve the overall project success.

After the introduction, the rest of the paper is structured as follows. First, we describe the research approach employed in this study. Next, in the main sections of the paper, we provide a detailed description and analysis of five ERP cases in China. Finally, we draw conclusions on how our paper addresses its goals.

2 RESEARCH DESIGN AND METHODS

The research for our study is best described as exploratory and descriptive of the emerging ERP market in China which aims to develop an understanding of the issues concerning BPR before ERP implementations. A case study approach is selected and is deemed appropriate because of the exploratory nature of this research (Walsham, 1993; 1995; Newman and Sabherwal, 1996). One advantage of using the case study method is its ability to explain what goes on in organisations and it is also particularly good for answering the 'how' and 'why' questions (Yin, 1994).

Table 1. ERP Case Studies in China

Case	Company A	Company B	Company C	Company D	Company E
Location	Suzhou China	Wuxi China	Suzhou China	Suzhou China	Suzhou China
Property of Organization	Foreign-funded Company	Sino Australian Joint Venture	Foreign-funded Company	Foreign-funded Company	Foreign-funded Company
Location of Headquarters	U.S.	Australia	U.S.	U.S.	Australia
Industry	Integrated Circuits	Beer & Beverage	Automotive Systems	Audio Products	Beer & Beverage
Number of employees worldwide	14,000	175,000	195,000	10,000	175,000
ERP systems Adopted	SAP	QAD	Fourth Shift	U8	QAD
No. Of interviewees	10	10	15	11	17

The data collection was undertaken between January 2002 and July 2002, through interviews and reviews of secondary data such as internal documents, minutes of meetings, company reports, numerous press releases and relevant websites. Table 1 above outlines the case companies and the number of interviewees from each company. Each interview lasted approximately 1.5 to 2 hours and was tape-recorded. Interview questions were semi-structured. In terms of data analysis, several concepts, themes and issues were identified and developed by the authors. The analysis went through several iterations to put together a logical overview of issues concerning BPR in those ERP implementation experiences in China.

3 CASE STUDIES

3.1 Company A

Company A is a subsidiary of a Fortune 500 company, which is a manufacturer of integrated circuits for personal and networked computers. The company's headquarters are based in the United States, with manufacturing facilities worldwide. Its China plant is in Suzhou and employs approximately 700 employees. The company had selected SAP as its ERP solution, at the request from their headquarters. Before the SAP implementation, Company A had legacy systems in all of its departments. The implementation process of SAP consisted of four phases where the legacy system was replaced by one ERP module in every phase. When asked about the project, the chairman of the company commented, "Before we kicked off the entire implementation, we had to review our existing business processes and identify some of our weaknesses in our processes."

The project manager who was in-charge of the ERP implementation also added, "From the beginning, we instructed our implementation team that the process should be done one step at a time because numerous 'mismatches' could occur and we should avoid speeding up any procedure irrationally." Employees in company A seemed optimistic about the project. For example, one of the end users commented, "Everyone in the company was fully aware of the system implementation and its importance. We were prepared to accept any change". Another informant also said, "I think innovation is encouraged in this company. If you can invent a feasible idea, your idea may be evaluated, adopted and rewarded by the company."

3.2 Company B

Company B was previously a Chinese state-owned beverage company, before being taken over by an Australian beer and beverage group in 1995.

In 1998, company B implemented most of the modules in the QAD systems-the ERP solution used by the headquarters in Australia. Being an ex-state enterprise, most of its employees were middle-aged and had rarely operated computers. This resulted in some difficulties in conducting user trainings. When asked to comment on the users' computer knowledge, one of the system analysts replied that, "Once I received a call inquiring about the operation of the system. Guess what? I even had to go over the basic computer knowledge by telling that person where the 'start' button was located."

Moreover, most middle-aged users studied Russian in school (in their early days, they were taught Russian rather than English as their foreign language). They did not understand English language at all. When asked how the users' limited knowledge of English had affected the implementation process, one of the system analysts spoke of his frustrations; "Once I asked one end user who had called for help, what system error message was on his screen, he answered me by spelling a whole sentence letter by letter."

These problems are often witnessed in state-owned enterprises or ex state-owned enterprises, which are traditionally managed with structured business processes and are less prepared and willing to go through radical changes. Such corporate cultures tend to be more conservative which imply that unknown or radical changes are often avoided whenever possible.

The other problem often mentioned by the respondents was the high turnover of personnel during the ERP implementation process. With process automation, job scopes were redesigned or positions were reshuffled. This set off a series of problems in terms of staff morale and productivity. When asked to highlight the problems that had surfaced, one manager commented, "For example, how to lay off redundant staffs without causing grievances among them? Who is supposed to be in charge of these changes? We didn't foresee such a great change involved in ERP implementation, we did not prepare before it occurred. We tried to lay off as few staff as possible since they all had worked for the company for a long time."

3.3 Company C

Company C is a subsidiary of an American conglomerate, which is a market leader in the mobile electronics and transportation system industry. It has a workforce of 300 in China and its headquarters are based in the US. While both its headquarters and the branch in Singapore are operating on SAP systems, Company C had instead opted for the Fourth-Shift ERP solution over other renowned ERP systems due to the issue of cost.

One problem that occurred during its implementation was the software customization. Having recognized many 'mismatches' between the functionalities in the ERP package and the business processes adopted in the financial department, the implementation team opted to customize the financial and accounting module by developing another fixed asset module during the implementation of the financial module.

However, some of the implementation team members had left the company soon after the customization. This had posed serious problems because the new staff had difficulty maintaining the customized module as they had little knowledge on how the former designers developed the module. This had resulted in the financial department encountering a lot of problems in operating their system. In the end, the financial module was abandoned. When asked to comment on the issue, the finance director said, "This customized module really drove me crazy. We wasted too much time and effort on it and now, nobody is able to take care of it any further".

3.4 Company D

Company D is a subsidiary of an American company that produces a full range of high-quality audio and video products, which employs approximately 400 employees in China. Company D had adopted a local ERP package named U8 and the entire implementation process lasted two months. There were no major "mismatches" in adopting the ERP solution. The company simply reengineered its business processes to be consistent with the standard models and configured the system according to its requirements. Company D also added some bolt-ons for additional functionalities.

On the whole, the company was satisfied with U8. As the financial director commented, "We are satisfied with this U8 ERP package in terms of its simplicity, low cost and vendor's prompt support. Moreover, we enjoy a very close relationship with the service engineers from U8's branch in Suzhou. We know that they will always be there while we adjust ourselves with new processes and environment."

In addition, the vendor offered strong customization support to help clients to resolve all kinds of problems according to the specific situation faced by each customer. The vendors also helped to resolve any bugs in the software and continuously provided training to new users.

3.5 Company E

Company E is an Australian owned beer and beverage plant that employs 400 staff in China. Company E had implemented the QAD ERP solution in

1997. All the employees, facilities and business processes were newly established and therefore, the implementation process was relatively smooth. When asked to comment on why the implementation was relatively successful, one of the IS analysts commented, "Good interdepartmental communication is the key factor, I must say. For example, frequent communication, lots of feedback on how to improve on our own processes to 'fit' the new ideas in the systems and total cooperation from members from various departments."

Another key success factor cited was a well-defined objective. One of the informants commented, "I think the clearly defined goals are important for us. We were clear and determined with what we need, with every department coping comfortably with the requirements."

When asked to comment about how the ERP system impacted their work, one project member of the ERP implementation team said, "We did not experience very dramatic change after we implemented ERP. Even if there were some changes, we were flexible and should be able to adapt easily."

4 FINDINGS

Since BPR is an important issue in ERP implementations (Bingi et al., 1999; Holland et al., 1999; Sumner, 1999; Holland and Light, 1999; Brown and Vessey, 1999; Parr et al., 1999), we argue that in the case of the Chinese organizations it is useful for ERP–adopting organizations to examine closely some of the issues related to bundling BPR with ERP. While our cases represent only a small number of ERP implementations in China, the cases still demonstrate an àdequate diversity in the patterns and characteristics of ERP implementations, which allow us to develop the findings below.

4.1 Selection of ERP Package Choice

Selecting an ERP package is a time-consuming task that needs a set of meticulous procedures. Even though it is well documented that choosing a package that best fits the companies' business processes is the most important criteria (Everdingen et al., 2000), however this is not always the case in China. There is a range of ERP packages available in the Chinese market, ranging from popular Western products such as 'SAP', to locally developed ones such as 'Yongyou'. Despite this, Companies A, B and E, for example, selected their ERP systems based on decisions made by their headquarters, rather than on cost or other criteria. This could be part of the companies' global IT strategy to consolidate their information systems for better communication and technical support. Furthermore, the vast presence of foreign ERP vendors in China, which accounts for more than 70% of Chinese ERP market (CCID, 2001), boosted these foreign companies' confidence to continue using these foreign ERP systems. Given the fact that foreign ERP packages continue to play a major role in the Chinese ERP industry, perhaps local ERP vendors could only seize a greater market share by offering quality but cheaper ERP packages with solid vendor support, as seen in the cases of companies C and D.

More research is needed to examine what the selection criteria are for choosing an ERP package for organizations in China. Specifically, to investigate the relative importance of various selection criteria adopted by the organizations and to validate the claim that choosing a package that best fits its business processes and functionalities is the most important criteria when the cost, risk of BPR and customization are being considered (Everdingen et al., 2000). A future study into how ERP packages are selected in a localized market like China may have implications for local companies as well as foreign vendors aiming to enter Chinese IT product market.

4.2 Readiness of BPR Process

Similar to Harrison et al's (2000) findings, we found in our cases that foreign companies are better prepared for the implementation of an ERP system as their structures and processes may already be based on modern Western management approaches, and so be more similar to the processes embedded in the ERP systems. On the other hand, Chinese companies, whether privately held or state owned, are likely to display more traditional management systems and business processes. Such traditional Chinese organizations carry with them years of cultural tradition and the far-reaching influence of Confucian dynamism, leading to a high uncertainty avoidance culture (Martinsons and Davison, 2000). People in high uncertainty avoidance cultures tend to have very structured activities and are wary of random events. Unknown or radical changes are avoided whenever possible and BPR may be too severe to be accepted by Chinese organizations.

Company B, which was an ex-state owned beverage company, had difficulties aligning its business processes with the standard procedures embedded in the ERP package. Most of the staff had been working for the company since it was still owned by the state, and many of them were in their middle age with low levels of English language and computer knowledge. They underestimated the radical changes required by the ERP systems. More importantly, they failed to analyze their businesses processes prior to the adoption and assess their readiness in adopting the new systems in terms of the quality and the level of available resources.

Conversely, for cases A and E, both reviewed and evaluated their internal processes prior to their ERP implementation. They identified their weaknesses and defined their objectives, which are often cited as the critical success factors for ERP implementations (Sumner, 1999; Holland *et al.*, 1999). Perhaps one of the reasons why foreign owned companies have an edge over the Chinese local companies in this respect could be due to their accumulated experiences in implementing ERP systems in their headquarters and other branches, which might have suggested BPR as a key step before ERP implementations.

Having a thorough knowledge of current business processes is the prerequisite for conducting BPR and matching its processes with the ERP package. However, it can be a complicated and demanding task to analyze business processes, and it is rarely conducted in real world situations (Scheer and Habermann, 2000). In some cases, some ERP vendors provide business process analysis model that facilitates better understanding of complex business processes (Holland et al., 1999). Further research should be carried out to examine the usefulness of these business process analysis models, which could be a potentially useful tool in helping organizations to conduct BPR before their ERP implementations (Newell et al., 2000).

4.3 Prefer "Matching" over "Customization"

Our study in China has shown that companies were more willing to reengineer their processes than customize the ERP software, since only one out of five companies performed customization during their ERP implementation process. Company C failed in their large-scale customization effort due to the departure of the software development engineers and poor documentation of amended source codes. This observation aligns with the results from the IS literature, which calls for organizations to change the business to fit the software with minimum customization (Mabert et al., 2001; Somers and Nelson, 2001). Conversely, Companies A, D and E were better prepared for the ERP implementation and adjusted their business processes to meet the systems requirements. Generally, the employees from these companies were more cooperative, adaptive, flexible and received solid support from ERP vendors to guide them through the transition period.

The adopting organizations should be aware of the business processes embedded in the ERP package before purchasing the systems. Prior to any implementation, they must thoroughly analyze these processes. If the processes embedded in the ERP package are inappropriate for the adopting organization, customization could be considered. Generally, the recommendation is that customization should be discouraged as far as possible (Sumner, 1999) to avoid errors and allow the adopting company to make the most of newer versions and releases (Nah et al., 2001). Future research must give more attention to how to conduct an analysis on these ERP packages in order to find the package with the best fit with the business processes.

4.4 Tendency to Over-Adjust Existing Business Processes

Company B, for example, conducted major process reengineering, which involved several job redesigns and shuffling of manpower. However the changes were far greater than what they had expected, which caused low staff morale and productivity. There must be a balance between what is being practiced and what is required from the systems. Furthermore, reengineering should take place iteratively to take advantage of improvements from the new system (Nah et al., 2001). On the other hand, Company D simply reengineered its business processes to be consistent with the standard models and customized the system according to its own requirements. As a result, it faced few 'mismatches' during the implementation process.

When bridging the 'gap' between the functionalities and business procedures implanted in the ERP package and those required by the adopting organizations, maintaining a balance between customization and BPR is essential. To reduce any potential risk and facilitate future maintenance and upgrading of the systems, customization should be independent from the original package. This would minimize making alterations in the modules offered by ERP vendors and instead, rely more on add-ons to achieve the required functionality. However, few companies prefer to make substantial changes to their business processes (Parr and Shanks, 2000) since BPR would also increase the complexity and the cost of ERP implementation (Gattiker and Goodhue, 2000). There must be a balance between reengineering the business processes and resolving any 'mismatches' through customization. More efforts should be made to explore how organizations could minimize the impacts on the existing staff from BPR. Instead of simply laying off surplus employees, training programs can be installed to re-skill these affected employees to prepare them for any forthcoming higher skilled jobs.

4.5 Continuous Improvement Efforts and Strong Vendor Supports

Management needs information on the result of ERP on business performance, which includes effective measurable project goals that meet business needs (Nah et al., 2001) and also as an authentication to silence any critics. To achieve success in an ERP implementation, continuous improvement efforts and strong vendor supports are important in the post-implementation period. Company A, for example, developed an environment which encouraged innovation and generation of new ideas. Also, Company D depended on vendors' support in terms of user training and minimum customization on the ERP systems. In addition, Company E attributed its implementation success to frequent communication and feedback among various departments.

Post-implementation review is critical as it provides a feedback channel for continuous improvement even after the new systems have been established. In any new systems implementation, there will usually be a temporary decline in productivity immediately after the implementation. However, this decline should last no more than 12 months, before it reaches the state where the ERP systems can be utilized more effectively by the adopting organizations (Mabert et al., 2001). This transition represents a period of instability with constant adjustments and adaptations to new processes. Standard activities usually involve providing continuous training for new users and cooperating with vendors to resolve bugs in the software. At this stage, preliminary benefits in adopting ERP may start to emerge (Ross and Vitale, 2000). Perhaps future research can explore more ways on shortening the transition period and to tackle the issues caused by the instability of the new systems.

5 CONCLUSIONS

Implementing a global product in a localized context has never been easy. Despite its potential, ERP is considered one of the most difficult systems to implement to date. The risk of implementing an ERP package in China is even higher due to the relatively backward, nonstandard management mechanisms and business processes in most Chinese companies. Taking into account these weaknesses, this paper argues that BPR could play a significant role in ERP deployment.

In this paper, we raised a number of issues associated with organizations implementing ERP systems in China. By developing an understanding of the importance of BPR as a preparatory-step to an ERP implementation, we derived five important findings. These comprise what literature suggests as key issues for an ERP implementation and integrate what our research revealed as essential issues for organizations implementing ERP systems in China. As the concept of ERP is still new to Chinese enterprises, it is natural that relatively few organizations were found to have carried out BPR as a pre-step for ERP implementation. As a famous Chinese saying, "*well begun is half done*", conducting BPR before ERP implementations should be of great help to potential Chinese ERP adopters.

The findings described in this paper are limited due to the lack of both locally developed ERP systems and locally owned company cases. This limitation was the result of the fact that it was extremely difficult to obtain access into organizations especially locally owned Chinese companies in China. The inclusion of these types of case should provide more insights into the difficulties of implementing ERP systems in a localized context, such as China with issues such as diversity in cultural background and economic conditions. Despite the limitation, we are convinced that this study is useful since there is very little research on ERP implementations in a developing country such as China (Reimers, 2002) and there can be no question about the importance of deeper understanding of the ERP implementation model for China.

Finally, we suggest that further research should include both locally developed ERP systems and locally owned company cases in a developing country to validate whether these findings apply. This research should investigate obstacles in the ERP implementations faced by foreign owned companies and locally owned companies and assess how this affects the integrated ERP implementation model for a developing country such as China.

REFERENCES

AMR 2002. Available at: www.amrresearch.com

Bingi, P., Sharma, M.K. and Godla, J. "Critical Issues Affecting an ERP Implementation", *Information Systems Management*, (16:3), 1999, pp. 7-14.

Brown, C. and Vessey, I. "ERP Implementation Approaches: Toward a Contingency Framework", *Proceedings of the 20th International Conference on Information Systems*, Charlotte, December 1999, pp. 411-416.

CCID (China Center of Information Industry Development), Available at: http://www.kingdee.com/news/zhuanti/erp/index.php

Everdingen, Y.V., Hillegersberg, J.V. and Waarts, E. "ERP Adoption by European Midsize Companies", *Communications of the ACM*, (43:4), April 2000, pp.27-31.

Gattiker, T.F. and Goodhue, D.L., "Understanding the Plant Level Costs and Benefits of ERP: Will the Ugly Duckling Always Turn Into a Swan?", *Proceedings of the 33rd Hawaii International Conference on System Sciences*, 2000.

Harrison, G.L., Mckinnon, A., Wu, A. and Chow, C. "Cultural Influences on Adaptation to Fluid Workgroups and Teams", *Journal of International Business Studies*, (31:3), 2000, pp. 489-505.

Holland, P. and Light, B. "A Critical Success Factors Model for ERP Implementation", *IEEE Software*, (16:3), May-June 1999, pp. 30-36.

Holland, P., Light, B. and Gibson, N. "A Critical Success Factors Model for Enterprise Resource Planning Implementation", *Proceedings of the 7th European Conference on Information Systems*, (1), 1999, pp. 273-297.

Huang, Z. and Palvia, P., "ERP Implementation Issues in Advanced and Developing Countries", *Business Process Management*, (7:3), 2001, pp. 276-284.

Jin, D. "ERP in China", *Computer Weekly*, 6 June, 2001.

Katzenstein, G. and Lerch, F.J., "Beneath the Surface of Organizational Processes: a Social Representation Framework for Business Process Redesign", *ACM Transactions on Information Systems*, (18:4), October 2000, pp. 383-422.

Krumbholz, M., Galliers, J., Coulianos, N. and Maiden, N.A.M., "Implementing Enterprise Resource Planning Packages in Different Corporate and National Cultures", *Journal of Information Technology*, (15:4), December 2000, pp. 267-279.

Lee, Z. and Lee, J., "An ERP implementation Case Study From a Knowledge Transfer Perspective", *Journal of Information Technology*, (15:4), December 2000, pp. 281–288.

Mabert, V.A., Soni, A. and Venkataramanan, M.A., "Enterprise Resource Planning: Common Myths Versus Evolving Reality", *Business Horizons*, (44:3), May-June 2001, pp. 69-76.

Martinsons, M.G. and Davison, R.M., "Cultural Considerations in Business Process Change", *Department of Information Systems, City University of Hong Kong*, 2000. Available at http://www.is.cityu.edu.hk/Research/Publication/working_paper00.htm

Nah, F. H., Lau, L. S. and Kuang, J. H., "Critical Factors for Successful Implementation of Enterprise Systems", *Business Process Management Journal*, (7:3), 2001, pp. 285-296.

Newell, S., Swan, J.A. and Galliers, R.D., "A Knowledge-Focused Perspective on the Diffusion and Adoption of Complex Information Technologies: the BPR Example", *Information Systems Journal*, (10:3), 2000, pp. 239-259.

Newell, S., Huang, J.C. Galliers, R.D. and Pan, S.L., "Implementing Enterprise Resource Planning and Knowledge Management Systems in Tandem: Fostering Efficiency and Innovation Complementarity", *Information & Organization*, (13:1), 2003, pp. 25-52.

Newman, M and Sabherwal, R., "Determinants of Commitment to Information Systems Development: a Longitudinal Investigation", *MIS Quarterly*, (20:1), March 1996, pp. 23-32.

Parr, A.N. and Shanks, G., "A Taxonomy of ERP Implementation Approaches", *Proceedings of the 33rd Hawaii International Conference on System Sciences*, (7), 2000.

Parr, A.N., Shanks, G. and Darke, P., "The Identification of Necessary Factors for Successful Implementation of ERP Systems", in: O. Ngwenyama et al. (eds.): *New Information Technologies in Organizational Processes*, Boston: Kluwer Academic, 1999, pp. 99-119.

Reimers, K., "Implementing ERP Systems in China", *Proceedings of the 35th Hawaii International Conference on System Sciences*, (8), 2002.

Ross, J.W. and Vitale, M.R. "The ERP Revolution: Surviving vs. Thriving", *Information Systems Frontiers*, (2:2), August 2000, pp. 233-241.

Scheer, W. and Habermann, F., "Making ERP a Success", *Communications of the ACM*, (43:4), April 2000, pp. 57-61.

Soh, C., Kien, S.S. and Tay-Yap, J., "Cultural Fits and Misfits: Is ERP a Universal Solution?", *Communications of the ACM*, (43:4), April 2000, pp. 47-51.

Somers, T.M. and Nelson, K., "The Impact of Critical Success Factors across the Stages of Enterprise Resource Planning Implementations", *Proceedings of the 34th Hawaii International Conference on System Sciences*, (8), 2001.

Sumner, M., "Critical Success Factors in Enterprise Wide Information Management Systems Projects", *Proceedings of the Americas Conference on Information Systems (AMCIS)*, 1999, pp. 232-234.

Walsham, G., *Interpreting Information Systems in Organizations*, Chichester, England: Wiley, 1993.

Walsham, G., "Interpretive Case Studies in IS Research: Nature and Method", *European Journal of Information Systems*, (4:2), 1995, pp. 74-81.

Yin, R. *Case Study Research: Design and Methods*, Second Edition, Thousand Oaks, CA: Sage, 1994.

About the Authors

Gary S C PAN is a PhD candidate in the Department of Computation, University of Manchester Institute of Science and Technology, UK. His research interests include implementation of enterprise systems and information systems project failure. He has presented in several major conferences and has won best track paper in the British Academy of Management conference in 2002. He can be reached by e-mail at s.pan@postgrad.umist.ac.uk

Shan L. PAN is Assistant Professor in the Department of Information Systems of School of Computing at the National University of Singapore. His primary research focuses on the recursive interaction of organizations and information technology (enterprise systems), with particular emphasis on issues related to work practices, cultures and structures from a knowledge perspective. He has published in IEEE Transactions on Engineering Management, Journal of Strategic Information Systems, Communications of ACM, European Journal of Information Systems, and Decision Support Systems. He can be reached by e-mail at pansl@comp.nus.edu.sg

Xin Xin CHEN is a research student in the Department of Information Systems of School of Computing at the National University of Singapore. Her primary research interests focuses on the strategic issues involved in implementing enterprise systems. She can be reached by e-mail at chenxinx@comp.nus.edu.sg

Donal FLYNN spent 11 years in the UK IT industry as programmer, database designer and project consultancy. He has degrees in Chemical Physics (BSc University of Kent), Sociology & Psychology (MSc Imperial College, University of London) and Conceptual Modelling (PhD University of East Anglia). His research interests are in the socio-organizational factors and processes that underlie the interaction between IT systems and their human and organizational contexts. He is currently Senior Lecturer in Information Systems, Department of Computation, University of Manchester Institute of Science and Technology. He can be reached by e-mail at donal.flynn@umist.ac.uk or visit his website at http://www.co.umist.ac.uk/department/staff_details_ac.php?staff_id=DJF

Part 3:

ICT Industries and Systems Development

13

THE PROCESS OF OFFSHORE SOFTWARE DEVELOPMENT

Preliminary Studies of UK Companies in Malaysia

Aini Aman and Brian Nicholson

School of Accounting and Finance, University of Manchester, UK

Abstract: There are many studies of information technology outsourcing but very few discuss in depth the process of offshore software development outsourcing. The aim of this research is to understand the important issues and strategies in managing activities and relationships among software development teams involved in offshore outsourcing. The study involves UK companies who are engaged in offshore outsourcing of software development to Malaysia either through joint ventures or fully owned subsidiaries. There are an increasing number of software firms choosing Malaysia as a venue for software development. The study contributes to the growing body of literature on offshore software development outsourcing by building a synthesized conceptual framework for global software outsourcing. This is derived from concepts found to be important in previous studies. The framework is illustrated using data from two ongoing case studies.

Keywords: Communication, Control, Culture, Globalization.

1 INTRODUCTION

Offshore software development outsourcing takes place when a customer uses a contractor in another to perform a software-related task, such as programming or data entry (Phillips, 2002). Amoribieta et. al, (2001) explains that the development of custom software is different from most other business activities since it requires a detailed understanding of business processes and the way information technology (IT) supports them. Managing offshore software development is highly complex because distance creates difficulties in coordination and control, knowledge transfer and communication (Carmel and Agarwal, 2001). Potential problems related to offshore software development include cross-cultural issues, language barriers, time zone differences, political instability and unreliable telecommunications infrastructure (Gao et. al, 2002; Walsham, 2002). Building an offshore partnership requires much effort and delicate handling by senior managers. The field of software engineering in this domain is relatively new and procedures for quality control and project management, though developing very fast, have yet to evolve fully.

Recent trends show that software developers are going offshore to develop and maintain their software as a strategy to improve cost control, product quality, product development, schedule reduction and focus on core business activities (Amoribieta et. al, 2001). Carmel and Agarwal (2001) outline some potential advantages of offshore software development over in country options as the ability to tap into specialized talent, facilitate a global presence and achieve cost saving of up to 60%. Other general reasons for adopting an offshore strategy include improved product quality, to engage in specialised product development, schedule reduction and to focus on core business activities (Amoribieta et. al, 2001). Among the favourite locations for offshore outsourcing are the first tier vendor nations often referred to as the "3Is" of India, Ireland and Israel. Popular second tier venues for a range of offshore software services are Singapore, Russia and Philippines. Other countries offering various types of IT services plus cheap labour costs are South Korea, Vietnam, Hungary, Czech Republic, Malaysia, Argentina, Brazil, Chile, South Africa, Poland, China and Mexico (Amoribieta et. Al., 2001; Heeks, 1999; Heeks and Nicholson, 2002).

There are very few in-depth longitudinal studies showing how companies achieve the benefits and overcome the problems of offshore outsourcing. The aim of this research is to understand how companies are managing activities and relationships in offshore software outsourcing. The study is concerned with UK companies engaged in offshore outsourcing of software development to Malaysia either through joint ventures or fully owned subsidiaries. There are an increasing number of software firms choosing Malaysia as a venue for software development. The study draws on a framework, which draws together concepts found to be important in previous empirical studies of offshore software outsourcing. The paper contributes to the growing theoretical body of literature on offshore software development outsourcing by building a synthesized conceptual framework for global software outsourcing. The practical contribution of this paper is in illustrating the framework using preliminary data from two ongoing case studies of offshore outsourcing.

The paper is organised as follows: in the next section the conceptual framework is presented which consists of a literature search drawing together literature on offshore software outsourcing. This is followed by a discussion of the research design for the case studies. Section four provides a description of the two cases followed by analysis drawing on the conceptual framework. The final section summarises the analysis and considers the contribution of the paper as well as outlining future work.

2 CONCEPTUAL FRAMEWORK

In order to select and develop an initial framework for the inquiry previous research was examined. Relevant frameworks and models related to outsourcing of software development are concerned with information technology outsourcing decisions (Willcocks and Fitzgerald, 1994; Apte and Mason, 1995); IT issues confronted by multinational corporations (Deans and Ricks, 1991); organizing existing knowledge about globalization of the software industry (Kim et al., 1989), division of labour in global software development (Meadows, 1995) and selection of projects and sites suitable for offshore software development (Ravichandran and Ahmed, 1993). Other frameworks are explicitly concerned with management of the process of offshore outsourcing of software development (Mitra and Narasimhan, 1996; Heeks et al., 2001).

Heeks et al. (2001) developed a dimensional framework on the basis of their empirical research and other studies of offshore software outsourcing. The COCPIT framework was chosen as a basis for conceptual frame development. The COCPIT categories were derived from empirical evidence from North America, UK, Korea and Japan based companies outsourcing to India involving over 200 interviews over a four-year period. The framework demonstrates that issues in offshore software development can be analyzed as the minimization of "synching gaps" between client and offshore vendor along six 'COCPIT' dimensions: **C**oordination/control systems (management coordination and control system), **O**bjectives and values (sharing the same objectives for their relationship and bringing the same values to that relationship), **C**apabilities (the developer's capabilities such as human capabilities matching the requirements of the customer); **P**rocesses (synchronization of work processes), **I**nformation, and **T**echnology (information completeness and technology usage). Within the COCPIT categories, successful relationships with a high degree of congruence achieved between vendor and client is termed 'synching'. The unsuccessful relationship with a low degree of congruence is termed 'sinking'. Using the framework Heeks et al. (2001) suggest that creating congruence or "synching" between client and developer along the COCPIT dimensions may lead to the higher project success rates in offshore software outsourcing. They explain that synching is also a means of reducing risks and costs. Their case studies also suggest that synching strategies have been poor at dealing with three overlapping issues; tacit knowledge, informal information and culture that indirectly affects all COCPIT dimensions.

COCPIT is useful to sensitise researchers and practitioners to the issues in offshore software development outsourcing. However, the framework does not explain each dimension in detail. For instance, Heeks et al. (2001) do not explain how and when each dimension influences the relationships between developers and clients. Furthermore, COCPIT does not explain other factors that might influence the existence of each dimension. Other weaknesses of this contingency framework are that the authors do not fully identify theoretical concepts within each dimension to facilitate in depth analysis. To overcome some of these limitations, the authors enhanced the COCPIT framework with other literature into each dimension and illustrate it using cases. The new framework may serve as a useful approach to systematically address the issues that arise in the context of offshore outsourcing of software development.

2.1 The Revised Framework

The revised framework is based on COCPIT with the extension on the explanation of each dimension based on existing literature. Table 1 lists current literature on each dimension of COCPIT. Although it is accepted that these categories overlap, the revised framework is used in the empirical study as a guide to data collection. Subsequently, the associated theory will be used to make sense and explore the empirical data and assist in explaining the issues encountered.

Table 1. Related Literature on each dimension of COCPIT

COCPIT	Related Literature
Coordination & control	Henderson and Lee (1992), Carmel and Agarwal (2001)
Objective & values	Hofstede (1991), Ramarapu and Parzinger (1997), Herbsleb and Moitra (2002), Walsham (2002)
Capabilities	Grimaldi and Torrisi (2001), Zhuge (2002), Gao, Itaru and Toyoshima (2002)
Process	Curtis, Krasner and Iscoe (1988),Rautiainen, Lasenius and Sulonen (2002)
Information	Heeks, Krishna, Nicholson and Sahay (2002), Herbsleb and Moitra (2002)
Technology	Whittaker and Schawarz (1999), Choi and Lee (2002)

In the *coordination and control* dimension of COCPIT congruence may be achieved when client and developer use corresponding management and control systems. Carmel and Agarwal (2001) explain that *coordination* is the act of integrating each task with each organizational unit, so the unit contributes to the overall objective, which requires intense and ongoing communication while *control* is the process of adhering to goals, policies and contracts, standards or quality levels. Carmel and Agarwal (2001) stress how distance can affect communication, which in turn can reduce coordination and control effectiveness. Developers not located together have very little informal, spontaneous conversation, which may help actors remain aware of dynamic progress. Failure in this respect might result in misalignment and rework. Synchronization requires commonly designed milestones and clear entry and exit criteria. Henderson and Lee (1992) combine research on managerial control and team-member control in order to explore a range of control behaviours that can affect the performance of an offshore software development team.

In the *objectives and values* dimension of COCPIT; Heeks et al. (2001) suggest that to achieve congruence, vendor and client should share the same objectives for their relationship and bring the same values and culture to that relationship. The issue of culture is particularly pertinent to offshore software development. Herbsleb and Moitra, (2002) explain that differences in cultural background sometimes lead to serious misunderstanding. Cultural barriers may hamper communication among the user, designer and programmer and can decrease productivity thereby affecting the development of quality software (Ramarapu and Parzinger, 1997). Some authors suggest different attitudes towards authority and hierarchy in the Eastern (less open and les confrontational) and Western culture (more open and extrovert) (Hofstede, 1991). This type of research is useful to sensitise the researcher to potential cross-cultural issues. However, more sophisticated analysis was performed by Walsham (2002) using concepts drawn from structuration theory. The theoretical approach is used to analyze cross-cultural conflict and contradiction, cultural heterogeneity, detailed work patterns and the dynamic nature of culture.

The COCPIT dimension of *capabilities* suggests that congruence can be achieved when human capabilities, skills and knowledge match the requirements of the clients. Grimaldi and Torrisi (2001) analyse the process of knowledge codification and the division of labour between software firms. They focus on the nature of tacit versus codified knowledge that firms share and exchange with other partners by means of collaborative agreements. Zhuge (2002) presents a notion of knowledge flow and the related management mechanism for realizing and ordered knowledge sharing and cognitive cooperation in a geographically distributed team software development process. Gao et al. (2002) suggest that to deal with complexities of knowledge, firms should try to simplify tasks and to coordinate activities of partners across time and space by standardizing various processes of knowledge transfer.

The *process* dimension in COCPIT; suggests that congruence is achieved when the vendor and client manage work processes and software development methodology effectively. Rautiainen et al. (2002) present a general framework for managing software product development. The framework combines business and process management through four cycles of control: (1) strategic release management; (2) release project management; (3) iteration management and (4) mini-milestones are used for daily or weekly task scheduling and monitoring to get an indication of system status during development. On the other hand, Curtis et al. (1988), used layered behavioural model in order to understand how different tool, methods, practices and other factors actually affect the processes that control software productivity and quality. The model emphasizes factors that affect psychological, social and organizational processes. The model focuses on the behavior of those creating the artifact, rather than on the evolutionary behavior of the artifact at each individual level, team level, project level, company level and business milieu.

The *information* dimension of COCPIT describes that congruence is achieved when the client and developer have access to the same information; for example, information relating to project requirements and timescales. The share of information requires both explicit and tacit knowledge. Herbsleb and Moitra (2002), suggest that without effective information and knowledge-sharing mechanisms, managers cannot exploit global software development benefits. Herbsleb and Moitra (2002), also suggest that to prevent assumptions and ambiguity and to support maintainability, documentation must be current and reflect what various teams are using and working on. In addition to documenting the various artefacts, updating and revising the documentation is equally important.

The COCPIT dimension of *technologies* is concerned with the software and hardware platforms for development work. Whittaker and Schwarz (1999) found that technology adept groups prefer to use what seem to be outmoded "material" tools in critical projects, despite a wealth of electronic group tools for coordinating the software development process. They also found that the medium of the schedule has a major impact on coordination problems. Choi and Lee (2002) suggest that a "system oriented" strategy and "human strategy" are important determinants of success of the offshore software development process. The system-oriented strategy emphasizes codified knowledge and focuses on codifying and storing knowledge via information technology to share, access and use knowledge formally. The human-oriented strategy emphasizes dialogue through social networks and person-to-person contacts focussing on acquitting knowledge via experienced and skilled people to share knowledge informally.

3 RESEARCH DESIGN

This research adopts a qualitative design as it aims to highlight the experiences of individual organizations in the process of offshore software development. The overall research strategy could be categorized as interpretivist guided by the knowledge of reality as socially constructed by individual human actors (Walsham, 1995). An interpretivist approach tries to understand the perspective of different actors involving multiple perspectives. This requires in-depth qualitative data including views of members in the software development teams concerning the process and management issues in the contexts within which the process is taking place. The methodology draws on contextualism (Pettigrew, 1987; 1990), which was found useful by Walsham and Waema (1994). The resulting interpretive case study (Walsham, 1995) is hoped will explain how and why contextual conditions and project management tactics and strategies interact. Pettigrew (1987) states that the researcher should come to the field situation equipped with a number of theoretical concepts, which could be used to analyse the data. The researchers using the revised COCPIT framework discussed in section 2, are equipped with range of theories to help and understand issues in offshore software development outsourcing.

The companies chosen for this study are currently engaging in offshore outsourcing software development. The sites were selected firstly through a literature search that indicated a number of UK companies who were engaged in outsourcing software development to Malaysia. A list of companies was contacted through telephone and fax. Data collection is through a variety of methods: unstructured and semi-structured interviewing, documentation review and observation. Triangulation or using various techniques of data collection is particularly beneficial as it provides multiple perspectives on issues, supplies more information on emerging concepts, allows for cross-checking and yields stronger substantiation of constructs (Eisenhardt, 1989; Glaser and Strauss, 1967; Pettigrew, 1990; Orlikowski, 1993).

Information was gathered on the software industry in Malaysia and two case companies known as Das Ltd and Ace Ltd. The real names of the companies are disguised for confidentiality. The study is ongoing and at this preliminary stage there have been a total of ten interviews with programmers and managers at both companies in UK and Malaysia. Interviews lasted for at least one and half-hours. The interviews were taped and transcribed and subsequently summarised Data from both companies were gathered and grouped into the dimensions of COCPIT using a data display method (Miles and Huberman 1994) which also enabling cross case analysis. Analysis of the data from interview transcripts in relation to the theory as suggested by Klein and Myers (1999) revealed the issues that exist in the process of offshore software development outsourcing and tentative explanation drawing on relevant theory.

4 CASE BACKGROUND: DAS AND ACE

Das and Ace are software development companies and have their headquarters in UK with wholly owned subsidiaries in Malaysia. Both companies are involved in offshore software development outsourcing. Das is a custom e-business software solution provider while Ace is a process plant engineering software services provider. In general, both companies are in the process of setting up software development teams in Malaysia for incoming projects. Both companies are currently recruiting software programmers in Malaysia for their incoming projects. Das is using Lotus and Domino while Ace is using Microsoft as a based for software development. Das is a medium-sized company established in 1994 with approximately fifty staff in three countries (UK, Malaysia and Bangladesh). They have a wholly owned research and development centre in Bangladesh. The development centre in Bangladesh serves both UK and Malaysia. Since 1997 Das has been using software development teams in Bangladesh. Das's office in Malaysia is currently tasked with client facing, consultancy and local project delivery. Ace is a large sized company established in 1983 with 340 staff in 21 locations across Europe and the Middle East, Asia Pacific and North America. Ace is currently using software development teams in UK and India. Ace's office in Malaysia is headquarters for administration in Asia Pacific with small project development capabilities to provide direct support for local clients.

The software industry in Malaysia to date has been import oriented and reliant on foreign base technologies. Overall, the information technology market is expected to grow by 10% to about RM 8 billion in 2003. The various Multimedia Super Corridor (MSC) flagship applications in Malaysia are expected to provide significant growth opportunities for the IT industry. Most of the MSC's companies are in the software development sector followed by Internet based business, content development, telecommunication, data centre, system security, consulting and incubators. The breakdown of European MSC approved companies' shows that most are from United Kingdom (Multimedia Development Corporation, 2002).

Likewise, the present size of software industry in Bangladesh is very small. Only a few firms are involved in the export of software and data entry business. The size of IT industry is estimated at around US$ 150 million, which is growing at more than 20% each year. In contrast, the Indian Software industry is the fastest growing industry in India. Indian software sector is forecast to continue growth at 50% per annum over the next few years.

5 ANALYSIS

In this section, selected preliminary empirical data will be explored using the revised framework outlined in section 2 above. The analysis will use the theoretical framework (COCPIT and associated literature) to illustrate the issues raised in offshore software development outsourcing.

5.1 Process, Coordination and Control

In the *process, coordination and control* dimension of COCPIT, a significant theme in the data collected so far is that of the pivotal importance and skills of those assigned to "straddler" roles. In both cases key individuals were seen to be instrumental in managing knowledge transfer and facilitating disembedding of coordination and control processes across time and space (Giddens, 1990) between on and offshore groups. Although this liaison person is regarded as important in managing the offshore process, little is understood about the skills and qualities of this person and the importance of the role in effective software development. A Das manager describes the importance of a "technical guide" in ensuring that specifications are understood and the process is controlled:

> *'In UK, we do all the upfront work with the customers over the analysis of the requirements. We manage the project here with our customers and with them (software development teams in Bangladesh). We will send a specification to one of the developers who act as a technical guide to them (the Bangladesh software development teams) to make sure we get the solution what we want. Then they will do all the programming and we will do the online testing with them. When they have completed it, they deliver back to us and we will deliver the final product to customers'.*

The liaison person or "technical guide" is a key personality in the "synching" between the onshore and offshore groups. This person is responsible for interpreting requirements and re interpreting and instituting control locally. Henderson and Lee (1992) stress that team member control (in this case; using technical guide) is necessary in software development practice because tasks in planning and design are difficult in nature. Henderson and Lee (1992) suggest that effective teams should have a manager with the skills and capabilities to influence how work is accomplished and to influence terms to behave in accordance with organizational gaols. In Ace, the role of project manager has full control on the project and he is a key person to liase with software engineers in offshore business units. In Ace, the project manager role is multifaceted as indicated by the company president:

> *'The way we manage the system is that we have a project manager. The project manager will define what we will develop. He gets into customers and brings back information about how the system will be defined, what improvement is needed for the future such as particular changes for certain industry, strategic program, define the kind of development project. Software development teams work close with the project manager. The project manager will allocate resources for them in UK and India. The software engineer cannot decide what to change but can decide the best way to do on the programming side and they can only make recommendation. The project manager is the key person who manages the project deliver the final product to business unit like us. The project managers are based in UK and they are in very close contact with project. One of the key tasks of project manager is to make sure that they deliver the product on time'.*

Despite the use of liaison persons in coordinating and controlling work offshore, distance can affect the way software is developed. This may reduce coordination and control effectiveness and might result in misalignment and rework (Carmel and Agarwal, 2001). The director of Das raised the issue as follows;

> *'Because we have a couple of project recently not going as well as we plan and it is painful to put it right, we recognize that a lot of the problems came from the project process. When we go to the next level down, we do not necessarily get a consistent delivery; every single project, every single time. We have been spending a lot of time helping them (software development teams in Bangladesh). We can't afford to have time helping them because that will delay the project here (in UK).*

5.2 Objectives and Values

In this dimension of COCPIT, the issues raised are concerned with cultural barriers. The director of Das compares the culture in UK and Bangladesh that sometimes leads to project delay:

'Here (in UK), the environment is very open where people have great trust on each other, ask questions very openly and freely. It is an informal collaboration where people do not think that asking question is a sign of failure. Bangladesh is a completely different situation partially because of culture and largely because of their education system. If we give so much freedom to the software development team in Bangladesh, they don't know what to do. The team wants specific instruction. The developers in UK have developed common sense of how things should work while people in Bangladesh are working in isolated world We notice that most of the time, humour is left out. We learn over the years that we have to develop specific techniques in dealing with them. Now, we have improved a lot because we have now got senior people there. What we know is that when we talk to the senior developer we notice that their knowledge and experience are very limited. We have to spend quiet a lot of time explaining to them'.

In the case of Das, our respondents told us that Bangladesh teams are not willing to express opinions overtly especially if contradictory to those of senior staff. Therefore, there appears to be a culture gap that is hard to bridge. The problem of culture is less intense in Ace because most of their software development teams are British. At Das, aspects of the reported need for specific instruction, education, humour and issues of common sense may be usefully explored drawing on Walsham (2002) structurational analysis once there is sufficient empirical data and multiple perspectives.

5.3 Capabilities, Information, Technology

The problems of capabilities and information are discussed here in terms of sharing knowledge and training new programmers. As in Das, knowledge sharing is important as they rely very much on Bangladesh expertise. Das use most of expertise in Bangladesh but sometimes using expertise from India. One of the managers explains how they build up their experts, which is problematic and also suffers from cultural difficulties:

'We have expertise in Bangladesh who can share knowledge with the new software teams in Malaysia. Sometimes, we go to India and get

people when we are short of workers. We realize that Indian developers bring severe problems. They interpret in their own way. The do not understand the business process very well. They have problems with UK staff too. The Bangladesh staff are used to UK culture while Indians are not used to it. When we first start a centre in Bangladesh, we used British staff to give training, set standards, procedures, coding etc. After two years the British staff were sent to Bangladesh to get training. We tend to do a lot of research. Once a week there will be staffs training. We come up with some topics that are beneficial to them and the company and do research on. Some times there is no time to train everybody. So, they just choose one small project say new language and study it on Saturday.

In contrast, Ace has more formal procedures in knowledge sharing and training new staffs. The project manger of Ace describes:

'The post of every new engineer and electrical staff in the company is made changeable. The purpose is to help he/she to be familiar with the company quickly. The development teams are going into intensive training, which is very different, and in timely change. Every year, we update training of new product for development teams because the development tools changes fast'.

To deal with complexities of knowledge, firms such as Ace try to simplify tasks by standardizing various processes of knowledge transfer (Gao et al., 2002). These standardized systems, often codified in manuals and databases, serve as points of reference to coordinate activities of partners across time and space. Such attempts to standardize are rarely unproblematic, and in tension with the need for flexibility at the 'local' level. As one of the managers in Ace indicates;

'We have concrete procedure and process on software development project because the software development task is very big. We need to be very rigid in testing, to make sure the system is stable, because we are the one that supposed to be doing it before we release it to our customer. We have a well-proven process anyway. We have it documented'.

While some degree of standardization is essential to enable global coordination, there is always the question of how much and what to standardize. A project manager in Das raises this issue;

'There is a lot of work to document what you have done. To document the work requires a discipline to ensure that the application is updated every

single time. We try to work out how to solve out most of this documentation problem'.

Documentation of systems by software developers is often regarded as a tedious task. In global software development, in addition to documenting the various artefacts, updating and revising the documentation is equally important. Herbsleb and Moitra, (2002) suggest that to prevent assumptions and ambiguity and to support maintainability, documentation must be current and reflect what various teams are using and working on.

The COCPIT dimension of *technologies* is concerned with the software and hardware platforms for development work. Both companies are using communication technology such as groupware applications in the software development process. The manager in Das describes the difficulties she has, especially on explaining certain procedures to the development teams in Bangladesh.

' I have been doing all of the work with the development team on line. They send me the application and I test it and I go online. If there is a problem we chat about it, doing it online and show the problems. We chat about it before we go to the customers to implement it. When you try to do things online and over the phone, you can make progress but ultimately teaching them down to absolute detail is a lot harder compared to doing it face to face'.

The biggest limitation Das face of using groupware is that they perceive it is not the same as face-to-face communication. Whittaker and Schwarz (1999) stress that electronic systems may reduce face-to-face communication, decrease awareness of the actions of other group members, and suffer from lack of visibility and performance.

6 CONCLUSION

The growth of offshore software development outsourcing indicates the need to better understand associated issues or problems. Literature reviews in the areas of offshore outsourcing indicate several frameworks to identify problems and issues in the area, which were discussed in section 2 above. Among the available frameworks, COCPIT was chosen as an all-embracing framework suitable for a processual study. However, this framework does provide sufficient detail or explanation of each dimension. Therefore, this paper has presented a revised framework drawing on other literature and illustrated it using preliminary data from two ongoing case studies. The new framework serves as a useful tool to systematically address the issues that arise in the context of offshore outsourcing of software development. This new framework was used as a guide to data collection and to help make sense and explore the empirical data as well as assist in explaining the issues encountered. The revised framework discussed in section 2 enables the researchers to be equipped with range of theories to help to understand issues in offshore software development outsourcing.

The contribution of this paper is two fold. For other researchers, this study contributes theoretically by drawing together relevant and current literature in each of COCPIT dimensions. If used as a contingency framework, researchers and practitioners can identify and "diagnose" issues in managing successful relationships in offshore outsourcing according to the specifics of a situation. For practitioners, this study illustrates selected aspects of the process of offshore software development outsourcing using two cases within the dimensions of COCPIT. The understanding of issues in the process of offshore software development is important for practitioners to manage successful relationships between clients and developers, which in turn will contribute to the success of offshore outsourcing.

REFERENCES

Amoribieta et al. "Using a Global Services Delivery Model to Rapidly Develop Quality E-Business Software Solutions", *The McKinsey Quarterly*, (2), 2001.

Apte, U. and Mason, R.O. "Global Disaggregation of Information-Intensive Services", *Management Science*, (41:7), July 1945.

Carmel, E. and Agarwal, R. "Tactical Approaches for Alleviating Distance in Global Software Development", *IEEE Software Journal*, March/April 2001.

Choi, B. and Lee, H. "Knowledge Management Strategy and its Link to Knowledge Creation Process", *Expert Systems with Application*, (23), 2002, pp. 173-187.

Curtis, B., Krasner, H. and Iscoe, N. "A Filed Study of the Software Design Process For Large Systems", *Communications of the ACM*, (31:11), November 1988.

Deans, P.C. and Ricks, D.A. "MIS Research. A Model for Incorporating the International Dimension", *The Journal of High Technology Management Research*, (2:1), 1991, pp. 57-81.

Eisenhardt, K., M. "Building Theories from Case Study Research", *Academy of Management Review*, (14:4), 1989, pp. 532-550.

Gao, J. Z., Itaru, F., Toyoshima, Y. "Managing Problems for Global Software Production – Experience and Lessons", *Information Technology and Management*, (3), 2002, pp. 85-112.

Giddens, A. "Structuration Theory and Sociological Analysis", in Anthony Giddens: *Consensus and Controversy*, J. Clark, C. Modgil, and J. Modgil (eds.) Brighton, England: Falmer Press, 1990, pp. 297-315.

Glaser, B. and Strauss, A. *The Discovery of Grounded Theory, Strategies for Qualitative Research*, Chicago: Aldine, 1967.

Grimaldi, R. and Torrisi, S. "Codified-Tacit and General-Specific Knowledge in the Division of Labour Among Firms: A Study of the Software Industry", *Research Policy*, (30), 2001, pp. 1425-1442.

Heeks, R. "Software Strategies in Developing Countries", *Communication of the ACM* (42:6), 1999, pp. 16-20.

Heeks, R., and Nicholson, B. "Software Export Success Factors and Strategies in Developing and Transitional Economies", *Development Informatics Working Paper Series*, Working Paper No. 12, Institute for Development Policy and Management, 2002.

Heeks, R., Krishna, S., Nicholson, B. and Sahay, S. "'Synching' or 'Sinking': Trajectories and Strategies in Global Software Outsourcing Relationships", *IEEE Software*, (18.2), 2001, pp. 54-62.

Henderson, J., C., and Lee, S. "Managing I/S Design Teams: A control Theories Perspective", *Management Science*, (38:6), June 1992.

Herbsleb, J.D. and Moitra, D. "Global Software Development", *IEEE Software*, March/April 2001, pp 16-20.

Hofstede, G. *Cultures and Organisations*, Maidenhead: McGraw Hill, 1991.

Kim, C., Westin, S., and Dholakia, N. "Globalization of the Software Industry: Trends and Strategies", *Information & Management*, (17), November 1, 1989, pp. 197-206.

Klein, H. K. and Myers, M. D. "A Set of Principles for Conducting and Evaluating Interpretive Field Studies in Information Systems", *MIS Quartely*, (23:1), 1999, pp. 67-93.

Meadows, C. J. "Globalizing Software Development", *Journal of Global Information Management*, (4:1), 1995.

Miles, M. and Huberman, A. *Qualitative Data Analysis*, California: Sage, 1994.

Mitra, S. and Narasimhan, S. "International Outsourcing of Software Development: A Framework for Issues", *Information and Management*, (31), December 1996, pp. 165-175.

Multimedia Development Corporation. Available at: http://www.mdc.com.my/

Orlikwoski, W. J. "CASE Tools as Organizational Change: Investigating Incremental and Radical Changes in Systems Development", *MIS Quarterly*, (17:3), September 1993.

Pettigrew, A. "Context and Action in the Transformation of the Firm", *Journal of Management Studies*, (24:6), 1987, pp. 649-670.

Pettigrew, A. "Longitudinal Field Research on Change: Theory and Practice", *Organization Science*, (1:3), 1990, pp. 267-292.

Phillips, D. E. "Selected Legal Issues in International Software Outsourcing, American Intellectual Property Law Association", *Spring Meeting*, April 19, 2002, New York.

Ramarapu, N. and Parzinger. M. J. "Issues in Foreign Outsourcing", *Information Systems Management*, (14:2), Spring 1997, pp. 27-31.

Rautiainen, K., Lasenius, C. and Sulonen, R. "4CC: A Framework for Managing Software Product Development", *Engineering Management Journal*, (14:2), June 2002.

Ravichandran, R., Ahmed, N. "Offshore Systems Development", *Information & Management*, (24), 1993, pp. 33-40.

Walsham, G. "Interpretive Case Studies in IS Research: Nature and Method", *European Journal of Information Systems*, (4:2), 1995, pp. 74-81.

Walsham, G. "Cross-Cultural Software Production and Use: A Structural Analysis", *MIS Quarterly*, (26:4), December 2002, pp. 359-380.

Walsham, G. and Waema, T. "Information Systems Strategy and Implementation: A Case Study of a Building Society", *ACM Transactions on Information Systems*, (12:2), 1994, pp. 150-173.

Whittaker, S. and Schwarz, H. "Meetings of the Board: The Impact of Scheduling Medium on Long Term Group Coordination in Software Development", *Computer Supported Cooperative Work*, (8), 1999, pp. 175-205.

Willcocks, L., and Fitzgerald, G. *Guide to Outsourcing Information Technology*, London: Business Intelligence, 1994.

Zhuge, H. "Knowledge Flow Management for Distributed Team Software Development", *Knowledge-Based System*, (15), 2002, pp. 465-471.

About the Authors

Aini Aman is a doctoral student in Information Systems at the School of Accounting and Finance, University of Manchester, England. Her research interest is in offshore software development outsourcing. For her PhD research, she is currently looking at the process of collaboration using groupware application in offshore software development outsourcing. She has a Master in Business Administration from Manchester Business School, University of Manchester. Her academic background is in Accounting Information Systems and Control and she is currently in a study leave as a university lecturer at Universiti Kebangsaan Malaysia, Malaysia. Aini can be reached by e-mail at ainiaman@hotmail.com.

Brian Nicholson is a lecturer in Information Systems at the School of Accounting and Finance, University of Manchester, England. His research into global software development and software export success for developing countries has involved several international collaborations involving fieldwork in India and recently Iran. Brian can be reached by e-mail at brian.nicholson@man.ac.uk.

14

CHINA TELECOMMUNICATIONS TRANSFORMATION IN GLOBALIZATION CONTEXT

A Structuration Perspective

Ping Gao[1] and Kalle Lyytinen[2]

[1]*Department of Informatics, Copenhagen Business School, Denmark;* [2]*Department of Information Systems, Weatherhead School of Management, Case Western Reserve University, USA*

Abstract: As globalization is becoming a prevailing topic, information technology and infrastructure in developing countries being non-ignorable forces of globalization has arisen the interests of scholars in information systems. Drawing on structuration theory this paper analyses China's strategy of telecommunications transformation.

Keywords: China, globalization, strategy, structuration theory, telecommunications.

1 INTRODUCTION

As now globalization is becoming a prevailing topic, information technology and infrastructure in developing countries being non-ignorable forces of globalization has arisen the interests of scholars of information systems. Academic researches on such a topic have begun to appear in mainstream information systems journals. As one significant example, the study about the application of geographical information systems (GIS) in India appears in MIS Quarterly (Walsham and Sahay, 1999). In this paper we study the Chinese telecommunications industry. In an international perspective, China presents an interesting case to study, being the largest potential market for telecommunications services and because of its unique economic and political environment for telecommunications reform and development. From the early 1980s, China's telecommunications industry has gone through a series of transformations. As a result, China's telecommunications industry achieves one the fastest developments in the globe scale. Now its subscriber numbers of mobile phone and fixed line rank the first in the world (Gao and Lyytinen, 2001; MII, 1999). But what a strategy has made China's telecommunications sector achieve such a big success, and why China has adopted such a strategy? These are the research questions of this paper.

There are many publications about China's telecommunications industry (Gao and Lyytinen, 2000; 2001; Mueller and Tan, 1997; Tan, 1994; 1999). But except for Zhang (2001; 2002) who uses North's (1990) institutional theory as an analytical tool, others focus on fact-finding and description. According to North, institutions are rules and constraints that structure individual interaction in political, economic and social affairs, like regulations and policies. Zhang argues that for the Chinese telecommunications industry there exist several institutions enacted by varied organizations, e.g. governmental branches, World Trade Organization (WTO), etc. Their enforcement is through a bargaining mechanism with strong political involvements. Different from Zhang who examines the implementation of different institutions or policies from the interaction between organizations as policy-makers, in this paper we investigate how the central government of China formulates its telecommunications transformation strategy. We challenge the existing position that the negotiating mechanism of decision-making characterized by China's political system has restricted the telecommunications reform (Mueller and Tan, 1997; Tan, 1994; 1999; Zhang, 2002). We insist that the centralized administrative system allows the central government to comprehensively consider the strategy environment, which enables it to adopt a proper strategy that is to gradually carry out telecommunications transformations by balancing requirements from technology advance and restrictions from telecommunications development and economic, political reform progress.

We address our arguments by drawing on the structuration theory of Giddens (1979; 1984). Structuration theory defines the structure of a social system and agency as a duality, which means "the structural properties of social systems are both the medium and the outcome of practices that constitute those systems" (Giddens, 1979: p.69). Structuration theory has been widely used in management studies to examine the change mechanism of social systems by taking these special social designs as structures, for example the inter-organization network (Windeler and Sydow, 2001). In this paper we define China's telecommunications market as a structure, and elaborate its transformation process which represents the reform strategy. Our idea is developed through six sections. Following this introduction section, the second section defines telecommunications reform strategy. The third section lays out theoretical grounding. The fourth section is methodology. The final two sessions describe analytical results, and derive conclusion.

2 CONCEPTUAL PRELIMINARIES

By examining the experiences of reforming the old PTT[1] systems in the developed world, Noam and Kramer (1994) conclude a telecommunications reform strategy has four aspects: liberalization, devolution, consolidation, and deregulation. Liberalization means introducing competition into a monopolized market. Devolution is a policy of splitting the monopoly structure. Consolidation is the opposite strategy to devolution. Deregulation means reducing government's intervention in market.

While deregulation is the main trend of telecommunications reform from long term of view, competition does not necessarily result in deregulation at all stages of reform. The role of regulation is even more crucial in the market liberalization phase than it was before. The experience has indicated that a wide change in the degree and a large difference in patterns of regulation have been presented with time evolving (Collins and Murroni, 1997). Also the institutional setting in regulation must change to adapt to market variation.

We revise Noam and Kramer's identification on national strategy for telecommunications reform, and define it as including three approaches: liberalization, reregulation and restructuring. Reregulation refers to the change in regulatory contents and patterns (deregulation, asymmetric regulation, etc), and in institutional settings. The restructuring includes devolution and consolidation.

3 THEORETICAL GROUNDING

3.1 Structuration Theory

The structuration theory defines the structure and agency as a duality. By definition of Giddens (1979; 1984), structure is the specific type of rules and resources or capacities recursively implicated in social reproduction, and human agency is the capacity to make a difference. For duality, Orlikowski and Robey (1991) interpret it as: the structure of social systems is created by human action, and then shapes future action. Agents in their actions constantly produce and reproduce the social constructs, which both constrain and enable them. The process that the duality of structure is reproduced over time and space is defined as structuration. Linked by modalities, which are the knowledge and resources accessible to actors, the social structure and human interaction are both broken down into three dimensions (Figure 1).

[1] PTT (Post, Telegraph and Telephone) is traditional state network operator. When regulatory functions are split up it is referred to as PTO (Public Telecommunications Operator).

When people act in organizations through modalities, they create and recreate three fundamental elements of social interactions and structures. As Walsham (1993, p.61) explains it:

Firstly, human communication involves the use of interpretive schemes which are stocks of knowledge that human actors draw upon in order to make sense of their own and others' actions. They thereby produce and reproduce structures of meanings which are termed structures of signification. Secondly, human agents utilize power in interaction by drawing on facilities such as the ability to allocate material and human resources; in so doing, they create, reinforce or change structures of domination. Finally, human agents sanction their actions by drawing on norms or standards of morality, and thus maintain or modify social structures of legitimation.

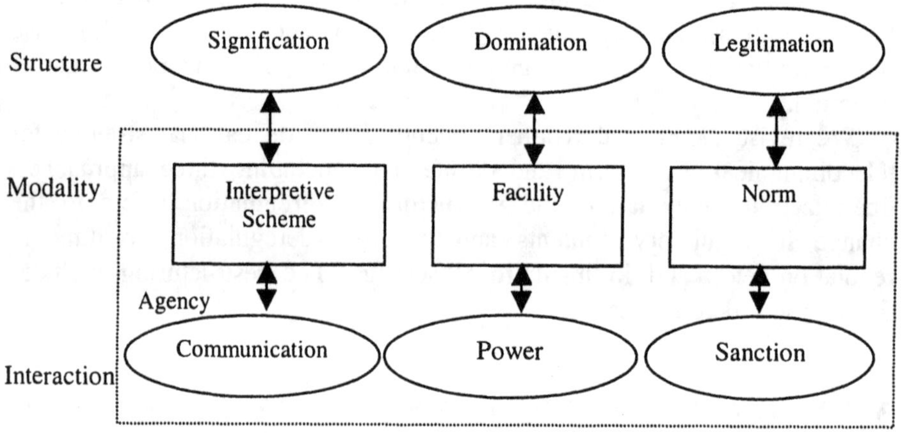

Figure 1. The duality of structure and action (Source: Giddens 1984)

3.2 Analytical framework

We accommodate the structuration theory to studying China's telecommunications reform strategy. The characteristics of a structurational model can be described in term of its definition on key concepts of structuration theory like structure, modality, actor and action (Walsham, 1993). We define the Chinese telecommunication market as the structure to be studied. The actor designing and then realizing the special structure is the state government. The structure and its formulation (action) constitute a duality: the structure characteristics affect the action, and in turn the structure can be modified through action which results in a new structure that is the basis of next step change. The strategy can be described by structure and action together, as shown in Figure 2. The distance denotes the change scale in ordinate, and time period in abscissa.

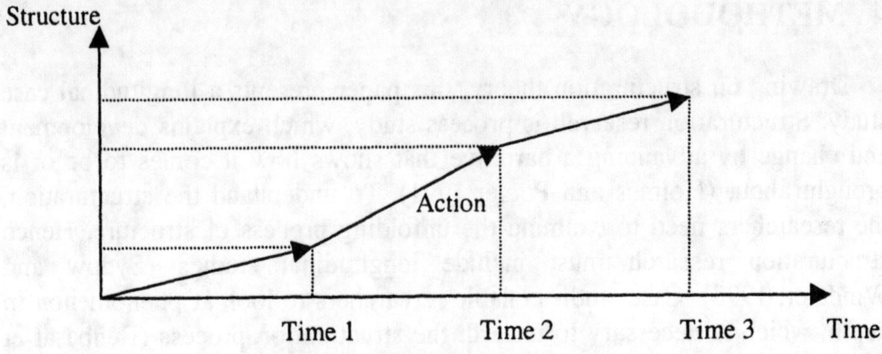

Figure 2. Strategy formulation process

The structure has three dimensions of signification, legitimation and domination. In the case of the Chinese telecommunications reform strategy, the signification is the meaning and characteristics of market structure. Legitimation is the generally accepted form of structure, or the opinion of what it should be. Domination is the economic, political, administrative, and technological ability to change or maintain a structure. The action is based on the interpretation on structure through modality. We assume the modality can be divided to several elements. From international experiences, we define them as the situations of technology advance, telecommunications development, macro economic and political environment, and international telecommunications reform (Gao and Lyytinen, 2001). Our structuration model of analysis is shown in Figure 3.

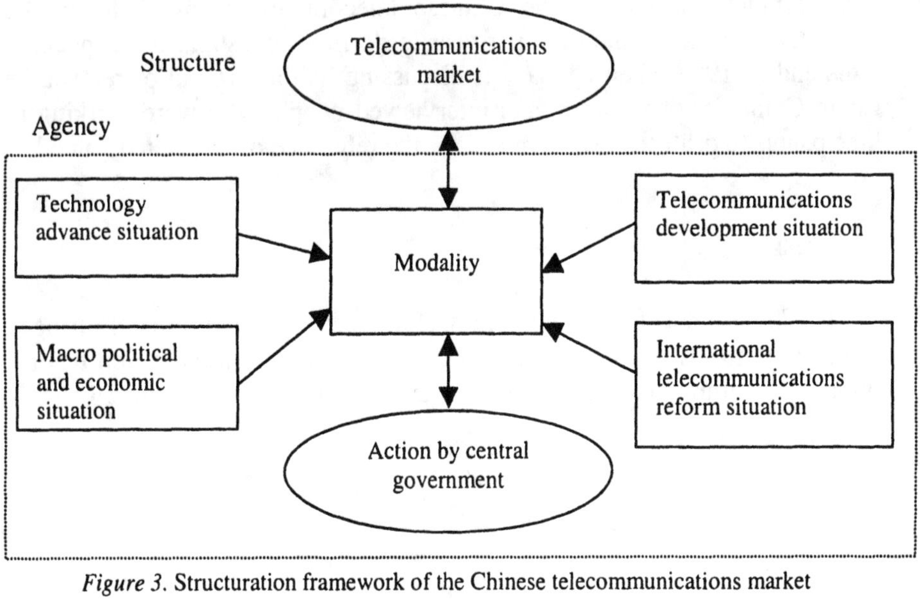

Figure 3. Structuration framework of the Chinese telecommunications market

4 METHODOLOGY

Drawing on structuration theory, this paper presents a longitudinal case study. Structuration research is process study, which explains development and change by advancing a narrative that shows how it comes to be or is brought about (Holmes and Poole, 1991). To understand the structuration, the researchers need to examine the unfolding process of structure. Hence structuration research must include longitudinal studies (Sydow and Windeler, 1998). Case studies enable researchers to look at phenomenon in depth, which is necessary to unearth the structuration process (Benbasat et al., 1987).

Data are collected from different ways. One option is field observation of structuration. Another method is to use archival documents that preserve time ordering of events to reconstruct a historical account of the structuration of technology. The third way is to conduct interviews so as to get the subjects' point of view concerning what structures and structuring mean from their perspectives. The first author of this paper had been a senior analyst in the affiliated consulting institute of China's Ministry of Posts & Telecommunications (MPT) for five years till the end of 1997. After 1998 when he moved to Europe, China's telecommunications reform has been his following-up project. He has kept a pragmatic network with the Chinese telecommunications industry, which enabled him to reach update data. Except browsing the official newspapers of China and having regular email and telephone communications with his previous colleagues who are still doing official consults to the Chinese telecommunications industry, he visited China at least one time per year to do field investigation. Especially at the end of 1998 when China was discussing its third round of reform, he was in China for one month, and interviewed people who were working in three panels to draft the reform plans for the reference of the State Council.

5 RESULTS

The Chinese telecommunications sector began to undergo major changes in the early 1980s by four stages. In the languages of structuration theory, China's telecommunications reform strategy is summarized in Table 1.

Table 1. Structure and reform strategy of China's telecommunications industry

	Signification	Domination	Legitimation	Action (Strategy)
1st stage	Monopoly	Poor economy and telecommunications basis. A user-pull market. Planned economy that rejected competition. Reform just began from the US, UK. Telecommunications reform was a restricted topic in China. State support by preferential policy was used	Natural monopoly was a deep-rooted norm on the globe. Monopoly was necessary to guarantee sovereignty, nation security, and socialism. Competition was acceptable for equipment and VAS	Sustain monopoly in network and basic services. Execute liberalization and deregulation on equipment and VAS
2nd stage	Monopoly in basic and mobile services and networks. Domestic non-private competition in VAS. Full competition in equipment. Multiple regulation system	The development level of telecommunications met demand of national economy growth. A user-push market. Technology advance required reform. Within a move of superficial macro reform the government got limited power to change telecommunications industry	Competition became a norm in international scope, and was practicing in global scale. A limited competition between state-owned firms as a test of future move was suitable for China	Introduce state-owned competitors to networks and basic services. Set a coordinating mechanism in regulatory regime. Liberalization and reregulation on PTT
3rd stage	Duopoly in mobile and basic services and networks. Limited, unfair competition. A non-neutral regulatory regime. Unfair regulation	Advanced, state-wide networks were formed. Deep macro reform was underway. Joining WTO in near future was a political task with the highest priority. The government had the power to enact a deep reform on telecommunications	Competition crossing telecommunications, televisions, and computer industries was accepted in the world. China constrained the reform not beyond the telecommunications	Restructuring on PTOs. Liberalization and reregulation
4th stage	Enforced, all-round competition. A neutral, coherent regulatory regime. An unreasonable market in that Unicom was weak	As a WTO member, China had to publish the timetable of full liberalization. To further adjust the market so as to meet the requirement of joining WTO has the highest priority.	Full competition with limitation on foreign capitals in a gradual way	Restructuring. Timetable of international liberalization was published

5.1 Up to 1994: Liberalization and Deregulation on Equipment and VAS

Due to its backward manufacturing industry, in 1982 China opened the equipment sector. China preferred a strategy of introducing production lines by joined-venture or foreign independent investments to directly purchase from abroad, which bounded the interests of foreign capitals with the Chinese market. Consequently some of domestic-made equipment has dominated China's network for its comparatively low price and competent quality (MII, 1999). China opened equipment market earlier and more intensively than most developed countries. Owning strong national manufacturing industries, they did not like foreigners to share their markets. For example, it was the late 1980s when most EU members opened terminal equipment sales to countries that have agreements of equal market access with them (European Commission, 1988).

In history the telecommunications network and services had been run under a monopoly in the global scope. This situation just began to change from the US and UK. In the US, the Modification of Final Judgment mandated AT&T to divest itself of its local service sector before January 1984, which was regrouped under seven regional holding companies. Meanwhile AT&T was allowed to enter into previously restricted markets like information services (Snow, 1995). In UK the market was fully liberalized in 1991 after a transition of duopoly in 1984 (Thatcher, 1999). Natural monopoly was still a general practice and legitimation in the world.

In macro economic and political context, reform on traditional socialism system and planned economy just began from non-dominant sectors. Telecommunications network was taken as an infrastructure and a basis of national economy, hence had to be under the absolute control of the state to guarantee national security and sovereignty. China's telecommunications network was extremely weak. In 1980, the telephone penetration rate was only 0.4%. From economic, political and ideological perspectives, the government had no power to reform telecommunications.

To promote China's telecommunications to develop from a very poor basis, the government granted preferential policies to telecommunications sector, which was possible under the state monopoly system. The State Council approved MPT to charge an installation fee from every subscriber that corresponded to the construction cost of one line. MPT had a privilege in advancing depreciation, using foreign currency and paying tax. More than half of MPT's investment was from these supports (MII, 1999).

New telecommunications services appeared from middle 1980s. Like in most other countries, in China value-added service (VAS) like paging without high network dependence was gradually opened to competition from the late 1980s. Currently VAS is a highly competitive market with thousands of operators (MII, 1999).

5.2 From 1994 to 1998: Liberalization and Reregulation on PTT System

The preferential policy and the strong market demand promoted the

Chinese telecommunications sector to enter to a new stage. In 1994, the telecommunications development level was already equivalent to the general growth level of the macro economy (Yang, 1997). The telecommunications market shifted from a buyer's pull to a seller's push. Therefore, unlike earlier times when the overriding objective was the development speed, currently service quality improvement and tariff reduction became the new focus, which could be best realized through a competitive mechanism.

In the US, the 1996 Telecommunications Act removed all remaining restrictions on the service scope of an operator and fully liberalized the American market (Baliga and Santalainen, 1999). For the liberalization on VAS in the global scale, and on basic telecommunications in the US and UK, generally an encouraging conclusion could be derived (Council of Economic Advisers, 1999; OECD, 1995). Hence competition was becoming a prevailing norm in international telecommunications world. Most developed countries were designing their reform plans. In China the support of preferential policy of the state met strong challenge from society and other industries. Also market competition was the demand of technological innovation. The appearance of Internet arose the topic of national information highway in the world, which required the convergence of media, banking industries etc with telecommunications. In macro aspect, China was initializing a statewide governmental reform to separate enterprise management functions from government branches. The state Council was empowered to reform the telecommunications industry. Consequently the market structure was transferred from monopoly supported by preferential policies to competition.

In 1994, as a part of macro reform, the State Council reformed the PTT system. Registered as China Telecom, the operating sector of MPT was changed from a functional department of MPT to an enterprise. Meanwhile Ministry of Electronic Industry (MEI) with some other state institutions formed Unicom to compete with China Telecom all-around, and Jitong that dedicated to Internet. Some ministries jointly set up the Sino Satellite Communications and China Orient to compete ChinaSat.

At the outset these changes were just like most international practices of telecommunications reforms, which are characterized by forming a coherent regulatory regime with an independent regulator and introducing new operators. But in China these changes were superficial. Both owned by the state, China Telecom and Unicom were the only two comprehensive public operators. They had a great disparity in strength. Unicom's income was less than only 1% of China Telecom's (Gao and Lyytinen, 2000). It was an unreasonable market situation in view of fair competition and should be improved. More importantly, China Telecom was still directly under MPT's control without independent rights in finance, investment, personnel, etc. Hence China Telecom was not a real enterprise, and the MPT was not a "pure" regulatory authority either. On the one hand MPT could not work fairly in executing regulation but might exercises bias toward its affiliation China Telecom. On the other hand it had no enough power to intervene in the domain of other ministries for example MEI that also participated in telecommunications market. The case was the same for the whole macro reform which was a superficial make-up. To prevent political disorder, China adopted an "act after trials" policy of transformation. At this time for the first move of reform, it was not intended to fully change the current system designed for the planned economy, but to perform a pilot test for future moves.

Prevented by the reform progress at the macro level, the 1994's reform on PTT failed. To meet the regulatory demand from market liberalization, the State Council set up a Joint Conference on National Economic Informatization in 1994 to coordinate the regulatory function distributed among different institutions. In 1996 it was substituted by the National Information Infrastructure Steering Committee (NIISC), which was responsible for the formulation and implementation of plans, policies, and regulations in information industry. But the NIISC was only an interim organization without legislative status, financial means, and administrative power to efficiently execute regulation. It still had to negotiate and cooperate with other powerful governmental agencies that were in charge of China's telecommunications industry (Tan, 1999).

5.3 From 1998 to 2000: PTOs Restructuring, Further Liberalization and Reregulation

From late 1990s China's telecommunications development entered a new phase marked by the forming of advanced, cross-country networks (MII, 1999). But the efficiency was low and service quality remained unsatisfactory. In international aspect, competition was not only a common-accepted norm but also put into practices. In EU, the January 1st of 1998 was the set deadline of full liberalization[2] (European Commission, 1997). Adapting to technology advance characterized by network convergence, a tide of enterprise mergers crossing national boundaries and spanning industries like telecommunications, computing and media emerged. A thorough reform on telecommunications industry was also a part of China's plan of joining WTO as soon as possible. The communist party of China with the supreme power in the country set it as a political task, which had the highest priority in national strategy. A deep reform was necessary to foster a fair competition market as required by joining WTO, and improve the

[2] Excluding Greece, Ireland, Portugal and Spain under derogation permission. In practice there are variances in different EU countries in their reform progress

competence of domestic operators in the forthcoming international competition.

In China, in 1998 a new round of governmental reforms started, which resulted in deep changes in the governmental system. As a result the Ministry of Information Industry (MII) was established based on MPT and MEI. MII was a coherent regulator by taking over the regulatory functions of MPT, MEI, NIISC, etc. MII was mandated with an exclusive power in administrating and regulating whole information industry.

For market transformation, the State Council set the principle as supporting the Unicom and protecting fair competition. China Telecom kept its fixed network and operations, and split mobile sector to form an independent body China Mobile. The paging sector of China Telecom whose turnover was five times of Unicom, and the mobile network of Great Wall as an army-run company were appropriated to Unicom (MII, 1999). The consolidation strategy was also applied to the satellite sector. ChinaSat, Sino Satellite Communications, and China Orient were merged to a group China Satellite. China Net was formed in 1999 by some state institutions, and the dedicated telecommunications network of the national railway sector was granted to enter into public telecommunications competition in 2000 by forming China Railway Communications. Consequently China has formed a "national fleet" composed by seven "carriers": China Telecom, Unicom, China Mobile, China Satellite, China Net, Jitong and China Railway Communications. This was the market prepared to meet for international competition (Gao and Lyytinen, 2001).

5.4 After 2000: Liberalization to Foreign Competition and Restructuring of Market

The fourth stage of reform was promoted by that China was formally adopted as a member of WTO at the beginning of 2001. For which China agreed to allow 49% foreign ownership in mobile communications within five years of accession; 49% in international and domestic services within six years upon accession; and 50% in VAS within two years. Foreign companies are going to enter into China within recent years, for which China believed it should re-arrange the market as preparations.

Another reason for the new round of reform was that the fair competition has not come as expected. Comparatively Unicom was still too weak to compete with China Telecom. Consequently, China Telecom was broken into two parts by areas. The northern part was composed by eleven northern provincial networks, China Net and Jitong, and took the name of China Net. The southern part was other provincial networks, and kept the name of China Telecom. Now Unicom is the only comprehensive telecommunications company in that its network spreads the whole country (though not strong), and it has licenses to provide all kinds of services.

6 DISCUSSION

6.1 Characteristics of China's Telecommunications Reform

In China the telecommunications reform has been carried out by governmental orders through applying administrative measures, which is different from other countries where the transformation is enacted by laws. By now the reform on China's telecommunications industry has gone through four stages and a full competition market is coming, a "Telecommunications Law" is still missing. This is the result of China's "after and trial" method in macro reform on the entrenched planned economy, which excludes the effects of laws.

In international experiences, telecommunications reform is major carried out through introducing new private competitors. In China PTO as the major reform target was split, by services firstly and by areas later. An explanation may be that this is in the consideration that the existence of a dominant operator will jeopardize the fair competition. A big firm has more power of influencing institutional environment of regulation enforcement (DiMaggio and Powell, 1983), for which China has a favourable cultural and political environment. Hence a devolution strategy was necessary.

China's ideological and political system favours that the state economy should control telecommunications. Hence in China the competition has been among state-owned operators, and a limitation is set on foreign capitals to let the Chinese companies take dominant shares in future cooperation.

Now China is still transforming towards a full competition. First, further deregulation is needed to remove the entrance barriers to allow more dedicated networks to enter into public telecommunications market. While the timetable of market opening to international capitals has been published, the restriction on domestic private capital should also be relaxed. Second, it is necessary to further deregulate the market, and change all operators to comprehensive ones like Unicom. The restriction on services for current operators should be released. The competition across areas should be encouraged. Third, the market should be restructured. Railway Communications is too weak to exist independently. Last, a convergence of telecommunications and other industries should be promoted.

6.2 The Application of Structuration Theory

This paper employs the structuration theory as the analytical tool in three aspects: defining market as a structure, describing the process of structure change as a strategy, and using modality to explain the decision maker's interpretation on structure and choice on strategy. The modality is defined as composed by four elements. Action or strategy includes reregulation, liberalization, and restructuring. The structuration theory dynamically connects different modality elements with strategy. As is summarized by Table 2, these elements have different influence on explaining the structure, and each element plays varied roles in different stages on formulating the strategy. Of all, the macro reform progress and politic consideration is decisive for strategy choice. In China, telecommunications reform has been a part of macro reform and national strategy of economy and politics, like joining WTO.

Table 2. Modality and strategy

	Technology advance	Telecommunications development	Macro context	International environment	Strategy
1st stage	Telephone network. Fibre was used	Poor networks. User-pull market	Planned system. Poor economy	VAS competition. US, UK led reform	Liberalization, deregulation on VAS and equipment
2nd stage	Internet, mobile network Internet appeared	The development level of telecommunications met demand of national economy growth. A user-push market	Superficial macro reform. Fast economic develop-ment	Reform PTT. Introduce competition	Reregulation on PTT. Liberalization
3rd stage	Network convergence	Advanced, state-wide networks were formed.	Deep macro reform. Fast economic develop-ment	Full competition	Restructuring on PTOs. Liberalization, reregulation
4th stage	Network convergence	A leading sector in national economy	Joining WTO	Full competition	Restructuring. Fixing timetable of international liberalization

For future research, one possibility of improvement on this paper is to incorporate other theories with structuration theory to provide a pluralistic account of strategies by different agents (Hung and Whittington 1997; Sahay and Walsham, 1997). For example by incorporating North's institutional theory as used by Zhang (2001; 2002), an extra institutional component can be added to modality. In this way the role of different institutes on strategy formulation can be explained to a better degree, and the question why some obviously better schemes were not accepted, e.g. to combine Railway Communications with other operators, can be answered clearly.

7 CONCLUSION

There is different ways to study market change. For the case of the Chinese telecommunications industry, most current studies investigate the influence of different institutes. Differently and as the first, we focus on the behaviour of the decision maker that designs the reform strategy. While previous studies have mainly focused on fact-finding and descriptions, here we demystify structuration theory and use it as the analytical basis for its strength in describing market transformation as a path dependent process. Our structuration model dynamically connects different elements of the modality with strategy formation. The concept of pluralistic components of modality can be generalized to studying other social systems, or policy-making in varied fields. We conclude that structuration theory is useful in describing the dynamic, varied influence of different elements on transformation of structure, as is telecommunications market in this paper.

We observe that varied modality elements have different influence on the reform strategy of China's telecommunications industry, and each element plays special roles in different stages. Of all, the macro reform process and national policy play a determinant role. Generally China's telecommunications transformation has been a part of the macro reform that abides by an "act after trial" motto, and greatly influenced by national policy like joining WTO. In the globalization context where a full competition is the main trend, China designs its own strategy major based on its own situation, instead of mechanically following others. China has cautiously promoted telecommunications market change in four stages.

Now China is still in the process of moving towards a full competition crossing telecommunications and other industries like media, televisions etc. According to previous experience, the advance will be determined to a large degree by macro reforms, e.g. to remove the tight control on media by the state.

ACKNOWLEDGEMENTS

The first author acknowledges the financial support of the PITNIT project led by professor Jan Damsgaard. The finial version of this paper is done when he was a researcher in PITNIT project. Please see: http://www.cs.auc.dk/research/ IS/PITNIT/.

REFERENCES

Baliga, R., Santalainen, T. *Telecommunications in Transition: The US Experience*, Helsinki. HeSE Print, 1999.

Benbasat, I, Goldstein, D.K., Mead, M. "The Case Research Strategy in Studies of Information Systems", *MIS Quarterly*, (11:3), 1987, pp.369-386.

Collins, R., Murroni, C. "Future Directions in Telecom Regulation: The Case of the United Kingdom", in *Telecom Reform: Principles, Policies and Regulatory Practices*, W.H. Melody (eds.), Lyngby: Technical University of Denmark, 1997.

Council of Economic Advisers. "Progress Report: Growth and Competition in U.S. Telecommunications 1993-1998", Washington DC, Feb. 8, 1999.

DiMaggio, P., Powell, W.W. "The Iron Cage Revisited: Institutional Isomorphism and Collective Rationality in Organizational Fields", *American Sociological Review*, (48), 1983, pp.147-160.

European Commission, "Directive on Competition in the Markets in Telecommunications Terminal Equipment", 88/301/EEC, 1988.

European Commission, "Status Report on EU Telecommunications Policy", DGXIII/A/1, 7 May, 1997.

European Commission, "First Monitoring Report on Universal Services in Telecommunications in the European Union", COM(1998)101, 1998.

Gao, P., Lyytinen, K. "Telecommunications in Transition: The Chinese Experience in the International Context", in *Managing Telecommunications and Networking Technologies in the 21st Century: Issues and Trends*, G. Grand (eds.), Idea Group Publishing, 2001.

Gao, P., Lyytinen, K. "Transformation of China's Telecommunications Sector: A Macro Perspective", *Telecommunications Policy*, (24:8-9), 2000, pp.719-730.

Giddens, A. *Central Problems in Social Theory*, Berkeley, CA: University of California Press, 1979.

Giddens, A. *The Constitution of Society*, Cambridge: Polity Press, 1984.

Holmes, M., Poole, M.S. "The Longitudinal Analysis of Interaction", in *Studying Interpersonal Interaction*, B. Montgomery and S. Duck (eds.), New York: Guilford, 1991.

Hung, S., Whittington, R. "Strategies and Institutions: A Pluralistic Account of Strategies in the Taiwanese Computer Industry", *Organization Studies*, (18:4), 1997, pp. 551-575.

MII, *China Telecommunications Over 50 Years*, Beijing. MII, 1999.

Mueller, M., Tan, Z. *China in the Information Age: Telecommunications and the Dilemmas of Reform*, Westport, CT: Praeger, 1997.

Noam, E., Kramer, R. "Telecommunications Strategies in the Developed World: a Hundred Flowers Blooming or Old Wine in New Bottles", in *Telecommunications in Transition: Polices, Services and Technologies in the European Community*, C. Steinfield, J. Bauer, and L. Caby (eds.), Sage, 1994.

North, D. *Institutions, Institutional Change, and Economic Performance*, Cambridge, NY: Cambridge University Press, 1990.

OECD, "Telecommunications Infrastructure: The Benefits of Competition. Information Computer Communication Policy", Paris. *OECD*, 1995

Orlikowski, W.J., Robey, D. "Information Technology and the Structuring of Organization", *Information Systems Research*, (2:2), 1991, pp.143-169.

Petrazzini, B., Krishnaswamy, G. "Socioeconomic Implications of Telecommunications Liberalisation: India in the International Context", *The Information Society*, (14:1), 1998, pp.3-18.

Petrazzini, B. "The Political Economy of Telecommunications Reform" in *Developing Countries: Privatisation and Liberalisation in Comparative Perspective*, Westport, CT: Praeger, 1995.

Sahay, S., Walsham, G. "Social Structure and Managerial Agency in India", *Organization Studies*, (18:3), 1997, pp.415-444.

Snow, M. "The AT&T Divestiture: A 10-year Retrospective", in *Beyond Competition: The Future of Telecommunications*, D. Lamberton (eds.): Elsevier Science, 1995.

Sydow, J., Windeler, A. "Organizing and Evaluating Interfirm Networks: A Structurationist Perspective on Network Processes and Effectiveness", *Organization Science*, (9:3), 1998, pp.265-284.

Tan Z. "Challenges to the MPT's Monopoly", *Telecommunications Policy*, (18:3), 1994, pp.174-181.

Tan, Z. "Regulating China's Internet: Convergence Toward a Coherent Regulatory Regime", *Telecommunications Policy*, (23:3-4), 1999, pp.261-276.

Thatcher, M. "Liberalization in Britain: From Monopoly to Regulation of Competition", in *European Telecommunications Liberalization*, K. Eliassen and M. Sjovaag (eds.), NY: Routledge, 1999.

Walsham, G. *Interpreting Information Systems*, Chichester: Wiley, 1993.

Walsham, G., Sahay, S. "GIS for District-Level Administration in India: Problems and Opportunities", *MIS Quarterly*, (23:1), 1999, pp.39-66.

Whittington, R. "Put Giddens into Action: Social Systems and Managerial Agency", *Journal of Management Studies*, (31), 1992, pp.829-845

Windeler, A., Sydow, J. "Project Network and Changing Industry Practices-Collaborative Content Production in the German Television Industry", *Organization Studies*, (22:6), 2001, pp.1035-1060.

Yang, P. "A Calculation of China Telecommunications Development Situation", *Posts and Telecommunications Soft Science Research*, (39), 1997, pp.3-7.

Zhang, B. "Assessing the WTO Agreements on China's Telecommunications Regulatory Reform and Industrial Liberalization", *Telecommunications Policy*, (25:7), 2001, pp.461-483.

Zhang, B. "Understanding China's Telecommunications Policymaking and Reforms: A Tale of Transition Toward Liberalization", *Telematics and Informatics*, (19:4), 2002, pp.331-349.

About the Authors

Ping Gao received a Doctor of Engineering degree from Xi'an Jiaotong University in 1992. Then he worked as a consultant for the Ministry of Posts and Telecommunications of China till 1997. From 1997 he was a researcher on mobile communications standardization at University of Jyväskylä, Finland. Now he is in Copenhagen Business School as a researcher of Mobiconomy project which studies mobile commerce implementation. His research interests include telecommunications regulation and strategy, technology diffusion, mobile commerce, digital economy and knowledge management. He has been following China's telecommunications industry, and taken it as his continuous study project. Ping can be reached by e-mail at pg.inf@cbs.dk.

Kalle Lyytinen is a professor at Case Western Reserve University and an adjunct professor at the University of Jyväskylä, Finland. He serves currently on the editorial boards of several leading IS journals including,

Senior Editor of AIS journal, Information Systems Research, EJIS, JSIS, Information & Organization, Requirements Engineering Journal, and Information Systems Journal. He has been the former secretary and chair of IFIP 8.2. He has published over 150 scientific articles and conference papers and edited or written eight books on topics related to system design, method engineering, implementation, software risk assessment, computer supported cooperative work, standardization, and ubiquitous computing. His research interests include information system theories, computer aided system design and method engineering, system failures and risk assessment, computer supported cooperative work, nomadic computing, and the innovation and diffusion of complex technologies and the role of institutions in such processes. Kalle can be reached by e-mail at kjl13@po.cwru.edu.

15

FACTORS INFLUENCING IRELAND'S SOFTWARE INDUSTRY

Lessons for Economic Development Through IT

Ciara Heavin[1], Brian Fitzgerald[2] and Eileen M. Trauth[3]

[1]Business Information Systems, University College Cork, Ireland; [2]College of Informatics and Electronics, University of Limerick, Ireland; [3]Pennsylvania State University, USA and Science Foundation Ireland.

Abstract: Although the Irish software industry has been in existence since the late 1960s, in the last ten years it has become an enormous success story and, by some estimates, is the largest exporter of software in the world. In this study, we derive a conceptual framework with which to investigate the emergence and evolution of the Irish software industry. In this framework endogenous factors – national, enterprise and individual – are examined against a backdrop of exogenous factors to explain Ireland's success in the software industry. Two important themes emerged from the study. One is the shift towards a risk-taking, entrepreneurial spirit in the software sector that had historically been absent in Ireland. The other is the long-standing government policy of intervention in the software sector, which became increasingly more focused throughout the 1990s. The framework and the lessons from the study should prove useful to researching other countries who are seeking to learn lessons from the Irish experience.

Keywords: Ireland's software industry, exogenous factors, endogenous factors, national, enterprise and individual.

1 INTRODUCTION

Ireland's software industry has been hailed in technology circles worldwide since the mid 1990s. It has emerged as a strong contender for multinational sites, along with Israel, India and Eastern Europe. Ireland's software sector employs 30,000 people in both indigenous and multinational operations and creates revenues in excess of €10 billion (Flood et al., 2002).

According to the OECD Information Technology Outlook 2000, Ireland has become the largest exporter of software goods in the world (IDA, 2002).

In the late 1990s Ireland emerged as an economic 'Celtic Tiger'. This has been the result of Ireland's openness to the global market coupled with the presence of heavy investment from multinational organizations. This rapid economic growth has enhanced employment levels and maintained low inflation rates (O'Riain, 1997). But this economic boom did not happen overnight. According to Trauth (2000: p.28), Ireland's development occurred in stages: the shift away from the protectionist policies of the 1950s, followed by recognition of the importance of information technology during the 1970s, and finally the realization of the information economy in the 1990s. Digital Equipment Corporation (DEC) was among the first multinational high tech companies to set up hardware operations in Ireland, doing so in Galway in 1971. While this government policy appeared to be the answer to Ireland's economic problems, O'Riain (1997: p.12) conveys the reality:

> *"This policy's success is questionable as companies created mainly low-skilled jobs, developed few linkages to the local economy, and often left once their tax breaks ended".*

The emergence of the software industry in Ireland as a key component of the economic "Celtic Tiger" parallels the larger journey that Ireland took from an impoverished agrarian society to a significant post-industrial society in the second half of the twentieth century. As documented by Trauth (2000: Chapter 2), this transformation occurred in phases as the economic development vision shifted from industrialization by invitation, to the recognition of economic development through the IT industry, to the creation of a diversified IT sector. A key component that has emerged within this diversified IT sector is the software industry.

After 1973 using Digital as an example, the Industrial Development Authority (IDA) began to attract investment from foreign multinationals to Ireland. Analog Devices, Amdahl and Apple mark the first high profile technology based (hardware) multinationals to open operations in Ireland. These companies produced mostly hardware in the early stages of the high technology industry in the country but this was to change. As IBM took the steps in the early 1970s to unbundle software and hardware costs, it realized that revenue could be generated from software development, something that could be located overseas (Cochran, 2001). Additionally, early indigenous software firms were established. In the late 1980s, key industry players such as Kindle and CBT Systems emerged, supplying products to vertical markets in banking and training. Since then, the Irish Software Industry has grown in prominence as Irish companies such as Iona Technologies, Baltimore and SmartForce Ltd. have become publicly quoted technical companies on the Nasdaq stock market.

Consistent with Trauth's broader analysis, contextual factors played a pivotal role in the evolution of the software industry in Ireland. Embedded in Ireland's transition from agricultural dependence through its quasi-industrial era of the 1970s, to the development of a software industry in the 1980s and 1990s is the picture of a rapidly changing societal context which facilitated this economic change and has felt its effects. This suggests the need for deeper investigation of this 'success story'. Such an investigation will contribute to a better understanding of the factors that have influenced the emergence of Ireland's software industry and can provide lessons for other countries/regions that wish to achieve a similar goal. Countries or regions that wish to pursue economic development through the creation of a software industry can benefit from a better understanding of the interplay between socio-cultural and policy factors. Ireland, itself, can benefit as well, from a better understanding of the uniquely Irish aspects that can be leveraged to ensure its continued success in this industry.

This paper is structured as follows. First, we present the conceptual framework that was employed in this research. This is followed by a discussion of the research methods, the research findings and the implications of this research.

2 RESEARCH FRAMEWORK

The purpose of this research was to investigate the emergence and evolution of Ireland's software industry since the 1960s and to identify the key contextual factors, which contributed to this growth. In order to do this, we first developed a research framework based on previous research. This framework is comprised of two parts; the exogenous factors that provide the background of societal factors in which the software sector was developed and the endogenous factors on which this study was based.

We drew upon three frameworks in the literature that consider the influence of societal context on IT industry success. These are: Ein-Dor's (1977) factors affecting IT industry success; Heeks' (1999) strategy for software success in developing countries; and Trauth's (2000) influence-impact model of society-technology interaction. The resulting conceptual framework accounts for contextual factors that could contribute to the development of a software industry. Our framework is presented in Figure 1, and its components and derivation are discussed below.

2.1 Exogenous Factors

Exogenous factors are those elements that may not be easily changed and that operate indirectly to influence IT success (Ein-Dor et al., 1997). They include demographic[4] factors and cultural factors. Demographic factors include a country's population, location, size and available natural resources. In this study we have also categorized culture as an exogenous factor that may contribute to the development or growth of a software industry. The term culture in this case refers to factors such as language, literacy, attitude to education and religion.

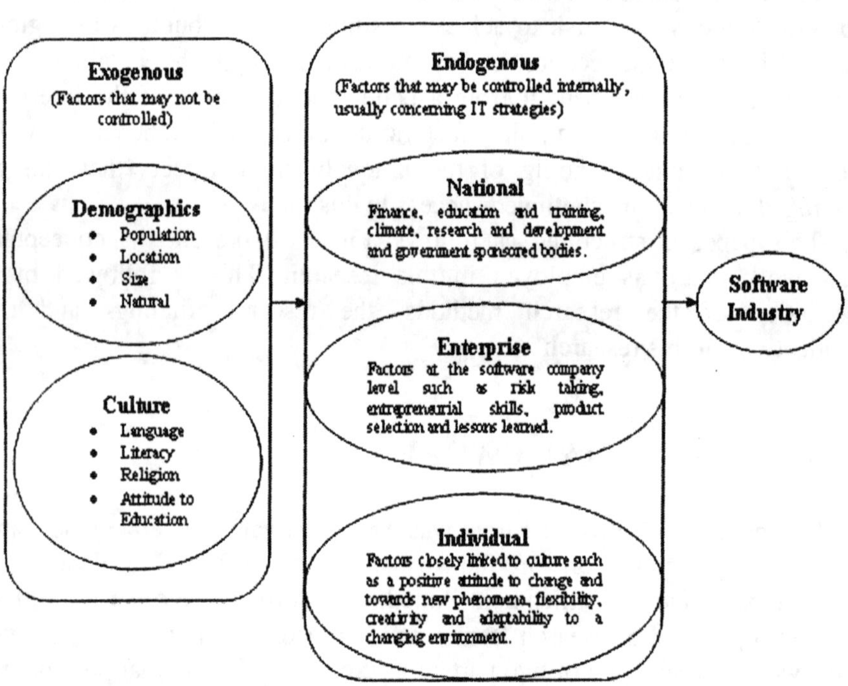

Figure 1. Contextual Factors Influencing the Development of a Software Industry
Adapted from Ein Dor et al. (1997), Heeks (1999), and Trauth (2000)

[4] The term demographics has both a narrow definition as having to do with the statistics of population, and also a broad definition as having to do with the general condition of life in a community. We use the term in its broader sense here.

According to the latest estimates from the Central Statistics Office, Ireland's population is 3.8 million with 38 per cent of the population under the age of 25 years, giving Ireland one of the youngest populations in Europe (IDA, 2002). Given this population profile, Ireland may be classified as a small economy, similar to Israel, New Zealand and Singapore. Whereas small economies had been considered at a disadvantage, they now are perceived to have the ability to compete with their competitors in high technology sectors disregarding the capacity of their mass markets (Ein-Dor et al., 1997). Two reasons are suggested as an explanation for this. The first is that a change in value creation has occurred. Value is no longer primarily created from industrial production. In the IT sector it is mainly created in analysis and problem solving tasks, identifying a customer's problem and offering solutions to those needs. Secondly, small countries no longer have to consider the issue of geographic location. Vast improvements in telecommunications mean that a manufacturer can carry out research and development in one country and production in another (Ein-Dor et al., 1997). While Ireland's small size may have been problematic during its industrial era, it is not a disadvantage for an information economy. Ireland's location has also acted in its favour. Being close to the European market, a member of the European Union and maintaining links with the US market has enabled Ireland to create an international software market.

Over the last twenty years Ireland's young well-educated population has acted as a huge incentive for multinational organisations to set up operations here (Trauth, 2000). In fact, Ireland has successfully educated more graduates than could be employed in the domestic sector. Given that almost all other countries worldwide have experienced shortages of engineers and computer science graduates, the surplus in Ireland has provided a unique advantage for the software industry here. Finally, Ireland's limited supply of natural resources has acted as a motivating factor in the development of a software industry.

For the purpose of this study, culture has been identified as an exogenous factor that may not be changed, but it is fair to say that in the long term this may not be true. When considering Irish culture as it relates to the software sector, the language spoken, literacy levels, attitude to education and religion have been found to be important factors (Trauth, 2000). Undoubtedly, the fact that Ireland is an English-speaking country is a major benefit to the software industry, above all with regard to doing business in the lucrative US market. It also has been a highly influential factor in attracting many leading US multinational technology companies to set up in Ireland. The improved literacy skills since the availability of free secondary education in 1968 and the positive attitude to education in Ireland have brought many changes to the country's economic position.

There are different views on the exogenous factor, religion. According to Ein-Dor et al (1997), religion does not directly affect the growth of an IT sector. Trauth (2000: p.197), on the other hand, did see an influence of religion in Ireland. "Religion is woven into the fabric of Irish culture". This factor was not considered in our research.

2.2 Endogenous Factors

Endogenous factors are those that can be developed and controlled and used in strategic IT industry development. National factors play a considerable role in the development of a technological sector. One of these national factors, government policy, has been shown to have a significant influence on the development of a technology sector (Ein-Dor et al., 1997, p85).

> *"Differences in the policies of governments seem to be major explanatory factors in differences in IT industry development".*

Government intervention may aid an industry expansion in areas that might be too difficult to deal with at an enterprise level.

With regard to enterprise factors, while the enterprise strategy and management tactics of software operations in developing countries may be unknown (Heeks, 1999), the evidence suggests that software companies are increasing their focus on choosing a niche product and market in order to avoid competing with Microsoft-type companies. The objective is to move away from the software localisation, software service, and traditional software application industry sectors. For example, in Israel software houses pursue the development of special purpose software such as computer systems management, data compression, encryption, virus detection and correction. They focus on particular product niches that, in turn, create a high level of value for the industry. In New Zealand software companies continue to produce a wide variety of software, yet focus their software domestically. In Australia the focus is on a specific market (Ein-Dor el al., 1997). These examples show that while it is difficult to pin down the various enterprise tactics, the enterprise factor plays a significant role in the development of the software industry.

At the individual level, the endogenous factors are closely linked to the exogenous factors. Trauth (2000) considers the Irish people involved in the information sector, their attitude to work, the quality of their work, the creativity that shapes the success of the sector and the society that surrounds the information economy. Undoubtedly, the individual element is an essential consideration when studying the development of a software industry. Ultimately, innovation, strategy and motivation are characteristics that act as key individual drivers for industry.

3 RESEARCH DESIGN AND METHODS

In order to investigate the role that these endogenous factors have played in the emergence of Ireland's software industry, a two-phased study of the Irish software industry was conducted. Initial data collection and analysis focused on document analysis of government documents, research papers and reports by the National Software Directorate and the Industrial Development Authority (IDA). This was used to inform the development of an interview guide that was developed in conjunction with the conceptual framework presented above. Interviews were chosen as a way of ensuring 'thick description' of the software industry's development since the late 1960s. In this way, firsthand insight was obtained about the facilitators and inhibitors as experienced by a cross-section of those involved in the emergence of the software sector.

Interviews with fifteen respondents from academe, industry and the government were carried out between March and May, 2002. During a taped interview of approximately one hour in duration the respondents discussed the interview questions related to three main endogenous factors: national, enterprise and individual. Interviewees were selected based upon their level of involvement with the software industry. Five of the interviewees represent academe, four represent government and six represent the software industry. In total, these interviewees offered key insights into the history of the industry, the public policies that have affected its development, and future plans that may be undertaken. The interviewees have been represented in this study with the use of pseudonyms to protect their privacy and opinions.

4 RESULTS

4.1 National

4.1.1 Government Intervention

There are conflicting views concerning the government's involvement in the development of Ireland's Software Industry. While the government undertook the policy of industrialization by invitation at the close of the

1950s, there is no clear evidence to support the deliberate strategy of intent for the cultivation of a software industry in the early years.

The academic and industry interviewees support the view that the development of the software industry is reflective of the key players that were involved in the early days. As one academic with considerable industry experience put it: "Software grew from the grassroots up". The government requested academic proposals with no real focus on software as an industry. During the late 1960s, it seems the government was interested in attracting any type of industry to Ireland, provided it was clean and created employment. While explicit foresight about the software industry was not in evidence, academics were confident in computers as the way forward.

> *"It was always believed that computers had a great future, people didn't exactly see where the future lay". [Pat, Computer Science Lecturer]*

Champions of the government's efforts believe that national policy for developing technology was conceived as far back as the 1960s through the industrialisation by invitation. This policy, then, became more focused as software began to naturally emerge as a potentially flourishing industry. The perception of a long-standing deliberate strategy by the Government for the software industry is not supported by the academic and industry-based interviewees, however. But it is acknowledged that the government's intervention and support for the industry has been successful over the last ten years. This turnabout in the early nineties is evidenced in the shift in Irish industrial policy in the 1990s as a result of the Culliton Report (National Software Directorate, 1992). The policy moved from a preoccupation with attracting foreign direct investment to the promotion of indigenous firms with a particular focus on software as a potentially flourishing sector in the world economy (Labour Commission, 1996).

4.1.2 Government Agencies

The Irish government has provided a significant support structure for the software industry. One of the many objectives of The National Software Directorate (NSD), set up in 1991, was to align industry with education, thereby creating niches in the software market and creating value from relevant research in the area of software technology. The Centre for Software Engineering (CSE) was also set up in 1991 as a support service for the software development community within Ireland, with the aim of helping companies improve in areas of quality control and productivity by implementing software engineering best practice, offering advice on company strategy and providing training.

Another important initiative has come from the Industrial Development Authority (IDA). The objective of this Government body is to create a dynamic by which Ireland increasingly sets the agenda and attracts other serious players in a small number of niche areas to emerge from clusters of technology based companies. The IDA is currently researching other areas that might create a new competitive advantage for the Irish software industry. In 2002, a Strategic Business Group (SBG) was set up within the (IDA) with the explicit remit of developing a new niche software position for Ireland "distinct from the nation of broad shoulders and smart people" [Andrew], as a government interviewee involved in developing the national software strategy put it.

4.1.3 Finance

One of the most limiting and contentious factors with regard to the development of an indigenous software company in Ireland, is that of funding. Ireland had virtually no venture capital until the 1990s. The National Software Directorate (1992) documented the problem at the time, suggesting that "the lack of available finance is proving to be a real barrier to growth and is the single greatest problem facing high-tech industries today".

The NSD established a state sponsored venture capital fund in 1996. It comprised 50 per cent Government money, with the remaining 50 per cent made up of investments acquired from a number of sources. Ironically when this fund became operational in 1996 other funds also appeared. After twenty years, private investment suddenly increased considerably. In trying to understand this phenomenon, one theory is that the mystery surrounding the software industry had dissipated even as investors were more confident backing software knowing that the government was also investing heavily in it (Keogh, 2000).

Undoubtedly, the availability of venture capital for software funding has acted as a significant enabler for the industry. While investment is more easily obtained, without a solid business plan and a quality product with a potential market, investment remains difficult to acquire.

4.1.4 Education

Ireland's education system has acted as a pivotal contributor to the development of the software industry. It has emerged as an undisputable factor influencing the growth of the industry. Since the availability of free secondary education under state legislation in 1968, a culture has developed in Ireland where the expected standard and quality of graduates each year remains high.

"The education system provided the seeds for the industry to grow". [Paul, Software CEO]

One of the key educational contributions to the development of Ireland's software industry, frequently cited during the course of the interviews, is the Computer Science department at Trinity College Dublin (TCD) and its inspirational head of department, Professor John Byrne. This department initiated the first Masters in Computer Science at Trinity in the late 1960s and had a significant impact on the embryonic software sector. This programme has acted as an incubator for young software companies and has been the origin of many successful software producers such as SoftTech, Mentec and Iona Technologies. These campus-led initiatives are commended by several of the interviewees as incubators for Ireland's software industry.

Examining the IDA statistics for 2002, it is evident that Ireland continues to improve its education system and it is clear that educational qualifications are valued highly. The number of students entering third level education in Ireland has increased by 25 per cent since 1992. There has as been a 35 per cent increase in the number studying engineering/technology courses. Irish education is considered quite high by international standards. According to the 2001 independent IMD World Competitiveness Report, Ireland boasts the highest public expenditure on education at 13.5 per cent, equaled in Europe only by Portugal (OECD, 2001). These data are consistent with interviewees' belief that the education system has significantly contributed to the growth of the software industry and on a larger scale the Irish economy as a whole.

4.1.5 The Climate for Development

With the closure of IBM operations in India in 1978, many ex–employees went on to set up their own indigenous software organizations (Heeks, 1996). In Ireland, there is a similar scenario whereby the software industry developed through the presence of multinational software companies. A number of the interviewees believe a turning point for the development of the industry came with the shutting down of a number of multinational operations. For example, while the closure of Digital Equipment Company (DEC) in 1993 may have been seen as disastrous at the time, the closure resulted in a new set of entrepreneurs who could use the software expertise they had developed to go into business for themselves. While there is no clear evidence to document this, Gallen (2001) suggests that the employees who worked with multinational companies have moved to work at indigenous operations.

> *"Staff who worked in multinational companies were, at one stage, more likely to have experienced structured approaches to software development, and this would be an asset to a small Irish company."*

4.2 Enterprise

Ireland has historically not been renowned for its entrepreneurial expertise and risk-taking ethos. A number of reasons have been suggested as to why this might be the case. According to Trauth (2000: p.321),

> *"Ireland did not have a significant economic infrastructure including venture funding...there wasn't economic payback for taking risks".*

There was also a belief among a number of the interviewees that Ireland is severely lacking in entrepreneurial history. This stems from the lack of industrialization experienced by the country up until very recently. It was not the done thing to start a business in Ireland during the 1970s and 1980s. It was considered "common" by the middle classes in Ireland to be involved in entrepreneurial business ventures. One software company founder, captured the essence of this feeling:

> *"Once upon a time in the early nineties if you were an entrepreneur you were a hustler". [Donal]*

Business failure is not readily accepted in Ireland and there tends to be a widespread feeling that when a business does fail the individuals involved tend not to be trusted to start again. The interviewees agree that Irish software companies need to accept failure before the industry can mature. The US software companies are more mature, and in US society, failure is not viewed entirely negatively, but rather as denoting experience in the industry and the ability to learn in order to succeed the next time. At the same time that business failure is not acceptable in Ireland, business success may also incur a negative reaction. Sometimes people who display initiative are resented for their success:

> *"People who go out and try to do things are knocked and begrudged for it, they are not complimented for it". [Donal]*

In the early years, software entrepreneurs in Ireland experienced considerable difficulty. That the software industry was poorly understood, contributed to a lack of confident investment.

"There was a mystery about software that would resist any investment by outsiders". [Dave, Software CEO]

The former Irish Prime Minister, Dr. Garrett Fitzgerald, expressed concern that since Ireland's culture is not traditionally entrepreneurial not enough people graduating university were prepared to set up in business. As one government based interviewee remarked, the Prime Minister felt that the young educated population lacked initiative and a sense of entrepreneurship. However, since the mid 1990s the opposite seems to be the case and now people are focused on setting up what are known as "High potential startups".

Irish software companies quoted on the Nasdaq have created increased confidence in the industry. The software entrepreneurs of the early to mid-nineties are credited for their confidence and entrepreneurial expertise that has aided the evolution of the strong software industry that Ireland boasts today. One government interviewee put it:

"It is individual, maybe it's luck, and maybe it's genius". [John]

Entrepreneurial Irish companies seem to have developed a winning formula: pursuing a market niche rather than competing with companies such as Microsoft, combined with the ability to develop a quality product that can be exported profitably. With the contribution of smart, enthusiastic individuals, a number of key companies such as Iona Technologies, SmartForce and Baltimore have emerged to challenge the non-entrepreneurial stereotype.

"It has spurred people on to see their next door neighbours doing all sorts of things, people are pushing themselves to keep up". [Donal]

The importance of a spirit of entrepreneurship was mentioned by all interviewees. They identified a combination of a well thought out business plan, a quality product with a market, a sensible business model, tight control on company growth and relating employee pay to shareholder wealth as a set of determinants for growing a successful software producing operation.

4.3 Individual

Ireland's technology industry has come a long way since the initial plan for industrialisation by invitation in the late 1950s. Irish culture has influenced this shift away from the land and towards the development of one of the most successful software industries in the world. One respondent with considerable experience in the software sector from both a government and industry perspective commented:

> *"People wanted to find their crock of gold at the end of the software rainbow". [Barry, Software Entrepreneur]*

Multitasking within an organisation has been identified as a predominant feature in smaller software development operations in Ireland. It seems that while other nationalities tend to adhere closely to their job descriptions, the Irish attitude "to get the job done" prevails when problems arise.

> *"A project won't stop dead if the operating system stops dead, they'll kick start it or something". [Sean, Software CEO]*

A number of the interviewees coupled flexibility with creativity as an individual contributor to the development of the Irish software industry. They believed that Irish people are imaginative software developers while other nationalities tend to be far more mathematical in their approach to software development. However, one industry interviewee felt that this creativity could lead to a tendency to be sloppy in an effort to succeed at a project. While the software product may be high in quality some developers do not appreciate the subtleties of jobs such as backing up work and security issues. One respondent drew a comparison between this 'wing and a prayer' mentality and that of an artist:

> *"It's a bit like Picasso, he paints a masterpiece and then forgets to put a frame on it". [Sean]*

However, this nonchalant attitude may have created considerable goodwill towards the country which has, in turn, benefited the software industry. The global recognition of the Irish culture means it is easy to attract well-educated foreign people to Ireland for the 'craic'[5] . Further, the progress of the Celtic Tiger, has encouraged many of those who emigrated during the slump times to consider returning to Ireland.

---

[5] This is an Irish word for 'good times'.

"The home call is quite strong, the result is experienced people returning". [Paul]

Trauth's (2000) findings reinforce the individual impact of the IT industry in Ireland. She considers Ireland's long-standing reputation as a laid-back, easy-going workforce and found that the information economy has altered this mentality, with a balance achieved between being relaxed and being productive. The quality of the individuals in Ireland has been earmarked as the key to its successful software industry. Beyond the Irish ability "to muddle through", the consensus is that the individual factor has provided a sizable contribution to the growth of the Irish software industry. The creativity, flexibility and innovativeness of the individual has driven the sector forward and inspired confidence among home-grown software entrepreneurs in the global software market.

5 CONCLUSION

The endogenous factors identified in the conceptual framework have been discussed and supported in varying degrees through the 15 interviews. While the interviewee opinions regarding each of these key issues differ in many aspects, there was consensus that, in general, the national, enterprise and individual factors have significantly influenced the development of Ireland's software industry since the late 1960s.

A number of suggestions were proposed by the interviewees as additional factors to consider. While each of these was carefully considered in relation to the model, the responses did not converge in a manner that strongly pointed to any significant alteration to the initial conceptual framework. Some of these suggestions included profit as a significant driver for the industry, and the presence of multinational organisations in Ireland. While we recognise that these factors have been important issues in the developing industry, they did not emerge strongly enough from the interviews to merit their addition to the model. As a result, we are confident that each of the factors identified merit their inclusion in the conceptual framework. Following analysis of the data in this study the model has been revised in order to allow for inter-dependency or overlapping between each of the endogenous factors (see Figure 2). While each of the factors has been confirmed as influencing the development of the software industry, it is evident that each of the factors does not fulfill its potential without the others.

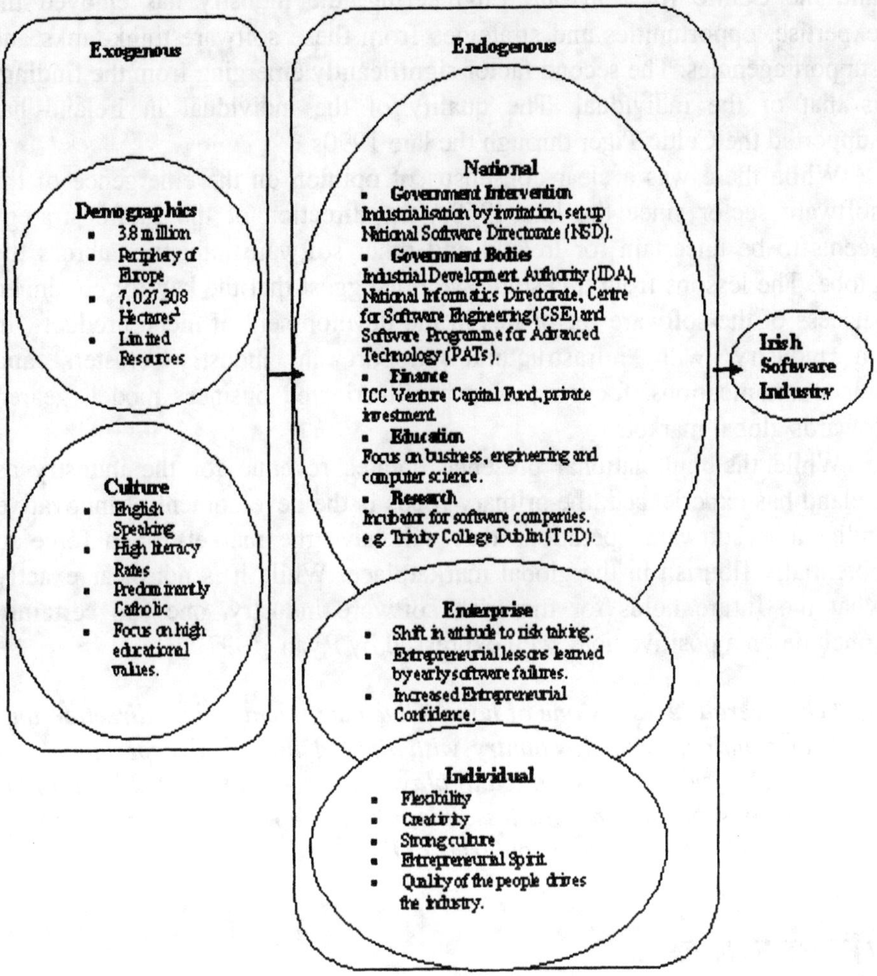

Figure 2: Summary of Factors Influencing the Development of Ireland's Software Industry

While the interviewees confirmed that these factors have driven the growth of Ireland's software sector, the degree to which they support these factors varies. Two areas of focus emerge from the interview analysis. These factors are the government's support for the industry over the last decade and the quality of the Irish individual, which has resulted in significant Irish entrepreneurial success in the software sector. The government's input since the early 1990s has shaped the way forward for the industry. With the provision of finance and extensive support from a variety of support agencies such as the National Software Directorate, the software Strategic Business Group at the IDA, the National Software Centre, Enterprise Ireland

and the Centre for Software Engineering, the industry has enjoyed the expertise, opportunities and strategies from these software think-tanks and support agencies. The second factor significantly emerging from the findings is that of the individual. The quality of the individual in Ireland has supported the Celtic Tiger through the late 1990s.

While there was a clear consensus of opinion on the emergence of the software sector since the late 1960s, the direction of the future strategy seems to be uncertain for Ireland and other software industries across the globe. The lessons from this study would suggest that the key for continued success of the software sector lies in the development of niche products, in an industry with infrastructural support in industry clusters and telecommunications, focusing on a profit-oriented business model, geared towards global markets.

While the multinational presence creates revenue for the industry as Ireland has experienced, the primary focus is the development of innovative indigenous software operations that will have the market-driven force to potentially flourish in the global marketplace. While it is not clear exactly what the future holds for the Irish software industry, one can certainly conclude on a positive note, well expressed by Ryan (1997):

> *"The overall lesson is one of hope. Given sufficient vision, direction and determination, a small country with limited industrial experience can establish itself as a significant player in a high-tech field within a relatively short period. This lesson is surely one worth learning by policy makers in all the aspiring countries of the world".*

REFERENCES

Central Statistics Office. "The Economy 2001", http://www.lrc.ie/annualreports/2001/2001-Report-3.pdf, Accessed at 2001.

Cochran, R. "Ireland: A Software Success Story", *IEEE Software*, March/April 2001, 87-89.

Condon, J. "The Irish Software Industry and Education", *National Software Directorate* (1998).

Dedrick, J. L., Goodman, S. E., and Kraemer, K. L. "Little Engines that Could: Computing in Small Energetic Countries", *Communications of the ACM*, (38:5), 1995, pp. 21-26.

Department of Education. "Key Statistics about the Department's Customers", http://www.education.ie/home/home.jsp?maincat=17216&category=17216&feature=Statistics&language=EN, Accessed at 2003.

Ein-Dor, P., Myers, M. D., and Raman, K.S. "Information Technology in Three Small Developed Countries", *Journal of Management Information Systems*, (13:4), 1997, pp. 61-89.

Flood, P., Heffernan, M., Farrell, J., MacCurtin, S., O'Hara, T., O'Regan, P. Carroll, C. *Managing Knowledge-Based Organisations: Top Management Teams and Innovation in the Indigenous Software Industry*, Dublin: Blackhall Publishing, 2002.

Gallen, S. "The Development of the Irish Software Industry". http://www.nsd.ie, Accessed at 2001

Heeks, Richard B. *India's Software Industry*, New Delhi: Sage Publications, 1996.

Heeks, Richard B. "Software Strategies in Developing Countries", *Communications of the ACM*, (42:6), 1999, pp. 15-20.

IDA. "Achieve European Competitive Advantage in Software", Dublin, *Industrial Development Authority*, 2001.

IDA. "Ireland: Vital Statistics", Dublin, *Industrial Development Authority*, April 2002.

Keogh, O. "Growth Explosion of Irish Software Sector", *The Irish Times*, November 20, 2000.

Labour Relations Commission. "Improving Industrial Relations: A Strategic Policy", 1996, http://www.lrc.ie/research_and_publications/improving_industrial_relations/29.htm

Myers, M. D. "Can Kiwis Fly? Computing in Finland", *Communications of the ACM*, (39:4), 1996, pp. 11-15.

National Software Directorate. "The Software Industry in Ireland: A Strategic Review", Dublin, *Industrial Development Authority*, 1992.

National Software Directorate. "The Software Industry: A Directory of Products and Services", Dublin, *Industrial Development Authority*, 1993.

National Software Directorate. http://www.nsd.ie, Accessed at 2002.

Organisation for Economic Cooperation Development (OECD). "Education At A Glance", 2001.

O'Riain, S. "The Birth of a Celtic Tiger", *Communications of the ACM*, (40:3), 1997, pp. 11-16.

Ryan, K. "The Irish Software Industry – Lessons Learned", *PICMET*, 1997.

Trauth, E. M. "Leapfrogging the IT Labor Force: Multinational and Indigenous Perspectives", *Journal of Global Information Management*, (7:2), 1999, pp. 22-31.

Trauth, E. M. *The Culture of an Information Economy*, Dordrecht: Kluwer Academic Publishers, 2000.

Watson, R. "The Great IT Race: New Zealand Lags Well Behind Finland", *Business Review*, (2:1), 2000, pp. 67-77.

About the Authors

Ciara Heavin, College Lecturer, University College Cork, Business Information Systems. She holds a BSc and MSc in Information Systems from UCC and is currently doing a PhD in IS. Her research to date includes the development of the ICT industry primarily focusing on Ireland's software industry. E-mail: c.heavin@ucc.ie

Professor Brian Fitzgerald holds the Frederick A Krehbiel II Chair in Innovation in Global Business and Technology at the University of Limerick, Ireland. He has a PhD from the University of London and has also held positions as Visiting Professor at Northern Illinois University in the US and the University of Gothenburg in Sweden, and was formerly at University College Cork, Ireland. His publications include six books and more than 60 papers, published in leading international conferences and journals. Having worked in industry prior to taking up an academic position, he has more than 20 years experience in the IS field. E-mail: bf@ul.ie

Eileen M. Trauth is Professor of Information Sciences and Technology at Pennsylvania State University, USA and Science Foundation Ireland Walton Visiting Professor at the University of Limerick. She conducts research on the influence of socio-cultural factors on IS, the IS profession, and underrepresented groups. Dr. Trauth has published several books including The Culture of an Information Economy: Influences and Impacts in the Republic of Ireland. She has taught and conducted research in Australia, Canada, France, Ireland, The Netherlands, New Zealand and the UK. She has written articles for the leading IS journals and serves on the editorial boards of several of them. Dr. Trauth received her PhD from the University of Pittsburgh. E-mail: etrauth@ist.psu.edu

16

RUSSIAN HIGH-SPEED SOFTWARE DEVELOPMENT

Overcoming the Challenges of Globalization

Jan Pries-Heje[1], Richard Baskerville[2], and Galina Ianshina Hansen[1]
[1]*The IT University of Copenhagen, Denmark;* [2]*Georgia State University USA*

Abstract: High-speed software development uses a number of techniques to move software quickly into production. Examples of these techniques include prototyping for fast requirement definition, release-oriented project management for scope control, and parallel development for rapid product completion and quality control. These techniques are spreading rapidly among software developers worldwide. Using the Kline model of innovation diffusion and the Greiner model of evolution and growth of organizations we analyze the enablers and barriers to diffusion of high-speed software development techniques in Russia. This analysis reveals a complex interaction of political, economic and technical elements that both enable and inhibit the development of knowledge necessary to support this case of innovation diffusion.

Keywords: Software Engineering, Information Systems Development, Agile Systems Development, Diffusion of Innovation.

1 INTRODUCTION

Since the commercialization of the Internet and the explosion of applications that followed the invention of the World Wide Web and browser technology researchers have studied this area intensively. One of the most interesting observations has been that there seems to have developed a practice of developing and launching software applications at high speed. In a study in U.S. of companies that believe they work at Internet speed (Ramesh et al., 2002) revealed a number of so-called Internet Speed practices. In the concrete the following nine High-Speed – or Internet Speed – practices were identified:

1. Parallel Development
2. Release Orientation
3. Tool Dependence
4. Customer Involvement
5. Prototyping
6. Fixed Architecture
7. Components
8. Maintenance Ignored
9. Tailored Methodology

The differences included the use of parallel development processes to speed up development. A new breed of tools that automated much of the development has emerged. Customers are being intensely involved in development, guiding the acquisition and prioritization of requirement chunks. Prototyping is being heavily used in understanding requirements as well as in developing throwaway and iterative releases. The critical role of good architectural design is well recognized. Developers are forced to reuse components as much as possible. Development has to take place in chunks of functionality, and an extreme release orientation is used to accommodate the need for speed and fixed delivery schedules. Maintenance issues often are ignored. The development process and methodology are tailored to match the needs for quality and speed for the next release. Further, they are also constantly changing and evolving as the products, markets and software development organizations mature.

These techniques are spreading rapidly among software developers worldwide. These are featured in popular new methodologies such as eXtreme Programming. The techniques are also visible in studies of both large and small organizations developing software for Internet applications (Baskerville et al., 2001). The benefits of these techniques, or at least the claimed benefits, are well known and this information alone provides a strong motive and credible explanation for *why* this rapid diffusion is occurring.

What is missing is an explanation of *how* these techniques are spreading in a borderless fashion. Our research was initiated by our curiosity towards whether Internet Speed practices had gone completely global, or were contained within developed consumer economies? In other words: Would we find the same practices in countries with emerging consumer economies? We were presented with an opportunity to study Internet Speed practices in one of the largest and most important emerging European consumer economies: Russia. This setting makes an ideal subject not only for reasons of size and importance, but also because of well-organized state programs to promote Internet and computing technology.

Internet Speed practices are relatively recent achievements in developed commercial economies. By examining these practices in Russia, we seek to discover if these techniques are diffusing to developing economies, and if so, how these are diffusing. In analyzing these phenomena, we use the Kline model of innovation diffusion. Analysis using this model enables us to discover the enablers and barriers to diffusion of high-speed software development techniques in Russia.

2 RESEARCH METHOD

We selected Russian software development as an arena of particular interest from the viewpoint of globalization. We can highlight three key reasons for this interest. First, we appreciate the sheer geographical size (12 time zones) and importance of Russia in the world. Second, in opposition to the geographical size, we discovered that the Russian Internet market is tiny when compared to countries in Western Europe or in North America. According to *PulseOnline,* only 1-2 percent of the population of 145 million people accesses the Internet regularly (2001). A third reason is the widely circulated prediction that this small percentage is expected to rise dramatically. Optimistic estimates predict upwards of 50% annual growth. Compared to the relatively low percentage of today's Internet access, Russia is characterized as a "developing country" in the arena of Internet software development with serious prospects of becoming a leading country within a relatively short period.

Having chosen Russia, in mid-October 2002 we conducted a number of interviews with three Russian software houses. Table 1 provides overview profiles of these companies. We also interviewed managers from a Danish software company with ten years of experience in outsourcing software development to a Russian software house in Saint Petersburg. Furthermore we carried out an extensive literature and internet search for documents in both English and Russian related to the issue of high-speed and internet software development in Russia.

Overall our research follows well-established research methods used in qualitative research. We carried out data collection through telephone interviews using a semi-structured interview guide.

3 CASE FINDINGS

We will discuss the case findings in two sections. First, we will describe the nature of software development at high-speed as found empirically in the

Russian cases. Second we will describe some of the Russian political initiatives described in the literature.

3.1 High-Speed Software Development

From a general perspective, we found a surprising degree of usage of high-speed software development techniques in Russian. All four of the software developing organizations in our study used some or several (but not all) of the techniques. Less surprisingly, the development of these techniques is not traced to training by universities or technical schools. Given the surprising degree of usage, we more expectedly found that the Internet itself has been the effective mechanism for spreading new knowledge about high-speed software development.

Table 1. Profiles of the Russian companies studied

Name (Pseudonym)[1]	Industry and What offered?	When Founded, and Size?	Interviewed	Organizational Roles represented
InterSoft	Intranet and internet systems, such as on-line catalogue shopping, content management, project planning and control system	Founded in 1999, 10-12 employees	One	Technical Director/ Chief of Maintenance Department
RusERP	Offers bookkeeping systems, ERP systems	Founded in 1991, 15 employees	Two	Two persons have been interviewed - developer and marketing chief
FinSoft	Applications to the Financial Sector	Founded in 1997, 53 employees	Two	Two persons have been interviewed - Analyst and Developer
BridgeOut	Decision support, Planning and scheduling applications for industrial, environmental and civil defense use	Founded in 1992, Subsidiary company (50%) of Danish Software House, 20 employees in Russia, 50 employees in Denmark (working together)	Three	CEO in Danish mother company, manager responsible for Russian development, manager responsible for outsourcing

In the remainder of this section, we provide a more detailed perspective on each case and how the techniques have evolved for these companies. We

[1] All the Russian Software companies requested anonymity.

will also give some specifics about where the development knowledge arises.

Intersoft didn't recognize the term "Internet Speed," but was nevertheless in no doubt about the content of the term. They were feeling time pressure in all their projects, whether these were for the Internet, for an intranet, or for some traditional software application. Intersoft was using parallel development heavily as their specific solution to cope with time pressure. They were also using a fixed architecture to save time, at least in some projects. They were tailoring their process – or their method – and this was justified because of the diversity of their customers and projects. Intersoft had looked at things such as eXtreme Programming (Beck, 2000), Rational Unified Process (Kruchten, 2000) and Agile methods (Aoyama, 1998). They had found the latter to be especially interesting and inspiring. However, we didn't find customer involvement to the degree most agile methods recommend. In fact Intersoft said that they didn't like it and only involved customers "if a customer insisted to be involved". We also did not find that their designs ignored maintenance. In fact, we found quite the opposite, since Intersoft members were quite proud of their ability to maintain their software products. Finally, when asked what their source of information and learning about these high-speed techniques had been, the answer was primarily the Internet plus a number of journals.

RusERP are developing both standard (ERP-like) products and totally new Internet-based products. They reported time pressure in producing these products, especially in the newer kind of projects. In response to this time pressure, RusERP now largely uses parallel development. To some extent they were working also with a fixed architecture and were dependent on tools for developing software. They agreed that they were often ignoring maintenance issues in their projects, but usage of this technique was dependent on the customer. The only learning source for new techniques they mentioned was the Internet.

FinSoft reported time pressure when developing the more important modules of a system, but not if it is just a "nice to have" module. Finsoft claims to use a fixed three-layer architecture. They also very often involve customers. Prototyping is a means for customer involvement, very commonly to communicate and get feedback from customers. Furthermore to cope with time pressure (for important modules/systems) parallel development and release orientation are applied. For those purposes, developers at FinSoft are dependent on tools. Again at Finsoft, the Internet was mentioned as a primary source of information and knowledge about high-speed techniques, together with different books, courses, and journals.

BridgeOut told us that "time to market is decisive, … to be first in a market means everything." This goal brings them time pressure. On the other hand, BridgeOut has also experience with launching a new product too early. In this case, "too early" means that, at launch, the product was of such inferior quality that it would forever destroy the product's potential. BridgeOut had worked especially on tailoring their methodology. This tailoring work had begun since their first prototyping development approach 10-15 years ago. After this initial work, they adopted a waterfall-influenced methodology, which was used to plan, and implement the cooperation between the Danish headquarters and the Russian subsidiary. However, they have since abandoned the waterfall-model and have now replaced it with an iterative methodology. This is an adaptive approach in which the number of iterations and the content of each are determined by those involved in each individual project. "I have so much confidence in iterative methods that I will use them in any situation" says the Manager responsible for Russian development. BridgeOut is also using parallel development in the sense that processes are taking place in Denmark and in Russia at the same time for the same project. In general, BridgeOut has a development process that includes several iterations with frequent deliveries (releases). BridgeOut has also looked at eXtreme Programming (Beck, 2000), trying, for example, Pair Programming. However they found Pair Programming was too stressful for their developers. Finally, BridgeOut is also using courses, journals, new books and the Internet to find new knowledge.

3.2 Russian Software Political Initiatives

The importance and growth of the Internet has not gone unnoticed in Russia. For example, they have launched a very ambitious federal program called "Electronic Russia 2002-2010" in January 2002. The program recognizes information and communication technologies (ICT) "as spurring economic change and development, boosting Russia's international competitiveness, improving the productivity and responsiveness of government, and creating a more educated, informed, and engaged citizenry" (Azrael and Peterson, 2002).

This program is not directly aimed at increasing the efficiency of the economy by developing a Russian high-tech or Internet marketplace. Rather it is aimed at improving management in the public sector, governmental performance, and transparency in decision-making. There are four concrete improvement goals comprising Electronic Russia 2002-2010 (Lakaeva, 2001; 2002; Hiltunen, 2002):

1. A more friendly environment in the form of effective legislation, and better communication between public institutions and private organizations such as the Internet software houses we interviewed
2. An Internet infrastructure in the form of better telecommunication networks, as well as access to electronic libraries, archives and

databases. More specifically, every city in Russia with more than 30,000 inhabitants should be connected.

3. E-Government through the establishment of e-commerce market places for state procurement and other commercial activities of the state. Thus in 2010 it is expected that 65% of all internal and up to 40% of external (across state organizations) communication is done in electronic form.
4. E-Education by providing computer training for education professionals and delivering a wide-range of distance learning packages.

Electronic Russia 2002-2010 aimed at creating a friendly environment (the first part of no. 1 above) in 2002, conducting feasibility studies and pilot projects in 2003-2004, and implementing the program at full scale 2005-2010 (Lakaeva, 2001). In general this Russian federal initiative has been received quite positively. For example Kimmo Sasi, the Finnish Minister of Transport and Communication, called it "a timely document that could increase the efficiency of the Russian economy" (Hiltunen, 2002).

However, criticism has also been raised. Azrael and Peterson (2002), for example, question whether resources will be available and state that "it is not clear that focusing on IT should be a priority at this time for Russian government or industry." And Hiltunen (2002) quotes Russian speakers at a conference in Helsinki as saying "that the programme is in many ways detached from what an average Russian person needs and can afford."

4 ANALYSIS

In this section we begin with an analysis of key enablers of the diffusion of these high-speed development techniques. For this purpose we will introduce the Kline (1985) linked-chain model of innovation diffusion. We selected the linked-chain model because the diffusion of high-speed software techniques across economies is undoubtedly knowledge intensive. It is a feature of post-industrial economics that recognizes how information industries have assumed a powerful, ideological role among consumer societies in knowledge economies. Kline's linked-chain model is a knowledge-centric process model that is not linear in nature. Its knowledge-centricity is important for modeling the origins and flow of knowledge related to the techniques. From the perspective of the Kline model, the Russian diffusion of these techniques appears to be well enabled.

After our discussion of the key enablers, we will turn our focus to the barriers confronting this diffusion. For this purpose we will introduce a model of organizational evolution and growth by Larry Greiner (1972;

1998). We selected Greiner's model of the evolution of organizations because it appears that the software houses in the Russian Internet development market are relatively small organizations. Given the key enablers suggested by an analysis using the Kline model, we can expect these firms to evolve rapidly the coming years. Greiner's model explicates organizational evolution and growth, and suggests the barriers the companies will be facing in this forthcoming evolution. From the perspective of the Greiner model, the Russian diffusion of these techniques faces some challenges. As a consequence, this model will also suggest enablers needed to overcome future barriers to the development of this high-speed software development industry.

4.1 Kline Linked Chain Model

Unlike linear models that define sequential patterns for idealized cases of the innovation process, five concurrent pathways or links characterize the linked chain model. In Figure 1, arrows denote these concurrent links within the other elements of the model. These elements are (1) market finding, an assessment of a product improvement or new product that meets an unfulfilled market; (2) analytical design, which is a preliminary design activity that establishes the scope of further design alternatives; (3) development, which includes detailed design, prototyping, and testing; (4) production, which includes redesign for manufacture and production; (5) marketing, which includes distribution as well as product marketing; (6) research and knowledge, which together constitute pure science. Knowledge is placed between the elements of the innovation chain and research as a buffer to imply that the store of human knowledge may often fuel the innovation chain without further research processes. According to this model, "research leads to product innovation only insofar as it stimulates a design via either invention or analytical design" (Kline, 1985: p.37). The concept of analytical design as distinct from design is not well-explored elsewhere in the literature, but might be characterized as the invention of a design through a complex intellectual trial-and-error process. The design component of development is also innovation, since it is an inductive and creative activity of the mind that synthesizes the old and new to satisfy its goals.

The links or pathways in the innovation chain include (1) central chain, which represents the long pathway of innovation through each element from market finding to marketing; (2) feedback links, which flow dynamically back-and-forth across the central chain boundaries especially product-improvement and new product innovation arising from market discovery process; (3) knowledge-linked research, innovation arising from the interaction of knowledge and the elements of analytic design, development, production and research; (4) invention-linked research, from the unstructured exploration of analytical designs; (5) product-linked research, regarding long-range product and "support for science" research.

The linked-chain model enables several major innovation diffusion implications important for understanding the Russian high-speed software development situation. First, knowledge is the base of innovation and its diffusion. Research alone is not the direct base of innovation, but is indirectly critical for creating the store of knowledge. Second, the systems and process research associated with product development and production are generally undervalued innovation elements. Third, a preoccupation with science has diminished our recognition of invention and analytic design as key elements of innovation. Fourth, our understanding of creativity and innovation in design is poorly developed, especially in terms of the interaction between the closely related activities of invention and analytic design.

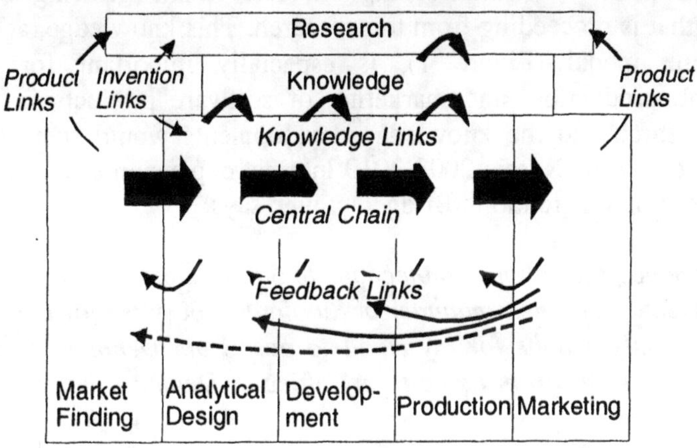

Figure 1. Kline's Linked-Chain Model

4.2 Analysis: Enablers of Diffusion of Practices in Russia

An analysis of the cases indicates four key Enablers of Diffusion that affect the transfer of these Internet Software Practices into and within Russia. These include (1) Internet-based research, (2) rapidly developed knowledge through the Russia 2002-2010 Internet expansion, and (3) the demands of rapidly developing markets through the Russoft Outsource marketing program. A fourth enabler (4) appears to be a developing social norm that software copyrights for high cost development tools can be disregarded unless they are generating revenue.

The first two key enablers are interlocking. Perhaps foremost is the availability of the Internet for purposes of research. This research, as shown in Figure 1, is important for all five phases of the product innovation cycle and for developing and maintaining the knowledge base required for innovation diffusion. It was thematic in our cases that the major research source for developing knowledge on Internet speed software development is the Internet itself.

> *The importance of the Internet as one important source of learning was thematic. InterSoft learned about Internet Speed development techniques "from different journals and the Internet", FinSoft and BridgeOut from "different books, courses, journals and the Internet", and at RusERP, the only learning source mentioned was - the INTERNET.*

The second of these interlocking enablers is the growing body of knowledge that is proceeding from this research. This knowledge, according to the Kline model (Figure 1), is especially important for design, development, production, and marketing of software products in Russia. Both the research and the knowledge developments would seem to have been enhanced by the Russia 2002-2010 Internet expansion project. A good illustration of this was found in BridgeOut who says:

> *"We have good telephone connections to Saint Petersburg and Moscow. But connections to the remainder of Russia are lousy. As for the Internet there are only 8 hubs for my email to pass from Denmark to Russia, whereas my email from home to my office in Denmark has to pass 12 hubs".*

The third key enabler of the diffusion of Internet Speed techniques is embodied in the efforts to develop a marketplace for Russian software products. The clearest embodiment of this enabler is in the Russoft outsource software marketplace being vigorously developed by Russian software industry groups. This project covers the marketing and market-finding elements in the Kline model.

> *"A number of associations (Rusoft, Inforus) similar in function to the Indian's NASSCOM was formed to promote Russian software development companies in the U.S., Europe, Asia and to improve Russia's image as that of a reliable center for offshore software development". (Luxoft, 2002: p.7).*

A fourth key enabler is especially representative of the knowledge links between research, knowledge, design and the processes of development and production. It regards a potential barrier to these links arising from the high cost and inaccessibility of many of the software development tools such as software development environments, object-oriented design packages, etc. These tools have been shown to be a feature of Internet speed development (Ramesh et al., 2002). Because access to these tools is difficult to acquire at low cost in Russia, some of these tools are being made accessible through back channels. The back channels are in no way planned or accommodated by any real organization, but rather have become enabled by a social network. The social network seems to be Internet enabled, and embodies some of the important knowledge links suggested by the Kline model (see Figure 1). The ability to acquire freely these tool packages enables developers and their organizations to research, experiment and build the minimum necessary knowledge and skills to compete with other Internet Speed software developers. Once the knowledge has been acquired, and serious production yields revenues, the software tools are subsequently acquired through more traditional front channels.

"Intellectual property and security are major concerns of companies considering offshore software development in Russia. Thankfully, the laws on intellectual property ownership in Russia are relatively clear and developed. While enforcement of these laws remain a problem for domestic consumer level products (music, movies, clothing), there have been no recorded violations of intellectual property rights in the offshore software development industry" (Lewin, 2001: p.5).

4.3 Greiner's Model of the Evolution of Organizations

As mentioned in the introduction to this section, the small-sized firm orientation of the Russian high-speed software development industry suggests a forthcoming field of organizational growth, adaptation, and development. A useful model for suggesting future barriers (and necessary future enablers) can be found in Greiner's (1972; 1988) evolutionary model. Figure 2 shows the five phases of organizational development and change as defined by Greiner. An organization starts its life in the "Creativity" phase shown in the lower left corner. In this phase, the founders of the company are typically entrepreneurs, communication among the people in the organization is informal, long work hours are normal, and the feedback from the market is immediate, as well as the reaction from management.

As the company grows in size and matures, it then reaches its first crisis, namely the "leadership crisis". Informal communication is no longer sufficient. The dedication, long hours and small salaries of the first hired "pioneers" are no longer sufficient motivation. Furthermore new procedures are needed to exploit efficiencies of size and to provide better financial control. To solve the leadership crisis a strong manager is needed. Often the owner/founders don't have the necessary skills and knowledge, and so will "hate to step aside even though they are ... unsuited to be managers" (Greiner, 1972; 1998).

When well through the leadership crisis the surviving organization will "embark on a period of sustained growth under able and directive leadership" (Greiner, 1972; 1998). In this second phase, communication becomes more formal, a hierarchy is built within the organization, and the upper levels take responsibility for the direction of the organization. It is also in this phase that formalized systems for accounting, incentives, work practice and job specialization will arise.

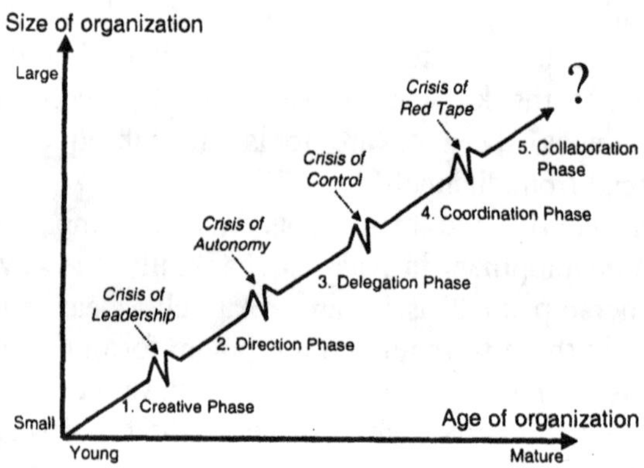

Figure 2. Greiner's model of organizational evolution and revolution

The second crisis then is the "autonomy crisis". Middle-level managers see the centralized decision structure of the second phase organization as a burden, and some of the more autonomous field unit middle managers will start running their own shows. Often the reaction by top management is an attempt to return to centralized management. To solve the second crisis a more decentralized organization structure is needed where middle management has greater responsibility and autonomy.

Phase three ends in the "control crisis" where top management realizes that they have lost "control over a highly diversified field operation." This crisis is overcome by the use of coordination techniques such as formal planning, creation of product groups treated as investment centers, and by initiating staff functions that controls and reviews for line managers.

The next crisis then is the crisis of "red tape" where the line is looking at staff functions with more suspiciousness, and distrustfulness evolves between headquarters and the field. Overcoming this crisis then leads the organization into what Greiner calls "the last observable phase in previous studies." In this last phase strong interpersonal collaborations are established to overcome the red-tape crisis. A more flexible and behavioral approach to management is implemented through the use of teams. The staff functions are reduced in number. And the motivational structure becomes more geared to team performance than to individual achievements.

4.4 Analysis: Barriers to Diffusion of Practices in Russia

An analysis of our four cases in relation to Greiner's model indicates a number of barriers for the future diffusion of high-speed software development practices in Russia. The model also suggests some potential enablers for overcoming these barriers.

First the prospects for establishing new companies seem quite bright. Thus new companies will enter phase 1 in Greiner's model where growth is primarily spurred by creativity. BridgeOut describes the market for qualified IT people in very positive terms:

> *"In Russia you have access to a pool of very strong resources ... highly motivated and well educated ... i.e. in Saint Petersburg you have about 50 technical schools and universities with more than 200,000 students to choose among ... financially attractive compared to salaries in Denmark. And the new generation is taught English as their second language so communication becomes easier in the future".*

This remark sheds light on both a barrier and an enabler. Language is a barrier. Russia doesn't use the same alphabet as in most Western countries and consequently communication problems are intensified between Russians and their foreign partners. It is, however, changing as also illustrated by the extensive use of the Internet as transfer mechanism for high-speed practices. Further, the remark above shows that continuation of the Russia 2002-2010 program's focus on bringing Internet access to Russia classrooms can be expected to be an enabler.

Second it is clear that many of the Internet software producing organizations in Russia are fairly young. They are also fairly small. These features suggest that the organizations are positioned right in the lower left corner of Greiner's model. These organizations are growing through creativity, and they can expect to meet the "crisis of leadership" in the nearer future. Consequently, a real barrier in a Russian context will be a sufficient supply of capable managers that can safely bring Russian software development companies forward into phase two. This is put in perspective by a remark from a respondent in BridgeOut:

> *"It is our experience in Russia that there is a lack of highly educated leaders".*

Unfortunately this barrier doesn't seem to be addressed at all in the Russia 2002-2010 program. A new key enabler will be needed to overcome this barrier in the near future.

Third, there are a larger number of foreign companies that have established themselves in outsourcing arrangements in Russia. One example is BridgeOut. In applying Greiner's model to the Russian companies it becomes clear that they will soon be growing through the direction phase. The quick growth of the Russian market – as aimed for by Russia 2002-2010 – will just as quickly lead to an autonomy crisis in many of the Russian companies. Again, the Russian organizations engaging in high-speed development of software lack key enablers to overcome this barrier. For example, carefully crafted legislation might enable a smooth transition from phase two to phase three in Greiner's model. Today, Russian contract law does not enable delegation of responsibility, especially in economic terms. A respondent at BridgeOut says:

> *"The Russian market is still immature. Of course you write a contract, but try to take that contract to court and you will find that your protection in a contract hardly can be trusted".*

The analysis using Greiner's model points to two near term barriers that lack key enablers for the future advance of companies developing software at high speed. First Russia is missing a source for better-prepared organizational managers for software companies. Second, Russia will need improvements, such as changes in their contract law, which enable companies to re-centralize their authority structures in the near future.

5 CONCLUSION AND FUTURE RESEARCH

High-speed software development uses a number of techniques to move software quickly into production. Clearly, many of these techniques were found in a case study of Russian Internet software houses. Using the Kline model of innovation diffusion and the Greiner model of evolution and growth of organizations we analyzed the enablers and barriers to diffusion of high-speed software development techniques in Russia, and found the following:

Enablers

1. Internet-based research
2. Rapidly developed knowledge through planned Internet expansion
3. Rapidly developing demands
4. A social norm against paying for copyrights unless there is associated revenue

5. Internet access in Russian classrooms can be expected to be an enabler

Barriers

1. Language is a barrier, slowly being overcome
2. Insufficient supply of capable software development organizational managers
3. Inability to centralize control over decentralized software development organizations

Future enablers needed

1. Expand Russia 2002-2010 initiative to include foreign language training
2. Improve training for software development organizational managers
3. Legislation is to enable a smooth transition from outsourcing agreements to more autonomous Russian software houses

Our study of the enablers and barriers to diffusion of high-speed software development techniques in Russia reveals a complex interaction of political, economic and technical elements that both enable and inhibit the development of knowledge necessary to support this case of innovation diffusion. This analysis permits us to understand how this diffusion occurs, and to suggest several future barriers and enablers that might be relevant in the near future.

REFERENCES

Aoyama, M. "Web-Based Agile Software Development", *IEEE Software*, November/December 1998, pp. 55-65.

Azrael, J. R. and Peterson D. J. "Russia and the Information Revolution", *RAND Center for Russia and Eurasia*, Issue paper IP-229-CRE, 2002, Available at: http://www.rand.org/publications/IP/IP229/IP229.pdf (Accessed at: November 8, 2002)

Baskerville, R., Levine, L., Pries-Heje, J., Ramesh, B., and Slaughter, S. "How Internet Software Companies Negotiate Quality", *IEEE Computer*, (34:5), 2001, pp.51-57.

Beck, K. *Extreme Programming Explained: Embrace Change,* Boston, MA, USA: Addison-Wesley, 2000.

Greiner, L. "Evolution and Revolution as Organizations Grow", *Harvard Business Review*, (50), 1972, pp. 37-46.

Greiner, L. "Evolution and Revolution as Organizations Grow", *Harvard Business Review*, May-June 1998, pp. 55-64.

Hiltunen, A. "Electronic Russia Gets Thumbs Up in the West", *Rosbalt News Agency*, Helsinki, 2002, Available at: http://rosbalt.com/2002/10/03/49224.html (Last Accessed: October 15, 2002).

Interactive Strana (2002). The citation of A.V. Korotkov is from an interactive Russian newspaper called Strana. Available at: http://interactive.strana.ru/24/ (Last Accessed: November 5, 2002).

Kline, S. J. "Innovation is not a linear process", *Research Management*, 1985, (28:4), pp. 36-45.

Kruchten, P. *The Rational Unified Process: An Introduction* (2nd ed.). Reading, Massachusetts: Addison-Wesley, 2000.

Lakaeva, I. "Electronic Russia Program", *U.S. & Foreign Commercial Service and U.S. Department of State*, 2001, Available at: http://www.bisnis.doc.gov/bisnis/country/011001E-Russia.htm (Accessed at: November 8).

Lakaeva, I. "Electronic Russia Update", *U.S. & Foreign Commercial Service and U.S. department of State*, 2002. Available at: http://tiaonline.org/policy/regional/nis/E-Russia.pdf (Last Accessed: November 8).

Lewin, R. "Whitepaper on Offshore Software Development in Russia", *The American Chamber of Commerce in Russia*, 21 March 2001. Available at: http://RussianSoftwareDev.eSolutions.ru (Last Accessed: October 15, 2002).

Luxoft. "White Paper on Software Outsourcing to Russia", *Luxoft/IBS Group*, 2 July 2002, Available at: http://www.luxoft.com/materials/wp.pdf (Accessed at: 15 October, 2002).

PulseOnline, "International Market Forecast: B2C E-Commerce in Russia" *PulseOnline*, 2001. Available at: http://pulse.tiaonline.org/article.cfm?ID=199 (Last Accessed: November 4, 2002).

Ramesh, B., Pries-Heje, J., and Baskerville, R. "Internet Software Engineering: A Different Class of Processes", in *Annals of Software Engineering: Process-Based Software Engineering*, Y. Wang and A. Bryant (Eds.), New York: Kluwer Academic Publishers, (14), 2002, pp. 169-195.

About the Authors

Jan Pries-Heje has a Ph.D. from Copenhagen Business School. He now works at The IT University of Copenhagen. From 2002 he is also Professor (part-time) at the IT-university in Gothenburg. His main research interests are information systems development, software engineering, and software process improvement. In particular he has carried out action research with industry on specific topics such as high-speed software development, IT project management, requirements specification, and successful organizational change with IT. He has published in these areas in journals like Journal of Accounting Management and Information Technology, IEEE Computer, The Data Base for Advances in Information Systems, European Journal of Information Systems, and Annals of Software Engineering. He can be reached by e-mail at jph@it-c.dk.

Richard L. Baskerville is professor of information systems and chairman in the Department of Computer Information Systems, College of Business Administration, Georgia State University. His research and authored works regard security of information systems, methods of information systems design and development, and the interaction of information systems and organizations. He is a Chartered Engineer, holds a B.S. *summa cum laude*, from The University of Maryland, and the M.Sc. and Ph.D. degrees from The London School of Economics, University of London. He can be reached by e-mail at baskerville@acm.org.

Galina Ianshina Hansen has a M.Sc. in economy from Moscow Technical University/State Institute of Roads and Automobile Transport.

She is presently a M.Sc. student at the Software Development Department at IT University, Copenhagen. Her research interests are in understanding software process maturity models as a change initiative that requires use of measurements of organizations ability to improve. She can be reached by e-mail at galina@it-c.dk.

17

INSTITUTIONAL STRUCTURES AND PARTICIPATION

Comparative Case Studies from India

S.K. Puri and Sundeep Sahay
Institute of Informatics, University of Oslo, Norway

Abstract: While participatory design in IS has found increasing acceptance, the role of participation in social arena has been lively debated in recent years. Development projects in the third world based on "western" assumptions have elicited sharp critiques, with consequent policy shift towards involving the poor people and local communities in these endeavours. In India, policies and programs aimed at reclamation of degraded lands, a priority thrust area which is directly linked to poverty alleviation, are typically controlled by central, state and local government bureaucracies, with little involvement, until recently, of the affected people. Some of these programs have attempted to use Geographic Information Systems (GIS) to model alternative development possibilities, and to then prepare action plans based on scientific knowledge for field implementation. These initiatives too have generally been bereft of end-user involvement. On the other hand, land development has been successfully carried out at several places in the country through community initiatives alone by harnessing indigenous knowledge. In this paper, we briefly present the continuum of participatory development through analysis of two contrasting case studies. The analysis leads to the challenge of integrating local knowledge within governmental institutional frameworks that can facilitate the larger spread of land reclamation efforts.

Keywords: Participatory development, information systems, community mobilization, degraded lands, GIS, structuration theory.

1 INTRODUCTION

End-user participation in design, development, and use of Information systems (IS) has now a reasonably well established tradition, having been promoted since the 1960s by a number of researchers and IS practitioners in the West, particularly in Scandinavia, USA and the UK (Ehn, 1993, Butler and Fitzgerald, 2001). Evolution of the ETHIC methodology is an illuminating example of this trend (Mumford, 1993). Participatory design (PD) has evoked varying and diverse trajectories in different societies impinging as it does on the "questions of democracy, power and control at the workplace" (Ehn, 1993: p.41). Many other contemporary disciplines, such as education, architecture and economics, have also embraced participatory approaches (Schuler and Namioka, 1993), the basic motivation being "people who are affected by a decision or event should have an opportunity to influence it" (ibid.: p.xii). In the last Participatory Design Conference in Malmo (2002), the emerging consensus was that participation should be viewed in broader contexts (outside of Western organizations), and should be analyzed within a broader process perspective that emphasizes the dynamic and political nature of participation.

Another domain of participation has emerged since about 1980s in the field of development studies involving third world countries, in which the involvement of end-user (i.e. the ultimate beneficiary of development efforts) is now increasingly being recognized as crucial to the success of such initiatives. We bring together ideas from these two domains of participation (i.e. IS and development studies) to examine the use of Geographic Information Systems (GIS) for land management in the overall context of rural development in India. We do so by presenting two contrasting case studies aimed at reclamation of degraded lands, viz. (i) community-driven without the use of GIS/computer technologies, and (ii) GIS-based (generally without end-user or community involvement). The primary aim of this comparison is to analyze the role of people's participation to develop implications for GIS-based approaches aimed at alleviating the crucial problem of land management.

However, there are at least three points of departure in our analysis with respect to the use of PD in IS, viz. (i) the 'system' development is embedded within broader economic and social development processes; (ii) end-users are not the traditional workforce of a typical organization, but subsume governments at central, district-, sub-district levels, as well as individual farmers and local communities that have deep stakes in the objective of development i.e. land use; (iii) while the role and importance of PD has been emphasized and reported in the mainstream IS research for at least over three decades, the import and significance of this approach has percolated relatively recently into the GIS literature (e.g. Craig et al., 2002).

The rest of the paper is structured as follows: In section 2, a theoretical perspective for analyzing the empirical findings, reported in section 3, is developed. The context of introducing use of GIS to the vexed problem of putting degraded lands to productive use in India is also provided along with the outcome of relevant research efforts in the past in this section. The analysis is taken up in section 4, while some concluding remarks to situate participation within land development efforts and the use of GIS are made in Section 5.

2 THEORETICAL FRAMEWORK

In this section, we first briefly discuss the evolution of people's participation in development projects. This leads to a discussion on how structuration theory can be drawn upon to examine the relationship between development and participation using the concepts of structure and agency. This linking provides the conceptual basis for our analysis.

2.1 Participation in Rural Development

Since the 1980s, a dominant critique of development initiated in the third world arose from the failure of its models enunciated, propagated, and implemented as per first world perspectives, including bilateral/multilateral aid agencies. This "impasse in development" (Kothari, 2002: p.35), has led to debates about the need to rethink development. Some of the criticisms that have triggered this rethinking are briefly discussed.

Harry S. Truman's (former US President) vision, expressed in 1949, was of poor countries becoming 'developed' by 'replicating' the economic models of 'advanced' societies, based on the application of capital, science and technology (Escobar, 1995: p.4). Escobar questions such 'invention' of development (ibid.: p.24) as an attempt by the first world to promote its strategic, geopolitical and commercial interests, leading to accentuation of poverty rather than its remission in developing countries.

The hypothesis of bilateral and multilateral organizations that aid can lead to development, especially poverty alleviation, has been closely questioned in the recent years. Despite substantial funding by donor agencies, their project-based approach has often ignored the realities of poor people and the underlying socio-cultural causes of poverty (Kothari et al., 2002). A top-down philosophy, bereft of stakeholder representation, has permeated the design and implementation of these development projects. As McGee (2002, p.93) argues, development organizations have typically perceived such projects "somewhere from outside its [project's] boundaries, and there were project beneficiaries – undifferentiated, passive recipients of its goods and services provided through the project channels ... (these were) created as a bureaucratic convenience, a manageable way of packaging assistance".

Chambers (1983) has argued that alternative development routes must be chalked out by people themselves rather than for them by outsiders. These critiques have contributed significantly "to a wave of thinking about the debate on human agency in development, which became loosely known as 'participation'" (ibid.). The World Bank's Learning Group on Popular Participation (later called participatory development) defined participation as a process by which people, especially disadvantaged people, influence decisions that affect them (Bhatnagar and Williams, 1992). Within this stakeholder perspective, the focus of participatory processes is to improve the dialogue among stakeholders, to chalk out the agenda in mutual consultation keeping in view local practices and indigenous knowledge. "This implies a process of negotiation rather than the dominance of an externally set project agenda" (Schneider and Libercier, 1995: p.30).

We thus see a clear shift towards attempts to increase participation of local communities in development policies and programs aimed at poverty alleviation and conservation of natural resources. In India, this shift is reflected in the promotion of Joint Forest Management (Kumar and Kaul, 1996), and the watershed committees (comprising officials and local people together) being made responsible to implement development programs for conservation of natural resources (Government of India, 2001).

Participation of people in IS and rural development takes place in a particular context set by governments, bureaucracies, organization structures and communities. In the following sub-section 2.2, we seek to understand the relationship between participatory processes and their context. To do so, we draw selectively upon the concept of structure and agency linkage proposed in structuration theory (Giddens, 1984).

2.2 Structure and (Participatory) Agency

Participation comes about from the knowledge and the capable agency of people attempting to support development processes, and in doing so creating the potential to reconstruct institutional structures which define, facilitate and foster development. Collective human agency, expressed as participation, has the potential to redefine existing structures and practices embedded in institutions of bureaucracy that typically implement development projects. Giddens discusses institutions as both product of and constraint on human action, where the institutional realm represents an existing framework of rules and typifications derived from a cumulative history of action and interaction. Actors' stock of practical knowledge influence "how people communicate, enact power, and determine what behaviours to sanction and reward" (Barley and Tolbert, 1997: p.98). We adopt a similar institutional nature of structure in this paper to examine mutual interaction between structure and participatory agency.

This institutional nature of structure is emphasized in the manner land management is organized in India. Land Development programs involve a network of government institutions like the central government ministries, scientific institutions, and district administration etc. These institutions have existed historically with a strong sense of bureaucracy, with rules and resources drawn from the British colonial rule, and later reinforced by the socialist agenda of post-independent India (Jain and Dwivedi, 1990; Saha, 1992). These institutional structures shape the manner in which GIS (and other technology development projects) are planned and implemented. In the next section, we describe the case studies that highlight some of these issues.

3 CASE DESCRIPTION

India has been called a biomass-based civilization (Gadgil, 1993) with a majority of rural population in the country subsisting primarily on resources produced or gathered from the land, for example fuelwood and fodder. Therefore, 'health' of land is closely linked to poverty mitigation and subsistence. Degradation of land has been caused by a variety of socio-economic-cultural reasons such as increasing human and cattle population, diversion of land in fragile eco-systems for other uses, indiscriminate deforestation, etc. Consequently, of the total land mass of the country, more than 40% "has productivity much below its potential" (Gadgil, 1993: p.168), making development of degraded lands a key concern of governmental programs. These efforts, whether supported from internal resources or by external donor agencies, have spawned, until recently, a bureaucratic, top-down approach, with minimum involvement of the local communities.

Globalization processes have escalated previously 'local' environmental concerns to a global stage, and the need for more wide spread use of information and communication technologies including GIS to address these concerns has become part of numerous international debates and accords such as the Agenda 21. The use of GIS is also being increasingly mandated in most externally-aided projects pertaining to the natural resource sector in India, for example in forestry management.

3.1 GIS for Land Management in India

India has attempted use of GIS in the management of wastelands since the early 90s. What distinguishes GIS from other systems like Computer Aided Design (CAD), and Database Management Systems (DBMS) is its ability to integrate geo-referenced data layers through operations like spatial search and overlay etc. to present the "original data in different ways and from different perspectives" (Aronoff, 1995: p.40). One of the first GIS initiatives was taken by the Ministry of Environment and Forests (MoEF) in 1991 through pilot projects in 10 districts, launched in collaboration with country's leading scientific institutions to explore whether land development programs could be designed on a more holistic, scientific basis by using the modelling power of GIS. These projects evoked several research studies, which concluded that full potential of technology had not been realized by the MoEF mainly due to socio-cultural rather than technical reasons. These studies also underscored lack of end-user participation. (Hutchinson and Toledano, 1993; Sahay, 1998; Sahay and Walsham, 1997; Walsham and Sahay, 1999; Walsham, 2000). Implementation of these projects was reported to have a strong technical focus, and the key motive for participating scientists was to have the opportunity to engage in research involving 'latest' technologies, the subsequent social acceptance and use of these systems in the field being accorded secondary importance (Sahay and Walsham, 1997, p.432).

Subsequently in 1993, GIS was sought to be institutionalized on a wider scale in reclamation of degraded lands under the Integrated Mission for Sustainable Development (IMSD), which we report in this paper. IMSD is jointly administered by the Department of Space, central government ministries, select scientific institutions and district administrations. GIS application software and database were typically developed for each district by one of the nominated scientific institutions, and then transferred to the district headquarters to help prepare land development action plans by using GIS. The scientific departments also developed user-friendly shells which "lay users who are not familiar with UNIX, GIS and remote sensing can use ..." (Dasgupta et al., 2000: p.32).

3.2 Empirical Work

For the present study, data was collected through semi-structured interviews conducted during February, April-May 2002, in two separate sites where land development projects were being implemented. The first site, located in Alwar district of Rajasthan state, concerns a land development effort realized through people's own initiative, without positive external support and without the use of GIS. The second site is a concerned District X (a pseudonym) that fell under the purview of the IMSD program, to develop action plans for land reclamation using GIS. In all, 24 participants were interviewed (14 in Alwar area, and 10 in IMSD district including four scientists from concerned remote sensing institutions).

These discussions were supplemented by the study of available material, and hands-on demonstration of GIS work at the IMSD district headquarters. The reason for selection of these two sites, disparate as these were in the philosophy and nature of work taken up, was the adoption of different models of participation and knowledge systems. In Alwar, water harvesting structures were constructed by the community without using computer technology. In the case of IMSD district, development plans, including location of such structures, were prepared by scientists using GIS. The aim of the interviews and analysis of secondary material was to broadly understand the usefulness of GIS in supporting development processes. We now discuss these two cases in some detail.

3.3 Rejuvenating a River through Indigenous Knowledge

Alwar district, a semi-arid region, lies in the north-east of Rajasthan state in Western India. The region is afflicted by severe water shortage, which has been exacerbated by recurrent droughts, and severe deforestation in recent years. 70% of the district's forests have been ravaged (Patel, 1997: pp.6-7) during the past three decades. The river Aravari had completely dried up many decades ago, primarily due to deforestation of its catchment, discarding of traditional conservation practices, encroachments, over-exploitation of natural resources etc. The main reason for abandoning traditional practices was "the new value system and enforced state control over land" (ibid.: p.7), which alienated the local communities over time. A villager stated:

> *Over the years, we had got used to government to do the work, since all forest and common lands were taken over by the government long back. As everyone knows, they do nothing for the poor except to give them bhashan[1]. TBS woke us up to be able to take our own decisions, and you can see the results. Earlier, our womenfolk had to walk up to 2 km to fetch water, but now there is sufficient water in johads and wells through the year.*

Tarun Bharat Sangh (TBS), an NGO with its origins in the University of Rajasthan, became involved in efforts to rejuvenate the river in 1985 mainly by reviving water conservation strategies using forgotten, discarded traditional practices. A small johad (johad is a semi-circular earthen water pond built along the contours of hill slopes for stopping and storing rain

[1] Lecture full of populist hyperbole

water. It has a wide base to collect runoff from tiny steams and rivulets. Since johads are made of earth, most of the construction cost is taken up by labor (Singh, undated: p.13)) was constructed by volunteers of TBS near the main tributary of the river Aravari. This improved water availability in the following months. Many villages, therefore, volunteered to build similar structures through community effort. TBS also contacted the local government functionaries to take up construction/repair of check dams in the area through government funds. The local officials expressed helplessness due to paucity of funds, but promised technical guidance if the villagers agreed to provide shramdan (voluntary labor). With assistance from government not forthcoming, the villagers, inspired by TBS, took up the work themselves, first desilting and deepening an old, abandoned johad. After the next year's monsoon, the water level stayed higher and was retained longer than in previous years. Thereafter, more than 200 johads were constructed in the catchment area. Patel, a researcher, reports (1997: p.36):

> *It was incredible for the villagers, because for generations, they had not witnessed such a transformation. It was further incredible that it had all been achieved due to their shramdan, and the catalytic help of TBS.*

This success motivated the other villages to come forward, resulting in over 1500 water harvesting structures being repaired or newly built during 1987-94. The community movement finally realized its dream when the Aravari turned into a perennial river in 1994. This led to a reversal of migration out of this area to the neighboring Delhi in search of livelihood. Foodgrain production increased four times, milk production doubled (despite no increase in cattle population), the water table rose significantly, forest cover was augmented from 5% to 40%. Besides improving the prosperity, quality of life, there was a resurgence of flora and fauna.

It is interesting to note that the revival of Aravari was not the end of troubles for the local people. As soon as this news spread, the local and state government officials descended on the scene in 1996, claiming ownership of the river, and plantations raised by the people. The state government awarded a fishing contract for the river to an outside contractor from Jaipur. TBS staff and the villagers were harassed and warned of 'dire consequences' by the local police and politicians. It was only people's organized resistance, culminating in a satyagrah (fight for truth) in February 1997, in which thousands of people from neighboring as well as distant places participated, that finally broke the bureaucrat-politician-contractor nexus. As a result of mass resistance, which became a media event, the government allowed the local people to be the custodians of these resources, and enjoy the benefits of their labor within communities' own arrangements (Patel, 1997).

3.4 Modeling Land Degradation: Application of Scientific Knowledge

Traditional approaches to combat land degradation in India have been implemented through various central and state ministries' schemes/projects that define degradation from sectoral perspectives driven by the respective mandate of these organizations, viz. to increase agriculture productivity, improving forest cover, augmenting employment opportunities for the rural poor, etc. An integrated approach that adopts watershed as the basic unit of land development has been largely absent until recently.

The IMSD program seeks to use remotely sensed satellite data and GIS to prepare land resources action plan for a selected watershed. One of the agencies responsible for the implementation of IMSD is the National Remote Sensing Agency (NRSA), an elite scientific institution focusing on space research and remote sensing. A component of IMSD has been NRSA's role in development of a GIS package since 1994, and its subsequent transfer in 1996 to the IMSD district on workstations in the office of the District Rural Development Agency (DRDA). Despite the local district staff having received training in GIS, they continued to depend on NRSA for modifications, updates, technical backup and 'handholding' through periodic visits of scientists to the district to resolve difficulties experienced by staff, and to help make enhancements. Several local DRDA staff after being trained on GIS typically left either through transfers or movement to the private sector. Two of the experienced staff who have been associated with the GIS work from its inception in the DRDA are now adept at operating 'normal' procedures around the GIS. Despite this, GIS is perceived as a 'black box' by the staff, albeit capable of generating alternative scenarios to quickly prepare projects aimed at development of degraded areas, for example where to locate water harvesting structures or to suggest appropriate land use. GIS was used to help in the prioritization of possible projects to be taken up for development. The staff indicated they were comfortable using GIS to visualize different development scenarios.

During a GIS demonstration, which we witnessed, the two experienced DRDA staff mentioned that they had other primary functions assigned to them (one was a civil engineer, and the other a project economist), and could spare time for GIS work only when a new project formulation, or related activity, was called for. The overall computer related work in the district was highly compartmentalized, with different groups working in isolation with limited interaction and learning from one another. The DRDA Director mentioned that while action plans prepared using GIS were discussed with the public, limited feedback was received. Modifications to action plans were based primarily on the criteria of 'scientific acceptability' rather than on public inputs.

The scientists noted that now the scientific institutions had formally 'adopted' specific districts to provide local support, training and hand-holding to the DRDA staff. This formalization was different from the earlier MoEF project in which the scientist involvement was primarily based on their volition. This adoption has led to a sense of greater ownership on behalf of the scientists.

4 DISCUSSION AND ANALYSIS

The two cases discussed provide contrasting but interesting examples of the nature of mediating institutions, participatory processes, and the application of knowledge. These are now discussed below.

Nature of mediating institutions: In the IMSD example, the development of GIS software in scientific institutions and its transfer to the districts is enabled primarily through bureaucratic structures, similar to the earlier MoEF attempts (Sahay and Walsham, 1997). Efforts to transfer technology to districts in the MoEF case were largely stifled because the scientists saw research into new technologies (like GIS) as their primary mandate. The scientists drew upon the notions of objectivity and superiority of science and technocrats, and were reluctant to involve the end-users in design processes. This dogmatic philosophy reflected the assumption that technological interventions using western models could be replicated to Indian conditions, based on the universality of scientific knowledge, and tended to ignore the significance of indigenous knowledge around land conservation.

Under the IMSD, the situation seems to have changed to some extent with the scientific institutions formally assuming greater 'ownership' of GIS activities in specific districts and thus transcending to some extent the earlier primary focus on technology development. As a result, institutions were seen to be providing more active technical support, and complemented ongoing efforts of district computing staff to support GIS work. Developing and providing user-friendly shells reflect further attempts to design the technology based on user capabilities. These efforts of the scientists have been reinforced by other central government initiatives requiring training programs for officials in project implementing agencies to include use of remote sensing data and GIS techniques in their curricula (Government of India, 2001). These multi-level capacity building efforts provide some evidence of “triggers to change structure” (Sahay and Walsham, 1997: p.436) through focused managerial agency.

In the case of the Alwar project, the mediating agency was the NGO TBS, but the impetus for change was driven primarily through the efforts of the community. The community members had a direct stake in the river restoration efforts, and frustrated by the inaction of the government, they decided to take the process into their own hands. These efforts, as the case description illustrates, has had significantly positive outcomes.

Participatory processes and application of knowledge: Attempts to facilitate participatory processes in IMSD are reflective of the Rapid Rural Appraisal (RRA) philosophy wherein while the project agenda is defined ‘externally’ (the scientific institutions in this case), the people (the district staff) are encouraged to participate. The involvement of the ‘end-users’ (the community members who are the land beneficiaries) is considered as not required under the IMSD. The merits of this thinking can be debated whether the community members need to understand the working of the GIS ‘black box’ to effectively participate in system design, or their point of entry could be an exposure to the outputs of GIS maps that depict the development plans which impact their land and livelihood.

In the Participatory Rural Appraisal (PRA) model, people are not seen as passive beneficiaries (Schneider and Libercier, 1995), and the development agenda, framework of participation and use of knowledge is expected to be decided in consultation with the community. However, PRA is initiated by outside ‘experts’, and the extent to which the people’s inputs are taken into account is still decided by the funding agency (e.g. bureaucracy). The Alwar example is seen to be different than the traditional PRA, as it is closer to the ideal of empowerment, wherein the people rather than the bureaucracy are driving the process. However, the efforts of the TBS cannot be discounted in facilitating these participatory processes, and contextualizing them within the framework of indigenous knowledge. The dynamics of PRA process, however, often lead to the articulation of new risks, as reflected in the conflict between the government and the local people noted at the end of the case. These risks are underscored by Mosse (1994: p.522):

> *[PRA techniques] have contributed significantly to the promotion of participatory development. But, while they offer new opportunities for the articulation of local knowledge ... they may also expose projects to new risks by creating public contexts and a new idiom in which dominant interests can gain legitimacy.*

In Table 1 below, we summarize the two approaches to land development that emphasize the contrasting nature of implementations, outcomes and key characteristics of the two cases. A further analysis of the case emphasizes the conceptualization of participatory processes ranging from a continuum where it is defined by the external agency to one in which it is driven for and by the people. This continuum reflects different kinds of relationship between structure and agency, and different forms of rules and resources that are drawn upon to exercise agency. We conceptualize and summarize this relationship in Table 2.

Table 1. Characteristics of the two approaches used for land regeneration

Community approach – Alwar	GIS approach – IMSD
Implementation based on:	
Intensive local participation; recourse to indigenous knowledge, local practices, use of appropriate technology; focus on tree-planting and protection.	Little local participation of end-users of technology and intended beneficiaries; normative approach based on scientific methods, computer-based modelling, imported technology.
Outcome:	
Successful regeneration of Aravari; immediate improvement in quality of life.	Action plans suggesting multifarious landuse in target areas; no clear assessment of the contribution of GIS.
Main features:	
TBS plays catalyzing role; people discover forgotten pride and value of mobilizing own resources.	Long gestation for technology to propagate; project-based funding as per pre-decided cost norms.

Table 2. Continuum of Participation in Development

Mediating Agency/Nature of Participatory Agency	Structure - agency interplay	Rules and resources mediating structure-agency relationship
External		
(i) *Govt./external donors*: orthodox approach [No participation] Designed and implemented by outsiders; local communities (intended beneficiaries) not consulted about their perception of development; monitoring and evaluation conducted by project designers themselves.	Existing structures dominate with little potential for agency to make change.	Prevailing hierarchies, bureaucratic norms, donor perceptions, pre-established funding patterns, power to allocate resources, superiority of external knowledge and technology.
(ii) *Govt./external donors*: RRA approach [Institution focused participation] Intended beneficiaries consulted in meetings organized by officials/donor agency experts; move towards eliciting participation; development agenda still externally-driven; aims to obtain community feedback into project formulation.	Possibility of creation of new structures to accommodate preliminary level of participation by people.	---- do ----- + local practices and indigenous knowledge.
(iii) *Govt./external donors*: PRA approach [User focused participation] Target group encouraged to define and spell out its aspirations; however, people's inputs not the entire basis for development – power exercised by bureaucracy/politicians still a major factor; relevance of local practices and knowledge more explicitly recognized; development agenda still externally-driven; faulted by development workers for treating community as a homogenous, non-problematic entity.	New structures to facilitate community participation, empowerment emerge; homogenous identity of community.	Community, indigenous knowledge, empowerment, bureaucratic/donor perceptions.
Internal: by, of, and for the community [Active participation] Direct involvement of local communities in development; impetus may be injected by an external agency; framework of development within the community, democratic and self-sustaining; indigenous knowledge fully exploitable.	Full potential for creation of new structures. Local power equations attempted to be rearticulated.	Trust, commitment, local power politics, indigenous knowledge, community resources.

5 CONCLUSION

We have conceptualized participation as a form and reflection of human agency that can be analytically viewed over four moments or levels of a continuum ranging from no participation, institution-focused, user-focused, to the participatory agenda being defined and enacted by the project beneficiaries themselves. We have described the first three levels to be mediated by governmental agencies, drawing upon institutional structures that emphasize hierarchy, a top-down philosophy, and the application of a project-based scientific knowledge. These structures tend to stifle participatory processes, or at best accept inputs from the end-users within the development framework defined by the bureaucracy. The fourth level of participation is described as being mediated 'internally' by the community members in partnership with a community-based organization, drawing upon contrasting structures reflecting trust, commitment, and respect for indigenous knowledge.

Through this analysis, we argue that the two cases provide elements of learning which together can help develop a more effective strategy to combat the larger problem of land degradation. While it is important to draw upon perspectives of local knowledge and conditions in the formulation of development projects, we must simultaneously also be concerned about the macro-problem of how learning and experiences that result can be drawn into efforts to address the complex problem of land degradation. At one level, land is a 'resource-in-use', inextricably related to people and society that use it, implying that degradation at one place and time will be situated and complex. At another level, there are patterns that repeat themselves in human-environment relations (Blaikie and Brookfield, 1987: p.16), implying some 'context-free' elements that can be 'modelled' using technologies like GIS. For example, land degradation and/or loss of vegetative cover can be predicted through modelling land use pattern and intensity of resource use.

The interesting challenge then is how to develop approaches that on one hand can take into account the capability of new technologies like GIS, which while potentially enabling the larger spread of projects, tend to come with government institutional structures that are not conducive to active participation, and to the contrary may even stifle such processes. On the other hand, initiatives like the Alwar case provide inspiration to draw upon the power of relevant local knowledge and the power of local agency. This predicament represents a dialectical relationship where one process can be seen to simultaneously both support and undermine the other. This dialectical relationship is captured neatly in Friedman's (1992: p.7) following argument:

> *Although an alternative development must begin locally, it cannot end there. Like it or not, the state continues to be a major player ... without the state's collaboration, the lot of the poor cannot be significantly improved.*

Castells (2000) has also referred to the above predicament as being a key challenge in creating new structures of governance in contemporary civil society. Castells would argue that movements like the Alwar case can be seen to reflect a 'resistance identity' that can help to prevent the potential 'disintegration of society' (through land degradation in our case). The challenge is to examine under what conditions this resistance identity can be converted into a 'project identity' where "social actors, based on whatever cultural materials that are available to them, build their position in society and, by doing so, seek the transformation of the overall social structure". (ibid.: p.120). Castells refers us to the challenge of sustaining resistance approaches like those offered by TBS, while simultaneously redefining governmental systems to support such approaches to blossom. Castells argues (ibid.: p.121) that if such transformations do not come about, movements like TBS remain largely symbolic, with impacts limited to the local domain. While resolving this predicament is beyond the scope of this paper, its identification helps to raise at least two key questions for future empirical and theoretical work. The first concerns the challenge of developing 'hybridized knowledge' that blends indigenous and scientific knowledge. The second relates to the development of institutional frameworks within which participatory processes are nurtured to facilitate application of this knowledge for the larger good of land conservation.

ACKNOWLEDGEMENTS

We thank Dr. Judith Gregory, IFI, University of Oslo, for reviewing an earlier draft, and for providing insightful comments, as also many useful suggestions. We also acknowledge with thanks the several useful suggestions made by two anonymous reviewers for restructuring the earlier draft to provide better clarity and focus to the participation debate.

REFERENCES

Aronoff, S. *Geographical Information Systems: A Managerial Perspective*, Ottawa, Canada: WDL Publications, 1995.

Barley, S.R. and Tolbert, P.S. "Institutionalization and Structuration: Studying the Links between Action and Institution", *Organization Studies*, (18:1), 1997, pp. 93-117.

Bhatnagar, B. and Williams, A. "Participatory Development and the World Bank: Potential Directions for Change", *The World Bank*, Washington D.C., 1992.

Binder, T., Gregory, J., Wagner, I. (eds.) *PDC'2002 Proceedings of the Participatory Design Conference*, Malmo, Sweden, June 2002.

Blaikie, P. and Brookfield, H. *Land Degradation and Society*, London: Methuen, 1987.

Butler, T. and Fitzgerald, B. "The Relationship Between User Participation and the Management of Change Surrounding the Development of Information Systems: A European Perspective", *Journal of End User Computing*, Jan-Mar 2001, pp.12-25.

Castells, M. "Globalisation and Identity in the Network Society: a Rejoinder to Calhoun, Lyon and Touraine", *Prometheus*, (4), 2000, pp. 109-123.

Chambers, R. *Rural Development: Putting the Last First*, London: Longman, 1983.

Craig, W.J., Harris, T.M., and Weiner, D. (eds.) *Community Participation and Geographical Information Systems*, London: Taylor and Francis, 2002.

Dasgupta, A.R., Rao, M., and Gopalan, A.K.S. "The National (Natural) Resource Management Information System", *IAPRS* (XXXIII), Amsterdam, 2000, pp. 27-34.

Ehn, P. "Scandinavian Design: On Participation and Skill", in *Participatory Design: Principles and Practice*, D. Schuler and A. Namioka (eds.), NJ: Lawrence Erlbaum, 1993, pp. 41-77.

Escobar, A. *Encountering Development: The Making and Unmaking of the Third World*, Princeton, NJ: Princeton University Press, 1995.

Friedmann, J. *Empowerment: The Politics of Alternative Development*, Oxford: Blackwell, 1992.

Gadgil, M. "Biodiversity and India's Degraded Lands", *Ambio*, (22: 2-3), 1993, pp. 167-172.

Giddens, A. *The Constitution of Society: An Outline of the Theory of Structuration*, Cambridge: Polity Press, 1984.

Government of India. "Guidelines for Watershed Development (Revised)", *Ministry of Rural Development*, New Delhi, 2001.

Hutchinson, C.F. and Toledano, J. "Guidelines for Demonstrating Geographical Information Systems based on Participatory Development", *International Journal of Geographical Information Systems*, (7:5), 1993, pp. 453-461.

Jain, R.B., and Dwivedi, O.P. "Administrative Culture and Bureaucratic Values in India", *Indian Journal of Public Administration*, (36:3), 1990, pp. 435-450.

Kothari, U. "Feminist and Postcolonial Challenges to Development", in *Development Theory and Practice*, U. Kothari and M. Minogue (eds.), Hampshire: Palgrave, 2002, pp. 35-51.

Kothari, U., Minogue, M., and DeJong, J. "The Political Economy of Globalization", in *Development Theory and Practice*, U. Kothari and M. Minogue (eds.), Hampshire: Palgrave, 2002, pp. 16-34.

Kumar, A., and Kaul, R. N. "Joint Forest Management in India: Points to Ponder", *Commonwealth Forestry Review*, (75:3), 1996, pp. 212-216.

McGee, R. "Participating in Development", in *Development Theory and Practice*, U. Kothari and M. Minogue (eds.), Hampshire: Palgrave, 2002, pp. 92-116.

Mosse, D. "Authority, Gender, and Knowledge: Theoretical Reflections on the Practice of Participatory Rural Appraisal", *Development and Change*, (25), 1994, pp. 497-526.

Mumford, E. "The Participation of Users in System Design: An Account of the Origin, Evolution and Use of the ETHICS Method", in *Participatory Design: Principles and Practice*, D. Schuler and A. Namioka (eds.), NJ: Lawrence Erlbaum, 1993, pp. 257-270.

Patel, J. *Story of a Rivulet Aravari: From Death to Rebirth*, Alwar, India: Tarun Bharat Sangh, August 1997.

Saha, A. "Basic Human Nature in Indian Tradition and its Economic Consequences," *International Journal of Sociology & Social Policy*, (12:1-2), 1992, pp. 1-50.

Sahay, S. "Implementing GIS Technology in India: Some Issues of Time and Space," *Accounting, Management and Information Technologies*, (8:2-3), 1998, pp. 147-188.

Sahay, S. and Walsham, G. "Social Structure and Managerial Agency in India", *Organization Studies*, (18:3), 1997, pp. 415-444.

Schneider, H. and Libercier, M. "Concepts, Issues and Experiences for Building Up Participation," in *Participatory Development: From Advocacy to Action,* H. Schneider and M. Libercier (eds.), Paris: Organization for Economic Co-operation and Development, 1995, pp. 29-64.

Singh, V. *Rejuvenating the Ruparel River*, Alwar, India: Tarun Bharat Sangh, undated, probably 1998.

Schuler, D. and Namioka, A. "Preface", in *Participatory Design: Principles and Practice,* D. Schuler and A. Namioka (eds.), NJ: Lawrence Erlbaum, 1993, pp. xi-xiii.

Walsham, G. *Making a World of Difference: IT in a Global Context*, John Wiley, 2000.

Walsham, G. and Sahay, S. "GIS for District-Level Administration in India: Problems and Opportunities", *MIS Quarterly*, (23:1), 1999, pp. 39-66.

About the Authors

S.K. Puri has worked in design of information systems for forestry sector, mapping of wastelands using remotely sensed satellite data, and applications of GIS for wasteland management. He is currently pursuing a doctoral program at the Department of Informatics, University of Oslo, and may be reached at satishp@ifi.uio.no.

Sundeep Sahay is professor at the Dept. of Informatics at the University of Oslo, Norway. Dr. Sahay has been actively involved in researching social, organizational and managerial implications of GIS systems, both in developed and developing countries. His current research activities include health Information systems, and issues relating to global software organizations. He has published extensively both in IS and organization studies. He may be contacted at: sundeeps@ifi.uio.no.

Part 4:

Societal Dynamics

18

SOCIAL EXCLUSION AND INFORMATION SYSTEMS IN COMMUNITY HEALTHCARE

Tony Cornford and Ela Klecun-Dabrowska
Department of Information Systems, London School of Economics and Political Science, UK

Abstract: This paper discusses information systems in community development contexts through the perspective of social exclusion. The discussion is situated within a particular geographical community in London and is based on initiatives to address healthcare needs within the context of a connected community initiative. The discussion considers the relation of information systems to social development and processes of inclusion (or exclusion).

Keywords: Information systems, community healthcare, social exclusion, digital divide, telehealth.

1 INTRODUCTION

Information systems (IS) research is biased towards business usage, serving the needs of late capitalism and focusing on a narrow and untypical sector's consumption of 'leading edge' technologies. There are of course exceptions, for example in literature discussing information systems in development and in the public sector (Avgerou, 2002; Heeks et al., 1999). Within the healthcare sector most attention is given to organisational information systems and to application of ICT to medicine and clinical practice. While some works do discuss community systems they tend to focus on individual projects and their evaluation (Brennan et al., 1997; Gustafson et al., 2001; Sosa-Iudicissa et al., 1997). In contrast, in this paper our aim is to explore how community health information systems might serve communities with high levels of deprivation, drawing from the theory of social exclusion and literature concerned with the relationship between an (information) society and the technology at its disposal (Dutton and Peltu, 1996; Kling, 1996).

Social exclusion provides a theoretical stance akin to the discourse of the digital divide, though we suggest that it is more encompassing and sharply focused, looking beyond isolated processes of digital exclusion, to a matrix of interlinked processes that contribute to the exclusion of people and groups within a society or community (Compaine, 2001; Haddon, 2000; Norris, 2001). Such exclusion processes span many mutually influencing levels, from the acts and attitudes of an individual through local contexts and communities to national and global structures (Burchardt et al., 2002).

This paper reports a local study, but the issues it addresses, of the relation of information systems to social development and processes of inclusion (or exclusion), have a wider significance. The study took place within the London Borough of Lewisham and in relation to its needs for health and social care as expressed by local people through discussions and initiatives they engaged in, and by local government bodies and health agencies through their strategies and programmes. The themes we identify - aspirations to 'harness' information and communication technology (ICT) to re-model health and social care delivery around patient needs and digital citizenship, and to enhance self-care and processes of community support, are relevant to many other poor or excluded communities in developed and developing countries as they encounter and assess the possibilities of ICTs. Equally, the potential problems such ICT-based services pose have a parallel in myriad other contexts across the developing and developed world. As Phipps (2000: p.64) argues "Applied to enhance access, choices, and social participation, new communications technologies can be a conduit for social inclusion – resting on our societal and strategic choices".

2 SOCIAL EXCLUSION AND INFORMATION SYSTEMS

The concept of social exclusion has gained popularity through the 1990s and it is now widely used within academic and political discourse across Europe in discussion of social policy, as well as in developing country contexts as an expression of a lack of civil and social rights (Gore and Figueiredo, 1998; Hills et al., 2002). Burchardt et al. (1999) describe social exclusion as a "contested term", tracing its European, in particular French, origins and contrasting it with the older and cruder concepts of poverty, or deprivation, or the more binary concepts of ghettoization and the underclass developed in American literature (Andersen, 1999). In policy debates the term expresses a dynamic and multifaceted understanding of the unfolding circumstances of people, groups or neighbourhoods at the margins of society; 'a short hand term for what can happen when people or areas suffer from a combination of linked problems such as unemployment, poor skills, low incomes, poor housing, high crime environments, bad heath, poverty and family breakdown' (SEU, 1999). Exclusion also features in debates about information society and innovative information systems in the public sector, often in terms of a scenario of information rich and information poor or the digital divide, framed in terms of a lack of ability to access information, including lack of resources or access to technology, and a lack of knowledge or understanding as to what to do with information. Social exclusion perspectives recognise change over time and dynamic processes of inclusion and exclusion at work within society, rather than identifying groups or individuals as being just excluded (or included) (Hills, 1997). Inclusion and exclusion is not absolute or binary but results from multiple circumstances, attitudes and dispositions, as well being relative to the wider society (Atkinson, 1998; Richardson and Grand, 2002). Milbourne (2002) illustrates how processes of exclusion can accrue over time and may lead to a gradual withdrawal from community networks and to decreasing access to social resources. Richardson and LeGrand (2002), based on discussions with focus groups from excluded communities in Britain, propose broad dimensions of exclusion in terms of participation in consumption activity, production activity, political activity and social activity, as well as use of publicly funded and provided services, such as health, education and transport, and of public goods such as a safe environment. Such a multi dimensional framework means that it is necessary to address the question of agency in such processes, as Burchardt et al. (2002) pose it, 'Who is doing the excluding?'. This might be answered in various ways ranging from accounts that focus on institutional structures, discrimination or lack of enforced rights, to seeing those excluded as responsible.

Social exclusion has been reflected in many contemporary health policy initiatives in the UK. 20 years ago the Black Report (DHSS, 1980) documented the way in which health inequalities had survived between the social classes in the UK even after 30 years of a comprehensive National Health Service (NHS) free at the point of delivery, findings confirmed by follow up studies (Whitehead, 1988; Acheson, 1998). The recent UK Government consultative green paper *'Our Healthier Nation'* (DOH, 1998), similarly expresses its aim, "To improve the health of the worst off in society and to narrow the health gap" (p.5). Thus social exclusion is identified as *both* one of the causes and the effects of ill health or disability. Growing out of this perspective has come a strategy for promotion of health information flows within the institutional world of healthcare provision and into the community, with UK initiatives including the web-site, *Wired for*

Health (DOH, 1999), and *NHSDirect*, a nurse-led telephone triage system handling 450,000 calls a month[1].

Attention to building stronger information flows within the health sector leads to consideration of the role that ICT might play in processes of inclusion and exclusion in relation to availability and accessibility of health services. Thus it may be connected to distance (which is relative – 3 miles in the City can be a very long journey, not least in cultural terms), to time (can a person use a service within their time constraints – for example as a single mother without child care), to mobility (ability to travel to access the services – for example as a disabled person), to cost (even a free service requires time off work), to 'understandibility' (having the language to understand information), to suitability and focus (is the service in some way 'personalised'), and to awareness (is a person aware of services on offer). We also understand that people can be excluded because of their attitudes and values (the agency dimension). All these characteristics are potentially interrelated in processes that reinforce exclusion, but are changeable as a person's or their community's circumstances and resources evolve. We note in particular the close link between individual and community exclusion, not only in terms of geographical constraints, for example lack of local health services or bad transport links, but also in terms of community-based culture, habits and shared attitudes.

Given the desire for an informational response to exclusion, ICTs are often proposed as 'solutions'. For patients, ICT-based services may offer better quality of care, for example improved access through teleconsultation facilities, convenience and cost-savings through local services and up-to-date information, better service or better trained and supported personnel able to provide better care (Yach, 1998). Others argue that telecare allows greater flexibility and responsiveness, for example, enabling the elderly to be cared for at home, providing safety, security and health benefits (Coyle et al., 1995; Sixsmith, 2000; Whitten et al., 1998), and a means of interacting in a client-centred manner, promoting patient autonomy through education and improved communication (Warner, 1997). We must also recognise that in the research literature these claims are not well documented (Collins et al., 2000; Hakansson and Gavelin, 2000; Mair and Whitten, 2000; Roine et al., 2001). Others note that such moves may change the nature of health services and the practice of medicine, resulting, for example, in a medicalisation of essentially social questions (Fisk, 1995; Gott, 1995).

[1] http://www.doh.gov.uk/newsdesk/inside/may2002/index.html Last accessed 08/09/02.

3 CASE STUDY OF TELEHEALTH IN LEWISHAM

The case study presented here, informed by interpretative and critical research modes, was conducted in 1998-2000 in the London Borough of Lewisham, a poor Borough in the south east of the city. A variety of data collection techniques were used, mostly un-structured interviews and when possible direct observation. Analysis of documentary sources offered different (generally official) viewpoints, additional details and background information. A number of councillors (elected politicians) and local government staff were interviewed about social development policies, IT strategies and their relation to national policies, as well as commitment to health related projects. Public forums organised to discuss issues of health, ICT and local needs, as well as local and national workshops and project specific meetings were attended. Following these meetings the researcher made contact with participants and subsequently was able to interview them and other people they suggested. In total 43 people were interviewed. The main themes discussed included the role of ICT in the provision of healthcare, local telehealth projects and initiatives, and the relationship between local strategies and national health policy. As a part of a larger study, the research data (documents, transcribed interviews and notes) were analysed from different perspectives using a broadly hermeneutic approach. In other work we have focused on an institutionalist perspective (Klecun-Dabrowska and Cornford, 2002), critical approach (Klecun-Dabrowska, 2003) and enacted technology (Cornford and Klecun-Dabrowska, 2001). This paper focuses on an analysis through social exclusion concepts.

Discussion with a variety of stakeholders who had institutional positions in healthcare and the local council revealed them as envisaging ICT as having an increasingly important role in enabling the delivery of services and changing the means of communication with citizens. This vision had led to the local council embarking on a *Connected Community* programme to harness ICT for the benefits of the community; in their terms, to regenerate it, to improve learning, employability and social inclusion, as well as to simplify government and strengthen democratic processes. For health and social care the *Connected Community* report (LBL, 1998) proposed, in partnership with community groups and health agencies, to enable people to engage with local clubs and voluntary groups through online connections, provide online community safety and health information, and offer support to homebound or fragile people, for example through Linkline facilities to emergency services, or by providing 'smart homes'. The programme was translated into concrete initiatives including the provision of terminals in libraries with free access to email and the Internet, multimedia kiosks in different locations including shops and pubs, and providing personal computers to schools, while IT skills were taught at local libraries and colleges. Other initiatives, included TellyTalk, One-Stop-Shop's and a community Intranet[2]. A One-Stop-Shop offers integrated and easy to access to multiple services at one point of contact, while TellyTalk is a free videoconferencing facility to connect people to local and national service providers[3].

Aiming to involve the community in shaping such local services (or being seen to do so) the Council organised a number of open one-day meetings. One of these, the *Connected Community Forum* provided an arena for discussions and workshops on how ICT could benefit people. This focus on technology rather than wider initiatives impelled one of the participants to ask: "Why are we talking so much about information technology?" Nevertheless, the Forum, did discuss many relevant issues including social cohesion and education, and was not overtly technologically deterministic, perhaps because local people from different professions, groups and ages led it. A number of initiatives were put forward, including re-designing Lewisham web-pages to centre them around local residents' needs rather than those of potential visitors and developing a degree of interactivity with links to local information and community projects. Other initiatives emerged and action groups were set up to work on different projects, not necessarily involving ICT, e.g. providing support to the elderly through home visits.

During one of the *Forum* workshops telehealth's potential to improve access to healthcare was the theme. Suggested services included televisits at home and teleconsultations in GP surgeries or local clinics and were seen as serving the socially excluded (either by their illness or fragility or by having little money to travel). It was also suggested that telehealth could offer busy people access to health advice outside their working hours (for example through *NHS Direct* or PolyClinics). Participants did express concerns about who would be able to take advantage of new digital services, recognising that while initiatives aiming to strengthen social inclusion and close the digital divide existed, focus on digital services could divert attention from other more valuable initiatives. The second *Forum* focused directly on health and healthcare. Here ICT played a small part in the local residents' vision of how the health of the local population might be improved. Instead, the participants established their sense of a clear link between health, health services and the local environment, demonstrating the necessity to retain a holistic understanding of community needs rather than isolated services.

Another initiative under the Institute Of Health Services Management (IHMS) Telemedicine and Telecare programme brought together

---

[2] see www.lewisham-visibledifference.org.uk Last accessed 29/10/2002

[3] see http://www.lewisham.gov.uk/NewsAndViews/tellytalk.asp Last accessed 29/10/2002

representatives of social services, health organisations and patients. The aim was to identify key healthcare related needs in Deptford (probably the most deprived area of Lewisham), discuss issues and problems facing health and social support services, and to explore the potential of ICT. Three broad aims were identified: enabling and supporting self-care, promoting inter-agency working, and improving efficiency. For each of these a number of potential ICT-based applications were identified and relevant services included Electronic Health Records, electronic referrals, more teleconferencing and teleconsultations between different health and care providers, and Internet-based information and training.

Taken all together we see wide-ranging needs expressed by citizens and healthcare providers, and a general ability to suggest possible ICT-based services that might help to address them, including dimensions of availability and accessibility as identified above. But ICTs were still largely seen as (fairly) unproblematic and such benefits were considered as either already proven in other settings or as just requiring further assessment. Little doubt was expressed about the viability of proposals while an integrated, 'strategic' approach and multi-use of equipment and resources was often proposed in order to reap benefits and reduce costs.

This programme of consultation, not withstanding its limitations, produced valuable suggestions regarding potential application areas for telehealth in Lewisham. It gave healthcare and social workers an opportunity to learn about ICTs, as well as to discuss potential reforms to the way their own services were run. It also provided a (limited) forum for citizens and patients to voice their opinions, and the approach of the workshops did encourage looking at contextualised problems first, rather than finding technological solutions. While the outcomes may appear to be 'wish lists', they did correspond quite closely with the national policy agenda (DOH, 2000; DOH, 2002; NHS Executive, 1998). Indeed, at the time many individual projects were under way in the Borough, including for example, a telepsychiatry service allowing consultations to take place in a family doctor's practice linked to a hospital-based consultant via teleconferencing equipment (McLaren et al., 1999). This project was described by the Development Manager at the Community Mental Health Trust as a part of a wider strategy : "We feel that we have a way of revolutionising the delivery of health care, making it more accessible and immediate. And particularly for disadvantaged communities, they may not have access to expert care – a way of bringing it to them". Yet one of the project's originators admitted: "We are not at all convinced that we are adding value". Another local project, Deptford's Women and Children's Centre, was an early pregnancy assessment unit in a very deprived neighbourhood with a telemedicine link for ultrasound scans and teleconferencing to hospital consultants.

Interestingly this telemedicine link was never used and was finally disconnected. However, the centre itself, although perhaps not effective in narrow cost terms, was popular locally and was seen as achieving its aim of providing new services and increased access to healthcare for a vulnerable group.

Two other indicative projects utilised the Web. EmpowerNet, offered information and limited interactive services for people with mental health problems and was developed by them as part of vocational training in web design. This project was given a clinical role, as the Director of the Community Mental Health Trust put it, "It is really important being honest with people about what information we have or don't have about the illness, effectiveness of treatment and about side effects. [...]And people find that very challenging because everyone always wanted to say that psychiatry is a science and it isn't". The SeaHorse project explored the potential of ICT for supporting people with HIV/AIDS and facilitating collaboration between carers (Cullen, 1997). This included assessment of information needs of self-help groups and development of computer-based applications, including interactive services and decision support tools, as part of a larger European funded project with partners from different countries. One of this projects co-ordinators emphasised its inclusionary potential: "Because we are offering services at home, we give access to information which otherwise the users wouldn't get if they weren't' well enough to get to a library or if they weren't earning enough money to have their own Internet connection.".

While most professional people interviewed – those involved in these projects and those working in local government – were enthusiastic about new technologies or at least believed that their potential could be harnessed to benefit citizens, some voiced concerns. For example, they worried that services may become depersonalised (e.g. caring aspects being replaced by technical intervention, leading to medicalisation of care), or be of lesser quality (e.g. provided by staff without adequate training in the use of a particular technology), and that patients might loose confidence in the services. As one interviewee noted, "I think we are giving people power to choose not to go to a doctor. I'm not certain that's actually good. [...] They would say 'but the user is in control because they can press the button'. But they don't want to press a button, they want help". A social worker too expressed uneasiness about sharing data between social and health services, foreseeing people withdrawing some information they would otherwise pass on to healthcare workers. She also questioned the viability of using teleconferencing equipment to communicate with the elderly and patients with hearing or sight problems and who find it difficult enough to use a telephone. More generally many staff felt unsure how to best utilise ICT and what the future would brings for them, "... we are moving too fast and some people will be left behind".

4 DISCUSSION

Social exclusion is a dynamic process, effecting different people in different ways at different points of time. Here we consider what this implies for ICT as a constitutive element of healthcare within a community – what sort of services it implies. Our understanding of social exclusion leads us to a paradox; that services supported by technology need to be both targeted and broad; meet individual needs but also foster positive community processes that promote use. One route is for these groups to be actively engaged in the design of the services (as in EmpowerNet). At the same time such services need to cater for people's changing circumstances over time. Conceptualising social exclusion as the interplay of multi-dimensional processes suggests that ICT-based health initiatives should focus less on consequences of exclusion, but rather on process that lead to it. For example, if the manifestation of social exclusion is through ill health related to poor diet and low income we may try to promote healthy eating, for example in schools, and offer skills training. Still, simply adding web-based information to this mix is unlikely to prove a powerful additional force unless it too is linked back into other community resources. As Milio (1992) points out, the value of health information goes beyond creating individual awareness to include educating and motivating for action. When shared (in two-way exchange), she suggests, it can also help to express and test ideas, bind people and communities and foster an environment conductive to well being.

In this case study we see national and local policies, new technologies and 'bottom-up' initiatives being translated into projects that attempted to use ICT to bring health and social services closer to people and their communities, to make them more 'customer-centred' and to provide support for self-help activities. Both central and local government sees such initiatives as linked and mutually reinforcing, evoking the concepts of 'joined-up thinking' and a multi-deprivation perspective on social exclusion. The Local Authority thus had policies based around ICT that aimed to address social exclusion by aiming at the 'regeneration' of the local community. The *Forums* indicate that local people see health, education and the environment as inter-connected, and ICT as playing a part in supporting social inclusion rather than having a central role. Similarly, the four projects briefly outlined illustrate the diversity of issues and multiple facets of exclusion that were being addressed in terms of different groups of socially excluded– the chronically ill, their carers, or a vulnerable population – but also in terms of the size and scope of projects and the technologies applied. Similarly, the extent to which they were shaped by local people varied considerably. What they share is an assumption that making service delivery local benefits the population; that it helps to target at local needs, gains the community's support, and can alleviate inequalities. This also reflects a more holistic approach to health, one that takes into account social and personal circumstances on health and addresses stress and stigma, as well as family and financial consequences, and the potential for peer support.

However, this research also indicates some problems encountered in addressing social exclusion through ICT. Critiques of the Lewisham policy documents (Sullivan, 1998; Sullivan and Quirk, 2000) suggest that they offer a simplistic portrayal of ICT-based reforms and unquestioning ascription to ICT of qualitative transformational potentials (Klecun-Dabrowska, 2002). A report assessing the *Connected Community* initiative highlights the unequal spread of resources that privilege some deprived areas over others - those with allocated special funds (Forbes, 2001). Many innovative ICT-based projects struggle during implementation and find it hard to sustain themselves once special funding is withdrawn. Equally expectations may be raised which may not be fulfilled, and while ICT have the potential to help to overcome some forms of exclusion, they may also create new forms or reinforce existing ones. The experience of the Deptford's Women and Children's Centre indicates that new technologies are not always needed and re-designing work practices or models of care delivery may be more beneficial. As Silverstone and Haddon (1998) suggest "While any benefits offered by new telematics may be welcomed, the new options and possibilities these technologies bring simultaneously mean that the mechanisms and nature of deprivation or inequality – what resources, competencies and cultural values are involved – are in a constant process of change." They note in particular the closing down of opportunities for interaction, involvement and the ability to influence ones environment through participation in political/economic life.

While in the initiatives we study here community involvement goes some way towards containing (if not reducing) such an emerging digital divide, as Van Dijk and Hacker (2000) (referenced in Kvasny and Truex (2001)) argue, structural forces beyond the physical access to technology act to perpetuate the divide, and increased material access may not necessarily lead to decreased structural divides. One dystopian vision of ICT in the UKs NHS is as providing an alternative to the established services, or as a 'fast track' that might undermine – not re-enforce – the inclusionary ethos. We also see some more specific dimensions of exclusion actually or potentially at work and enhanced through ICT in the allocation of scarce resources and acceptance of opportunity costs, and in the medicalisation of care. Although projects such as SeaHorse focus specifically on empowering socially excluded groups, they may also serve to increase the isolation of their target communities if they do not promote a sense of involvement in the wider community. For carers too the solidarity they can develop, or the sense of power they feel in decision-making, may come to reinforce their isolation. In summary, health services delivered through ICT, particularly personal telecare systems, might positively contribute to social isolation and be exaggerated by displacement of other relationships. For example, substitution of day centres, where older or disabled people meet and engage in varied activities, by discussion groups over the Internet, or the substitution of a visit to Grandma with a call to the call centre to receive a report on her vital signs and her mobility index in the past 24 hours.

5 CONCLUSIONS

How to deploy ICT to reform and restructure the delivery of care in order to achieve better, more equitable healthcare is a wide spread concern for communities, governments and international organisations as they struggle to define (and enact) new conceptions of citizenship and belonging within an illusive 'global information society'. This paper has set out to consider these issues and to present community view(s) on ICT-supported health services through a social exclusion perspective. Social exclusion, as we argue in the introduction, provides a robust and theoretically sound basis for such analysis and the paper seeks to add to the qualitative based empirical work in a field where the weight of literature is either theoretical or based on statistical analysis.

We argue here that the relationship between information systems, health services and social exclusion is complex, and cannot be approached through a narrow understanding of access (access to ICT and access to health services). Thus, section 2 develops descriptions of different dimensions of social exclusion in health and relates them to the potentials of ICT as seen in the IS literature. Our study suggests that ICT-based services are seen as potentially benefiting the community, but there is a strong concern that this must not lead to further exclusion of some groups. The *Forum* meetings and workshops often suggested that such services should be provided *in addition* to conventional ones, and initiatives should be based on active participation of community members. Our interviews also indicate that some health and social workers themselves feel excluded by technology and are afraid of being 'passed by'. Finally, both the *Forums* and workshops saw the potential of information systems and ICT-based services only as part of and within a wider program of regeneration and reform of health service. The indicative projects we describe (telepsychiatry, telelink at the Deptford Centre, SeaHorse and EmpowerNet) show that such initiatives do have potential to address some dimensions of exclusion (e.g. distance, time and awareness), but they may have little effect on agency aspects if not embedded in local contexts and institutions, and receiving long-term support.

We suggest that viewing ICT through the social exclusion perspective helps us to go beyond medical, technological or even organisational frameworks and to link to concepts of citizenship and belonging in an Information Society. Still, in this paper, we have told an incomplete story. We have no space here to discuss the different and often ambiguous meanings that ICT acquire in policy and strategy, and the meanings that are forged during the development, implementation and use of each such service. To do so we would have to draw explicitly on the more substantial theoretical approach of critical theorists (Feenberg, 1991; Habermas, 1979; Horkheimer and Adorno, 1972 [1944]), where social exclusion can embody one dimension of a critical study of ICT. But then it would another paper.....

REFERENCES

Andersen, J. "Social and System Integration and the Underclass", In *Capitalism and Social Cohesion: Essays on Exclusion and Integration*, I. Gough and G. Olofsson (Ed.), London: Macmillan, 1999, pp. 127-148.

Atkinson, T. "Social Exclusion, Poverty and Unemployment", London School of Economics, *Centre for Analysis of Social Exclusion*, 1998.

Avgerou, C. *Information Systems and Organizational Diversity: The Articulation of Local and Global Rationalities*, Oxford: OUP, 2002.

Brennan, P.F., Schneider, S.J. and Tornquist, E. (Ed.) *Information Networks for Community Health,* New York: Springer-Verlag, 1997.

Burchardt, T., Grand, J.L. and Piachaud, D. "Social Exclusion in Britain 1991-1995", *Social Policy and Administration*, (33:3), 1999, pp. 227-244.

Burchart, T., Le Grand, J. and Piachaud, D. "Introduction", In *Understanding Social Exclusion*, J. Hills, J. Le Grand and D. Piachaud (Ed.), Oxford: OUP, 2002, pp. 1-12.

Collins, K., Nicolson, P. and Bowns, I. "Patient Satisfaction in Telemedicine", *Health Informatics Journal*, (6), 2000, pp. 81-85.

Compaine, B.M. (Ed.) *The Digital Divide. Facing a Crisis or Creating a Myth?,* Cambridge, Ma.: MIT Press, 2001.

Cornford, T. and Klecun-Dabrowska, E. "Telehealth Technology: Consequences for Structure through Use", *Proceedings of the 10th World Congress on Medical Informatics*, London, 2001

Coyle, G., Boydell, L. and Brown, L. "Home Telecare for the Elderly", *Journal of Telemedicine and Telecare*, (1), 1995, pp. 183-184.

Cullen, J. "Final Report on the Feasibility of SEAHORSE", *Telematics Application Programme* (EC DGXIII); Project: HC 1111; SEAHORSE Support and empowerment for AIDS and HIV: the on-line research and self-help exchange, February 1997.

DHSS "Inequalities in Health: The Black Report", *Department of Health and Social Services*, London, 1980.

DOH "Our Healthier Nation: A Contract for Health", *HMSO*, London, 1998.

DOH "Saving Lives: Our Healthier Nation", *HMSO*, London, 1999.

DOH "The NHS Plan -The Command Paper 4818-1", *HMSO*, London, July 2000.

DOH "Delivering the 21st Century IT Support for the NHS: National Strategic Programme", Department of Health, 2002.

Dutton, W.H. and Peltu, M. (Ed.) *Information and Communication Technologies: Visions and Realities*, Oxford: OUP, 1996.

Feenberg, A. *Critical Theory of Technology*, New York: Oxford University Press, 1991.

Fisk, M.J. "A Comparison of Personal Response Services in Canada and the UK", *Journal of Telemedicine and Telecare*, (1:3), 1995, pp. 145-156.

Forbes, M. "Implementing a Connected Community Strategy for Lewisham: A Framework for Action", Doc Ver. 2.0, *London Borough of Lewisham*, 2001.

Gore, C. and Figueiredo, J. (Ed.) *Social Exclusion and Anti-Poverty Policy: A Debate*, Geneva: ILO, 1998.

Gott, M. *Telematics for Health: The Role of Telehealth and Telemedicine in Home and Communities*, Oxford: Radcliffe Medical Press, 1995.

Gustafson, D.H., Hawkins, R.P., Boberg, E.W., et al. "CHESS: Ten Years of Research and Development in Consumer Health Informatics for Broad Populations, Including the Underserved", *Proceedings of the 10th World Congress on Medical Informatics*, London, 2001, pp. 1459-1463.

Habermas, J. *Communication and the Evolution of Society*, Boston: Beacon Press, 1979.

Haddon, L. "Social Exclusion and Information and Communication Technologies", *New Media and Society*, (2:4), 2000, pp. 387-406.

Hakansson, S. and Gavelin, C. "What do we *Really* Know About the Cost-Effectiveness of Telemedicine?," *Journal of Telemedicine and Telecare*, (6:1), 2000, pp. S1:133-136.

Heeks, R., Mundy, D. and Salazar, A. "Information Systems for Public Sector Management", Working paper 9, *Institute for Development Policy and Management*, Manchester, 1999.

Hills, J. "A CASE for Investigation", *LSE Magazine* (Winter), 1997, pp. 7-9.

Hills, J., Le Grand, J. and Piachaud, D. (Ed.) *Understanding Social Exclusion*, Oxford: OUP, 2002.

Horkheimer, M. and Adorno, T.W. *Dialectic of Enlightenment*, New York: Herder, 1972 [1944].

Klecun-Dabrowska, E. "Telehealth and Information Society: a Critical Study of Emerging Concepts in Policy and Practice", *London School of Economics and Political Science*, 2002.

Klecun-Dabrowska, E. "Telehealth in the UK: a Critical Perspective", *Proceedings of the 2nd European Conference on Research Methodology for Business and Management Studies*, Reading, UK, 2003,

Klecun-Dabrowska, E. and Cornford, T. "Organising vision of telehealth," *Proceedings of the 10th European Conference on Information Systems*, Gdansk, Poland, 2002, pp. 1206-1217.

Kling, R. "Beyond Outlaws, Hackers, and Pirates: Ethical Issues in the Work of Information and Computer Science Professionals", In *Computerization and Controversy: Value Conflicts and Social Choices*, R. Kling (Ed.), San Diego: Academic Press, 1996.

Kvasny, L. and Truex, D. "Defining Away the Digital Divide: A Content Analysis of Institutional Influences on Popular Presentations of Technology", *Proceedings of the IFIP Working Group 8.2*, Boise, Idaho, USA, 2001,

LBL "Connected Community", *London Borough of Lewisham*, Policy and Resources Committee, London, 1998.

Mair, F. and Whitten, P. "Systematic Review of Studies of Patient Satisfaction with Telemedicine", *British Medical Journal*, (320:7248), 2000, pp. 1517-1520.

McLaren, P., Mohammedali, A., Riley, A., et al. "Integrating Interactive Television-Based Psychiatric Consultation into an Urban Community Mental Health Service", *Journal of Telemedicine and Telecare*, (5:1), 1999, pp. S1:100 - 102.

Milbourne, L. "Unspoken Exclusion: Experiences of Continued Marginalisation from Education Among 'Hard to Reach' Groups of Adults and Children in the UK", *British Journal of Sociology of Education*, (23:2), 2002, pp. 287-305.

Milio, N. "New Tools for Community Involvement in Health", *Health Promotion International*, (7:3), 1992, pp. 209-217.

NHS Executive "Information for Health: An Information Strategy for the Modern NHS 1998-2005", Leeds, 1998.

Norris, P. *Digital Divide: Civic Engagement, Information Poverty and the Internet Worldwide*, Cambridge: Cambridge University Press, 2001.

Phipps, L. "New Communications Technologies: A Conduit for Social Inclusion", *Information, Communication & Society*, (3:1), 2000, pp. 39-68.

Richardson, L. and Grand, J.L. "Outsider and Insider Expertise: The Response of Residents of Deprived Neighbourhoods to an Academic Definition of Social Exclusion", *London School of School of Economics*, Centre for Analysis of Social Exclusion, 2002.

Roine, R., Ohinmaa, A. and Hailey, D. "Assessing Telemedicine: A Systematic Review of the Literature", *Canadian Medical Association Journal*, (165:6), 2001, pp. 765-71.

SEU "So you Would Like to Know More", *Social Exclusion Unit*, Cabinet Office 1999. Last Accessed 10 February 2003, http://www.cabinet-office.gov.uk/seu/index/mre.html

Silverstone, R. and Haddon, L. "New Dimension of Social Exclusion in a Telematic Society, Working paper 45", *Graduate Research Centre for Culture and Communication (CULCOM)*, University of Sussex 1998.

Sixsmith, A.J. "An Evaluation of an Intelligent Home Monitoring System", *Journal of Telemedicine and Telecare*, (6), 2000, pp. 63-72.

Sosa-Iudicissa, M., Oliveri, N., Gamboa, C.A., et al. (Ed.) *Internet, Telematics and Health*, Amsterdam, Berlin, Oxford, Tokyo, Washington, DC: IOS Press, 1997.

Sullivan, D. "Modernising Lewisham", *London Borough of Lewisham*, London, 1998.

Sullivan, D. and Quirk, B. "Where its @: Lewisham's Strategy for Getting Connected", *London Borough of Lewisham*, February 2000.

Van Dijk, J. and Hacker, K. "The Digital Divide as a Complex and Dynamic Phenomenon", *Proceedings of the International Communication Association*, Acupulco, Mexico, 2000,

Warner, I. "Telemedicine Applications for Home Health Care", *Journal of Telemedicine and Telecare*, (3:1), 1997, pp. S65-S66.

Whitten, P., Collins, B. and Mair, F. "Nurse and Patient Reactions to a Developmental Home Telecare System", *Journal of Telemedicine and Telecare*, (4), 1998, pp. 152-160.

Yach, D. "Telecommunications for Health - New Opportunities for Action", *Health Promotion International*, (13:4), 1998, pp. 339-347.

About the Authors

Tony Cornford is a senior lecturer in Information Systems at the London School of Economics and Political Science. He graduated in Economics, obtained a Masters degree in operational research and a PhD in information systems, all from the LSE. His research interests include the dubious concept of implementation. Tony can be reached by e-mail at t.cornford@lse.ac.uk. or through his home page at http://is.lse.ac.uk/staff/cornford

Ela Klecun-Dabrowska is a lecturer in Information Systems at the London School of Economics and Political Science. She was born in Warsaw, Poland where she obtained her matura. She completed her first degree, masters degree and PhD in the UK. She holds a PhD in the area of telehealth from the London School of Economics. Her research interests include health information systems and telehealth, evaluation of information systems, and the application of critical theory in the field of information systems. Ela can be reached by e-mail at e.Klecun@lse.ac.uk

19

KNOWLEDGE, NGOS AND NETWORKS

Applying Epistemology To The Work Of Development

Justine Johnstone
London School of Economics; London Metropolitan University, UK

Abstract: Non-governmental development organisations (NGDOs) are increasingly adopting knowledge-based roles such as lobbying, advocacy, policy formulation, research and dialogue facilitation. At the same time, they have on the one hand been criticised for knowledge errors and failures, and on the other been urged to adopt knowledge networking technologies. Building on earlier work which developed a three-dimensional model of knowledge based on contemporary epistemology and cognitive science, this paper develops a knowledge-based view of NGDO activity, arguing that there is evidence of a complex range knowledge roles, involving some very demanding informational, epistemic and conceptual challenges. The expertise, skills and resources required for effective knowledge work can be supported by networks and networking technology, but only if deficits as well as benefits are recognised and if informational, epistemic and conceptual capacities are developed alongside technical expertise.

Keywords: Knowledge, development, NGOs, NGDOs, Internet, networks, epistemology, information, conceptual development.

1 INTRODUCTION

It is widely recognised that the roles of non-governmental organisations (NGOs) are changing (Murphy and Bendell, 1997; Rawcliffe, 1998), partly as a result of previous successes and partly because of the failings of the public and private sectors in dealing with complex problems of poverty, underdevelopment and environmental damage (Heap, 2000). In the development sector specifically there has been a change from operational work to international advocacy (Madon, 1999), and towards the building of civil society and democracy (Clark, 1995; Fox, 1992). Such activities, alongside emerging concepts of development as public action and policy formation, and the proliferation of information and communication technology (ICT), have caused non-governmental development organisations (NGDOs) to recognise the fundamental importance of knowledge and communication activities such as research and lobbying (Thomas et al., 1998).

In a paper to a previous IFIP conference, I reviewed some of the large and growing literature advocating a knowledge-based approach to development and technology (Johnstone, 2002) and argued the need for philosophical grounding. This paper extends the three-dimensional analysis of knowledge developed there, based on the 'externalist' strand in contemporary epistemology (Putnam, 1988; Crane, 1995; Goldman, 1999). Where the previous paper presented a very general framework, this paper applies the framework specifically to NGDOs. Using existing literature as evidence, it indicates firstly how certain activities identified as crucial to the functioning of NGDOs can be seen as exemplifying one or more of the framework dimensions. The second section argues that a three-dimensional approach can help to diagnose knowledge problems and perhaps to suggest solutions. The final section considers critically some knowledge claims made for network technology.

The view of knowledge adopted here is based on recent epistemology which stresses the importance of how beliefs come about rather than their intrinsic properties or relationships to other beliefs. Different philosophers offer slightly different formulations; the framework applied here is based on that developed by the American epistemologist Alvin Goldman, whose definition of knowledge can be summed up as 'true belief formed by a reliable process'. This 'process reliabilism' shifts epistemology towards the empirical sciences since a wide range of processes can be reliable, including vision, memory, introspection, good reasoning and social processes such as communication and argumentation. Knowledge processes, however, do not occur in a vacuum or yield knowledge on their own. They depend crucially both informational inputs and preexisting conceptual structures.

These three requirements – information, reliable process and conceptual structures – give rise to a three-dimensional framework for analysing activities and interventions in terms of knowledge. At the informational level, a knowledge analysis must consider factors such as the relevance, accuracy, timeliness, inclusiveness and organisation of content, whether derived from direct experience or through communication. At the conceptual level, the analysis must focus on the ways in which knowers categorise, make sense of information, and form ideas. At the epistemic level, it must look at how knowers judge the degree of confidence to place in the ideas they form – treating a new idea with little evidential backing as a hypothesis, for example, but accepting a long-standing, tested belief as a near certainty.

Although no one of the three factors can on its own yield knowledge, and any individual act of knowing requires all three, any improvement in informational, epistemic or conceptual status may contribute to overall knowledge status. Whether it does or not, will depend on whether the other two factors can operate on it – for example, access to a document may provide information but unless it is intelligible (conceptual dimension) and reliable (epistemic dimension) will not amount to a gain in knowledge.

The three-dimensional analysis treats knowledge as a fundamentally individual and cognitive phenomenon, something that exists only within the mind of a knower. On the other hand, it acknowledges an important social component at each level, other people being among our most important sources of information and concepts, and communication being central to many epistemic processes. In considering knowledge in organisations – in this case NGDOs – the social aspects are crucially important.

2 KNOWLEDGE IN NGDOS

Development NGOs engage in a wide range of activities and processes aimed at generating and supporting knowledge both within the organisation itself and externally in wider society. Some NGDO activities such as advocacy and research are obviously directly linked to knowledge, but even where goals are not related to knowledge, it is frequently the case that knowledge is instrumentally essential to attaining them.

In each case, informational, epistemic and conceptual factors are involved, although the three dimensions are not necessarily of equal significance. Any actual activity will inevitably focus explicitly on just one or perhaps two dimensions, depending on the purpose of the activity and assumptions about what is problematic and what is not. If an AIDS NGO wants to communicate with others about a training workshop for home-based carers they will probably focus on information provision (when, where, who, how much etc), assuming that the concepts of training, workshop and home-based care are relatively unproblematic. If they want to open the workshop to the community to recruit new carers such assumptions may not be valid and effort may have to be put into explaining what home-based care is, why it matters, and perhaps what happens in a workshop.

Knowledge failures and problems often arise when assumptions are made that turn out to be wrong about which dimensions are unproblematic – using terminology that the intended audience does not understand (failure through assuming unproblematic sharing of concepts), or sending 'uppers' to ask 'lowers' their views (failure through assuming reliable epistemic process).

The following sections analyse some central informational, epistemic and conceptual roles highlighted in recent work on NGDOs. It needs to be stressed that all real-life activity involves all three dimensions to a greater or lesser degree. The categorisation employed below is based on the ultimate purpose of the activity, but is not intended to suggest that information-directed actions, for example, may not also involve conceptual development or epistemic judgement; indeed, they are bound to.

2.1 Informational Roles

The informational dimension is concerned with the nature, quality and properties of the raw material upon which knowledge processes operate. Information activities and their inputs and outputs occupy a central place in the new roles being adopted by NGDOs (Meyer, 1997), as project-based work is increasingly displaced by information-based work (Edwards and Hulme, 1992). NGDOs are information-rich environments (Powell, 1999), having access to sources that are seldom brought together anywhere else. They may have close ties to local communities and to grassroots groups while at the same time being well-informed about the functioning of government and even international organisations and decision-making processes. At the same time cheap, prolific information and communication technology (ICT) has broken governments' informational monopolies and realigned relations with non-state actors including NGDOs, for example, by enabling them to 'reach behind' national borders and force governments to become responsive to international public opinion (Mathews, 1997, p.51).

Information is also a strategic resource for NGOs aiming at empowerment of the poor (Grimwood-Jones and Simmons, 1998). To be effective, informational flows need to be a two-way process: on the one hand ensuring the voices of the poor are heard by policy-makers, and on the other that information about government, policies, services and benefits reach the poor (Madon and Sahay, 2002). Similarly, NGDOs increasingly participate in two-way exchanges with business, acting as indicators of public opinion while in return receiving insight into business perspectives (Heap, 2000).

NGDOs' diverse information linkages can make them a valuable resource for local people, as communities benefit from being able to access a wide range of different information sources (Brown, 1991). NGOs increasingly pay attention to informal and locally derived sources such as folk media, drama, storytelling and voice recording (Edwards, 1994; Mundy and Compton, 1995) and informal channels (Madon and Sahay, 2002). The mixture of formal and informal sources and communication channels can promote the inclusion and participation of marginalised groups such as slum dwellers in consultations with government (Madon and Sahay 2002). At the other end of the spectrum NGOs which operate internationally or which have links at supranational level can access play a crucial role in local – global information transfers (Madon, 1999).

2.2 Epistemic Roles

The epistemic dimension of knowledge is concerned with the individual, organisational and social processes by which beliefs are formed. Perhaps the most obvious and ubiquitous epistemic demands on NGDOs are in the processes of planning, executing and evaluating action. All groups have to have methods for deciding what to do, how to do it and whether it is working. Often the methods remain implicit – at least until problems arise. However, demands from donors and partners, and general requirements for accountability in the sector, are increasing the need for explicit methods. This is most advanced in the sphere of monitoring and evaluation (M&E), often perceived as problematic and a source of criticism for NGDOs, which are generally seen as not being very good at it (Brett, 1993). Although M&E has traditionally been seen – and sometimes resented – as a donor imposition, sustainable development work needs local management and M&E is therefore as important for recipients as for donors (Mikkelsen, 1995). The area remains fraught, with many commentators pointing out methodological problems, weaknesses of NGDO evaluation and the difficulties inherent in epistemic decisions such as what information to collect, how to collect it, who should collect it, how and by whom it should be processed. During the 1990s a shift occurred away from progress reviews and impact assessment by 'neutral' outsiders' (Mikkelsen, 1995) to more process-focused, participatory and interpretive techniques. These have been defended on epistemic grounds, Chambers (1994), for example, arguing that they can be more reliable than surveys in delineating dimensions of poverty.

Participatory approaches have however recently come in for significant criticism, for ignoring vital aspects of development such as political and empowerment dimensions. Clearly there is a need for ongoing critical and creative research into evaluation techniques. Fowler (2000) argues against the establishment of a development 'mono-culture' and stresses the key role of NGOs working with local agents to develop innovative cross-cutting approaches rather than narrow conformity of methods.

Epistemic roles extend beyond the internal needs of the organisation and its partners. NGDOs are also often important providers of knowledge tools and techniques to the community. These organizations' proximity to and experience with those they serve, contributes to their valuable – yet often overlooked – roles as community facilitators. In that role, they serve in an important position of early adopters and arbiters of tools, resources, and practices most likely to succeed in addressing individual and community needs. More resources (technological and otherwise) should be deployed to places where the need is greatest for innovation to address growing numbers of underserved populations in an effective and efficient manner, yet where demand is suppressed due to lack of awareness and understanding as to their availability or relevance (Turner, 2002).

Another epistemic role for NGDOs is facilitating testimony, dialogue and dissent at grassroots level (Edwards et al., 1999). Processes of public dialogue are one of the main ways in which social interactions can facilitate knowledge (Goldman, 1999) and NGOs are often key contributors in civic participation and democratic, dialogic cooperation among parties involved in development (Edwards, 1999; Fowler, 2000). These processes however, are not unproblematic and NGOs have themselves been criticised for 'an eclectic outpouring of ideas and views, without organised and coherent debate' and for inability to ensure equitable participation (Pearce, 2000).

Dialogue processes are seldom quick and decisive, particularly in political discussions and where there are numerous stakeholders. The existence of mechanisms for ongoing political exchange both within communities and between communities and policy-makers has been considered a key development indicator, and NGDOs key players in these processes (Brown, 1991; Madon and Sahay, 2002). Such roles depend crucially on epistemic virtues such as reputation, trust, integrity, which in turn depend on high standards of truthfulness, error recognition, open-mindedness to new ideas and ability to learn (Chambers, 1994; Edwards, 1997; Edwards et al., 1999; Heap, 2000). Typically third sector organisations have trust advantages over other organisations but they can be discredited by poor research (Heap, 2000) or come to be viewed with suspicion by the poor owing to their perceived wealth (Holloway, 1999) or closeness to government, business and funders (Pearce, 2000).

2.3 Conceptual Roles

At the conceptual level, knowledge analysis is concerned with categorising and sense-making activities, for example the development of causal, ethical or structural understandings of the world. This level presents probably the greatest challenges of all knowledge work since it is here that high-level cognitive skills are most required, such as imagination, synthesis and the generation of new and possibly contradictory or controversial ideas.

One key area of conceptual activity in NGDOs is advocacy work. Activities such as lobbying, research, publishing, policy input and helping to draft ethical guidelines and codes of conduct involve exchanges and networking across boundaries (Edwards and Hulme, 1992; Meyer, 1997; Madon and Sahay, 2002; Thomas et al., 1998) and in some cases explicit attempts to change others' existing theories of the domain. On the three-dimensional analysis presented above, advocacy requires efforts on all three dimensions. Campaigns have to be backed with high quality, relevant information, both to plan effectively and to support the arguments being made. At an epistemic level, advocacy organisations must be trustworthy and seen to be so. They must present high-quality evidence and argument, and ensure methods are reliable and transparent. Fundamentally, however, advocacy is about changing conceptual structures such as underlying values and outlooks, sometimes in ways that run counter to vested interests or long-standing prejudices. Given these demands it is not surprising that many NGDOs have relatively weak advocacy skills (Manji et al., 1999)

The same factors that give NGDOs informational advantages can also be a source of conceptual wealth, for instance the simultaneous attachment to the local and engagement with the global that characterises many international NGOs (Madon, 1999). Exploiting these resources can allow NGDOs to bridge conceptual divides and act as 'translators' between different groups. In theory NGDOs should be fertile generators of new conceptual frameworks, both for internal consumption – reconceptualising their aims and activities for instance – and for ideas to feed into the wider development community, public understanding and government policy.

In practice, NGDOs do not always deal well with conceptual issues, often failing to clarify their concepts and make meanings clear when this would result in controversy (Pearce, 2000). At a macro level, there is also the need to question the fundamental assumptions of development itself and to look for new conceptions (Pearce, 2000). Edwards (1999) speaks about the need for a 'paradigm shift' in development work, and the need for NGDOs to develop 'strategic understanding' of global issues (Edwards et al., 1999). To deal with this NGDOs need to develop stronger theoretical bases on which to ground their identity and actions (Pearce, 2000).

Developing local theoretical capacity is important outside the development community too. Escobar, for example, argues that 'the belief that theory is produced in one place and applied in another is no longer acceptable practice. There are multiple sites of production and multiple mediations in the generation and production of theory' (Escobar, 1995: p.221). NGDOs are obvious candidates for participating in such developments yet, for all the talk of new knowledge and informational roles, little attention has been paid so far to their potential as agents of social-theoretical change.

3 DIAGNOSING PROBLEMS, IMPROVING PERFORMANCE

Knowledge failures and errors in NGDOs have recently come in for comment. The reasons for error are complex, and one application of the three-dimensional framework is in generating more systematic and epistemically grounded analyses of knowledge problems. Chambers (1994), for example, analyses knowledge failures in terms of the distorting effects of power relationships. Subordinate people ('lowers') distort their words and behaviour in the presence of powerful 'uppers' such as agency workers; uppers accept the distortion at face value since it conforms to their expectations and interests. In terms of the three-dimensional model, the process by which uppers move from receiving information (what they see and hear from lowers) to forming beliefs is highly unreliable. Since they lack (or fail to look for) meta-information about the motivations and outlook of informants, their processes of judgement and inference are faulty: in terms of the three-dimensional view, the epistemic status of uppers is poor. But so is their informational status, since although they do receive information from and about lowers, it is information about lowers *in the presence of uppers*. What uppers need – but cannot access – is information about the lives of lowers undistorted by the presence of uppers. The information uppers have fails to meet the criterion of relevance and, worse than this, it looks very like relevant information, making it easy for uppers to draw the wrong conclusions. In this way uppers and lowers perpetuate self-sustaining 'systems of misinformation'. If Chambers is right, uppers are probably conceptually challenged as well, since they fail to interpret information correctly partly because of fallacious preconceptions (conceptual structures).

Similarly, a three-dimensional analysis can help to categorise and ground the analysis of dangers inherent in what Thomas et al. (1998) term the 'finding out fast' research of NGDOs with restricted time and resources. Problems such as static conceptualising (preconceptions about the nature of the situation), assuming the problem is already known, and not involving participants in defining the issues are aspects of the conceptual dimension of knowledge. Routinised (often inappropriate) choice of techniques is an epistemological matter. Ignoring literature, secondary data and difficult sources are informational failures.

A three-dimensional approach can also be used to improve practice. As argued above, NGDOs are often well placed to access different conceptual perspectives and different understandings of a situation, having links at many different levels, from grassroots to international institutions, and across many sectors (policy, scientific, religious, academic, etc.). These relationships are powerful knowledge resources. Equally important for conceptual development is the need for self-awareness and a critical attitude. All institutions can succumb to groupthink, and the values and convictions of third-sector organisations can, as in the public or private sector, be antithetical to knowledge.

At an epistemic level, too, NGDOs can benefit from a more critical approach. Limited time, skills and resources mean that academic standards of data collection, analysis and corroboration may be unattainable (Thomas et al., 1998). Nevertheless, there is a need for rigour in obtaining and interpreting evidence, and for fully-fledged investigations into causal relationships ('how and why things happen as well as what has happened') (Thomas et al., 1998). Methodological diversity can be a strength of NGDOs (Fowler, 2000) but researchers under pressure may apply techniques unthinkingly and in over-standardised ways, yielding superficial results (Thomas et al., 1998).

Compared with achieving high conceptual and epistemic standards, improving informational status is generally relatively straightforward. Of course, there may be crucial information that an organisation cannot get hold of or information that it would like to disseminate but lacks resources to publish, but these are the kinds of problems third-sector organisations typically excel at resolving, requiring practical if unconventional solutions.

Knowledge is not just a resource within NGDOs; for many it is part of the aim as well as the method, product as well as process – for example, when the organisational aims include educating beneficiaries, the public or policy makers. Improving knowledge status in such cases does not mean just achieving internal conceptual, epistemic and informational development; it may mean a vast range of activities, including providing information to outsiders such as clients, the media or other NGDOs; facilitating dialogue in the local community or among development agencies; or bringing about conceptual shifts in policy-makers, the public or international agencies such as the World Bank. A number of writers have pointed out the limited successes actually achieved by NGDOs in influencing policy as opposed to carrying out projects (Edwards and Hulme, 1995; Edwards, 1997; Madon, 1999). Such activities themselves demand specialised knowledge and skills.

For a variety of reasons, then, if NGDOs are to be effective knowledge organisations far more attention needs to be paid to developing core informational, epistemic and conceptual capabilities. It is often suggested that ICT and, particularly in the context of globalisation, computer networks, have a crucial role to play. The final section of this paper shows how here too the three-dimensional approach can help to delineate some key issues.

4 NETWORKS FOR KNOWLEDGE

The use of ICT to support knowledge activities in NGDOs has been much discussed, particularly as professional management practices have entered the sector, bringing with them the language and tools of knowledge management, knowledge sharing, communities of practice and learning organisations (see for example Hunt (2000) and Van der Velden (2002)). All the associated technologies have knowledge strengths and weaknesses that a three-dimensional analysis can help to identify: databases, for example, may be informationally effective, but they assume preexisting and unproblematic conceptual categories. They also suffer from a serious epistemic defect in that originators and receivers may know virtually nothing about each other (see Goldman (1999) for a philosophical account of the importance of mutual knowledge). Originators may have little incentive to produce high-quality content since features such as anonymity, separation in time and place from receivers, lack of emotional and social contact with receivers all remove important incentives. This paper is not concerned with ICT generally, however, but specifically with networks.

Even before Internet access became widely available to southern NGDOs writers such as Brown, started to show an increasing awareness of the importance of networks, describing 'linkage indicators' as a crucial aspect of evaluating community development (Brown, 1991; Madon and Sahay, 2002). From the early 1990s researchers began to recognise the networking potential of the Internet and other ICT for facilitating connections between organisations and for organising collective action (Annis, 1992; Salamon, 1995). Whether electronically based or not, networking aspects of NGDO activity are now stressed by many researchers: Madon and Sahay, for example, speak of intermediary NGDOs as information 'hubs' having both vertical linkages to government and grassroots, and lateral connections with other NGDOs and the media (Madon and Sahay, 2002). Madon highlights the potential for learning in networks, which enable organisations to benefit from outsiders and to draw on extended resources beyond themselves (Madon, 1999). Nath has developed what he calls a 'knowledge networking' approach to development founded on an Internet-based model (Nath, 2000). And Castells has proposed a view of networks as inherently democratic, transferring information, knowledge and power from the centre to the periphery, and bringing about structural social change (Castells, 1997) – as when the Zapatistas were able to use the Internet to mobilise worldwide support against repression by the Mexican government. Other authors identify networks and the Internet as having been instrumental in the growing importance of NGOs and a 'levelling factor' in relations between NGOs and the business and state sectors (Mathews, 1997; Bray, 1998; Winter and Steger, 1998; Heap, 2000).

At the same time, a number of intermediary and capacity-building organisations have emerged offering specialist Internet services and products to NGDOs. The Association for Progressive Communications, one of the first and still one of the largest, has been joined by many others such as One World and Kabissa, and even some very small-scale local initiatives such as Kznaidslink which offers a web presence to AIDS groups in the KwaZulu-Natal province of South Africa. Organisations such as IDRC and Bellanet have long hosted electronic discussion groups on development topics, and most international development conferences now run parallel electronic conferences, permitting otherwise excluded voices to be heard. Extensive technical training is also available online and, to a lesser extent, training in more abstract informational skills, such as the Distance-Learning Course for Online Efficiency run for NGOs by Dr Burkhard Luber of The Threshold Foundation in January-April 2002.

Clearly the Internet is vitally important to NGDOs for many reasons, and it might seem as if there were a relatively straightforward equation between networking technology and knowledge gains[1]. In fact, the situation is more complex, and there are many factors that compromise the Internet's knowledge-supporting capabilities. Goldman (1999) diagnoses weaknesses at informational, epistemic and conceptual levels that many users will recognise: vast, uncatalogued resources; unreliable and even deliberately deceptive communications (witness the rapidity with which false virus warning circulate); a tendency for networks and electronic forums to become introverted, reinforcing prejudices rather than opening up spaces for critical argumentation; loosening of the social constraints on truth-telling that operate in face-to-face communication; anonymity, leading to reduced accountability; lack of meta-information by which to judge accuracy and integrity of information; hidden commercial – and other – interests and motivations; lack of editorial controls and low barriers to entry; hidden selectivity of hypertext linkages.

Many of Goldman's theoretically derived concerns are reflected in empirical studies of Internet use among NGOs. Madon diagnoses widespread problems of information overload and underdeveloped systems for information seeking, storing, transferring and disseminating (Madon, 1999), and Edwards (1994) points to the negative impact of large quantities of email adding to the already severe workload of hard-pushed NGO employees. Surman (2001) finds NGOs experiencing multiple difficulties in making effective use of the Internet, and Song (1999) argues that while the Internet increases research efficiency and collaboration it also increases workload and resource requirements.

Of course, the fact that a technology is not perfect is not an argument for rejection but rather for users acquiring greater expertise and skill. In the case of knowledge and networking tools, the expertise and skills required are complex – partly technical but to a large extent also cognitive and communicative. There is evidence, however, that even on a technical level many NGDOs are only gradually moving beyond basic tools such as email and static web pages and very few, even among northern agencies, are making extensive use of more strategic applications such as online publishing, lobbying and discussion forums (Saxton and Game, 2001). Knowledge about the technology and how to use it is lacking, and even

[1] Edwards (1994), for example, speaks of the Internet bringing about 'a paradigm shift in storage, retrieval, handling and dissemination of information.'

when it exists there is often a lack of technical skill to implement it (Surman, 2001). We have already seen that higher-level skills such as research, analysis and conceptual development are extremely demanding, even from a conventional perspective. Manji et al. (1999) found that many NGOs recognise a need for training in knowledge aspects of networking technology such as using the Internet for research, and they diagnosed a need for applications for democratic participation. However, with a few exceptions such as information seeking and research networks we have hardly begun to discuss the use of Internet technology in supporting the knowledge-based activities that lie at the heart of the new roles being adopted by NGDOs.

Goldman (1999) suggests some starting places: better use of search engines and intelligent agents; reliance on named and accredited sources and websites; more use of email and other interactive applications (allowing epistemically valuable contact between communicators and receivers); hypertext to help structure an argument or link to supporting evidence; promoting open and active discussion groups where the seeds of conceptual change, not just information exchange, may be sown[2]. These ideas of course need specialist application to make them relevant to the work of NGDOs, a task that is likely to require collaboration between academic researchers, capacity-building intermediaries, and NGDOs working in the field. It is perhaps asking a lot, but without integrating informational, epistemic and conceptual capacities into technological programmes NGDOs risk duplicating familiar errors and failures, albeit with advanced technology.

Perhaps the most fundamental problem in developing an integrated approach to networks (as knowledge and as technology) is lack of recognition that there *is* a problem: in many organisations the connection between networking and achieving social goals remains (Surman, 1999). Making those connections explicit is a job for technologists, epistemologists and development theorists to work on together.

5 CONCLUSION

The knowledge and informational aspects of development work, and specifically of activities undertaken by NGDOs are receiving unprecedented attention. In the realm of service provision but even more so in their growing advocacy and capacity-building roles, NGDOs depend vitally on information

[2] In similar vein Bray (1998) gives the example of an Internet discussion group on NGO-business relations hosted by Newcastle University www.mailbase.ac.uk/lists/business-ngo-relations, and Shell, which has used its website to acknowledge and openly discuss controversial issues.

and knowledge, both for their own internal processes and functioning, and as part of their explicit external goals. This poses a challenge – the need for new skills, for research and sometimes for high-level conceptual thought. In addition, knowledge goals may be external in that the NGDO may aim to change the external information environment, to bring about a shift in someone else's conceptual framework, or to change donor-driven methods of evaluation. It is not enough for NGDOs to be good at finding things out themselves; they have to be able to communicate insights to others and sometimes to persuade others to change their views – perhaps in opposition to vested interests and long-standing assumptions. Sometimes this may be a relatively unproblematic matter of information provision within existing categories and arrived at by well laid down methods. In other cases, it may be much more demanding, and the categories and methods may themselves be under investigation. A whole range of new skills and expertise may be required, spanning everything from information acquisition, storage and dissemination, to methodological evaluation and conceptual development.

At the same time that attention is focusing on their knowledge roles, NGDOs are coming under criticism for failures and errors. It is vitally important to understand why such failures happen, and to recognise the enormous demands that knowledge roles impose on organisations and individuals. NGDOs generally work under severe internal and external constraints that can mitigate against effective knowledge development. I have tried to show how an analytical approach that separates out various key dimensions of knowledge can help us to diagnose failings and pressure points – and, I hope, suggest ways of improvement. Not all the remedies lie in NGDOs themselves. Donors and partner organisations also have a role to play – for example by being more open to alternative, locally generated and locally meaningful forms of monitoring and evaluation. This is not an argument to dispense with rigour or objectivity but rather to look at different methods and evaluate them open-mindedly as possible routes to knowledge.

As well as taking on new roles, NGDOs are finding their structures and forms of organisation and association changing. Networking models are gaining increasing recognition, driven partly by Internet use and partly by the growing emphasis on information and communication work. A whole layer of intermediary and northern NGOs now exists with the aim of enabling southern, community and grassroots groups to make more sophisticated use of networking technologies and applications. Networks, however, are defined by users and their actions, not by technology alone. A three-dimensional analysis helps to show why and highlights the need to integrate technological capacity-building into a broader knowledge project involving informational, epistemic and conceptual development. So far, the amount of research, help and support available for this is negligible by comparison with the attention paid to technology itself. It is time to change this, and to start paying more detailed attention to the complex intangible dimensions of knowledge.

REFERENCES

Annis, S. "Evolving Connectedness among Environmental Groups and Grassroots Organisations in Protected Areas of Central America", *World Development* (20:4), 1992, pp. 587-95.

Bray, J. "Web Wars; NGOs, Companies and Governments in an Internet-connected World", *Greener Management International*, (24), 1998, pp. 115-29.

Brett, E. A. "Voluntary Agencies as Development Organisations: Theorizing the Problems of Efficiency and Accountability", *Development and Change*, (24), 1993, pp. 269-303.

Brown,. D. "Methodological Considerations in the Evaluation of Social Development Programmes - an Alternative Approach", *Community Development Journal*, (26:4), 1991, pp. 259-265.

Castells, M. *The Power of Identity*, Oxford: Blackwell, 1997.

Chambers, R. "All Power Deceives", *IDS Bulletin*, (25:2), 1994, pp. 14-26.

Chambers, R. *Whose Reality Counts? Putting the First Last?* London: Intermediate Technology Publications, 1997.

Clark, J. "The State, Public Participation and the Voluntary Sector", *World Development* (23:4), 1995, pp. 593-601.

Crane, T. *The Mechanical Mind: A Philosophical Introduction to Minds, Machines And Mental Representation*, London: Penguin, 1995.

Edwards, M. "NGOs in the Age of Information", *IDS Bulletin*, (25:2), 1994, pp. 117-125.

Edwards, M. "Organisational Learning in NGOs: What Have We Learned?", *Public Administration and Development*, (17:2), 1997, pp. 235-50.

Edwards, M. *Future Positive: International Cooperation in the Twenty-First Century*, London: Earthscan, 1999.

Edwards, M. and Hulme, D. "Scaling up the Development Impact of NGOs: Concepts and Experiences", *Making a Difference: NGOs and Development in a Changing World*. M. Edwards and D. Hulme. London: Earthscan, 1992.

Edwards, M. and Hulme, D. "Non-Governmental Organizations: Performance and Accountability – Introduction and Overview", *Non-Governmental Organizations: Performance and Accountability - Beyond the Magic Bullet*. M. Edwards and D. Hulme. London: Earthscan, 1995.

Edwards, M., Hulme, D., et al. *NGOs in a Global Future: Marrying Local Delivery to Worldwide Leverage*. NGOs in a Global Future, University of Birmingham. (1999).

Escobar, A. "Imagining a Post-Development Era", *Power of Development*. J. Crush. London: Routledge, 1995.

Fleming, S. "E-Government Models and NGO Cooperation: Civil Society's Use of Information and Communication Technologies to Strengthen Public Participation", Available at: http://www.bytesforall.org/Egovernance/html/egov_models.htm

Fowler, A. "Civil Society, NGDOs and Social Development: Changing the Rules of the Game", *Geneva 2000 Occasional Papers*, (1), Geneva, UNRISD, 2000.

Fox, J. "Democratic Rural Development: Leadership Accountability in Regional Peasant Organizations", *Development and Change*, (23:2), 1992, pp. 1-36.

Goldman, A. *Knowledge in a Social World*. Oxford: Oxford University Press, 1999.

Grimwood-Jones, D. and Simmons, S. Eds. *Information Management in the Voluntary Sector*, London: ASLIB, 1998.

Heap, S. *NGOs Engaging with Business: A World of Difference and a Difference to the World*, Oxford: INTRAC, 2000.

Holloway, R. *Freeing the Citizen's Sector from Global Paradigms. And Trying to Get a Grip on the Moral High Ground*, NGOs in a Global Future, University of Birmingham, 1999.

Hunt. P., "Knowledge Management: Implications and Applications for Development Organisations." *Report of Workshop Co-Organised by Bellanet International Secretariat, the Benton Foundation, the Canadian Development Agency*, The International Development Research Centre, Benton Foundation: Washington. 2000. Available at: http://www.bellanet.org/km.

Johnstone, J. "Knowledge Perspectives on ICT and Development: what the theory of knowledge can add", paper delivered at *ICTs and Development: New Opportunities, Perspectives & Challenges*, conference of International Federation of Information Processing Working Group 9.4, Bangalore, India, May 2002. Forthcoming publication in *The Digital Challenge: Information Technology in the Development Context*, S. Krishna, S. and Madon, S. (Eds.) Ashgate.

Madon, S. "International NGOs: Networking, Information Flows and Learning", *Journal of Strategic Information Systems*, (8:3), 1999, pp. 251-261.

Madon, S. and Sahay, S. "An Information-Based Model of NGO Mediation for the Empowerment of Slum Dwellers in Bangalore", *The Information Society*, (18:1), 2002.

Manji, F., Jaffer, M. and Njuguna, E. "Enhancing the Capacity of Human Rights and Advocacy Organisations in Southern Africa", IDRC, 1999.

Mathews, J. T. "The Age of Nonstate Actors", *Foreign Affairs*, (76:1), 1997, pp. 50-66.

Meyer, C. "The Political Economy of NGOs and Information Sharing", *World Development*, (25:7), 1997, pp. 1127-1140.

Mikkelsen, B. *Methods for Development Work and Research*, London: Sage Publications, 1995.

Mundy, P. and Compton, L. "Indigenous Communication and Indigenous Knowledge", *The Cultural Dimension of Development*, D. Warren, L. Slikkerveer and D. Brokensha. London: Intermediate Technology Publications, 1995.

Murphy, D. F. and Bendell, J. "The Politics of Corporate Environmentalism: Civil, Legal or Self-Compliance for Sustainable Development?", *Business Responsibility for Environmental Protection in Developing Countries*, Heredia, Costa Rica, UNRISD and Universidad Nacional, 1997.

Nath, V. "Heralding ICT Enabled Knowledge Societies", *Sustainable Development Networking Programme*, 2000.

Pearce, J. "Development, NGOs, and Civil Society: The Debate and its Future", *Development, NGOs, and Civil Society*, D. Eade. Oxford: Oxfam, 2000.

Powell, M. *Information Management for Development Organisations*, Oxford: Oxfam, 1999.

Putnam, H. *Representation and reality*, Cambridge, Mass.: MIT Press, 1988.

Rawcliffe, P. *Environmental pressure Groups in Transition*. Manchester: Manchester University Press, 1998.

Salamon, L. M. *Partners in Public Service*. Baltimore: Johns Hopkins University Press, 1995.

Saxton, J. and Game, S. "Virtual Promise: Are Charities Making the Most of the Internet Revolution?", *London: Third Sector*, 2001.

Song, S. "Guidelines on the Use of Electronic Networking to Facilitate Regional or Global Research Networks", IDRC, 1999.

Surman, M. "Making the Internet Work for Civil Society: A Product and Service Manifesto for APC", APC, 1999.

Surman, M. "From Access to Applications: How the Voluntary Sector is Using the Internet." *Government of Ontario*, 2001.

Thomas, A., Chataway, J. and Wuyts, M. (Eds.) *Finding Out Fast*, London: Sage Publications in Association with The Open University, 1998.

Turner, R. "IT Utilisation by Non-Profits and Community Groups", Jarboe, K. Extending the information revolution: a white paper on policies for prosperity and security, 2002, *Athena Alliance*. Available at: www.athenaalliance.org/pdf/AA202_WhitePaper.pdf.

Van der Velden, M. "Knowledge Facts, Knowledge Fiction: The Role of ICTs in Knowledge Management for Development", *Journal of International Development*, (14), 2002, pp. 25-37.

Winter, M. and Steger, U. *Managing Outside Pressure. Strategies for Preventing Corporate Disasters*, Chichester: John Wiley & Sons, 1998.

About the Author

Justine Johnstone lectures Humanities Information Technology at London Metropolitan University and is completing a PhD in Information Systems at the London School of Economics. She holds a Masters in Philosophy from London University. She grew up in South Africa, where she now conducts research into Internet use among AIDS NGOs. Her theoretical research interests focus on the application of contemporary analytical philosophy to issues in information systems, particularly in the fields of knowledge, ethics and justice. She can be contacted by e-mail at j.e.johnstone@lse.ac.uk, or through her home page at http://is.lse.ac.uk/homepages/johnstone.

20

LESS CYBER, MORE CAFÉ

Design Implications for Easing the Digital Divide with Locally Social Cyber Cafes

Tony Salvador, John W. Sherry and Alvaro E. Urrutia
Intel Corporation, USA

Abstract: Shared models access to information and communication technologies (ICTs), e.g. telecenters and cyber cafes, have been considered as one means to reduce the digital divide. Cyber cafes in particular have proliferated in some locales yet not in others with apparently similar characteristics. This paper questions the prevailing emphasis on the "cyber" characteristics of access, e.g. computing and internet access as is currently known, and attempts to refocus the conversation by considering computing and access in the context of the "café", e.g. as public life in the sense of Habermas. This analysis is based on extant literature and direct ethnographic research in several public places in six countries. We offer design perspectives based on a reflection of "third places" as inspiration for appropriate innovation in the provision of computing and communications.

Keywords: Cyber café, digital divide, third place, ICT

1 INTRODUCTION AND JUSTIFICATION

While it is estimated that about 10% of the planet has access to information and communication technologies (ICTs), the remaining 90% do not, creating what's been commonly called the "digital divide". Despite the unifying label, the digital divide is not a uniform population; rather, there are various gradations of both access and the capability to derive benefits from such access that cross a wide swath of economic and social aspects (Castells, 2001). It seems that "desperately poor people" (often in rural villages) exemplify the digital divide to the exclusion of others, for example: "The 'digital divide' is the subject of almost daily reports… But since when have desperately poor people had an 'urgent need' for a computer?" (Keniston,

2003). While there is certainly a large population of the desperately poor, there also appears to be a significant population not quite so poor, yet without ICT access – a population that might be considered the "world's middle class" (Milanovic and Yitzhaki, 2001). The current paper considers possibilities for ICTs access within this latter context with the purpose of easing the digital divide.

Shared models of ICT access are often imagined when considering ICT access for the digital divide. In contrast to personal ownership of equipment and personal access to infrastructure, shared access models, e.g. telecenters, variously shift the physical, economic and social burdens of access from the individual to businesses, communities and governments (Pedrelli, 2001; Proenza et al., 2001). "Telecenter" is a general term applied to any establishment that provides public access to ICTs. However, as the term "digital divide" obscures difference within it, "telecenter" also obscures critical differences among the various forms of shared access. For example, at a high level, there appears to be a growing, albeit still rough, consensus around three categories of shared access: Telecenters, Cyber Cafes and Information Access Points (Colle and Yonggong, 2001).

Telecenters, per se, take as defining characteristics an emphasis on community service and development in addition to the provision of ICT access (Proenza, 2001; Gáspár, 2001). Globally, there are numerous attempts to establish telecenters as a means of erasing the digital divide, often but not always, in the most economically disadvantaged and geographically rural areas. Telecenters of this ilk operate mostly as not-for-profit organizations relying on various sources of funds, including governments, foundations, non-governmental organizations in addition to fees for use. Some examples that have met with varying levels of robustness include, Australia (Short, 2001), Brazil (Kyle, 2001), Egypt (Hashem, 2001), Hungary (Gáspár, 2001), India (Thamizoli and Balasubramanian, 2001) and South Africa (Benjamin, 2001).

Cyber cafes represent a second common model of shared access that may have relevance for easing the digital divide and which have, in recent years, expanded from (the almost mythical) first cyber café[1] to the uncounted thousands that are reputed to exist today (Ivala, 1999). A typical cyber café is composed of between five and twenty computers with varying levels of internet access, e.g. dial-up or broadband, in addition to providing other business services such as internet telephony, copying, faxing, etc. (Liff et al., 1998). They are particularly attractive in that they represent economic opportunity for the café owner/operator (Holmes, 2001).

---

[1] Reported to be in 1984: http://www.theiac.org/. It could be Electronic Café International: http://www.ecafe.com/about.html.

In addition to cyber cafes that support people temporarily disconnected from their home computers, e.g. London commuters, many cyber cafes serve local populations in developing nations. For example, they are popular in places like Ecuador and Peru. However, they are considerably less prevalent in Brazil or Mexico, where those that do exist seem more for tourists and professionals (cf., Kelly et al., 2001). Furthermore, for a variety of reasons, there are very few outside of urban environments in town or rural settings.

As currently conceived, it's not difficult to understand key reasons why cyber cafes have remained tethered to their urban, this-side-of-the-divide environs. Among these are available power and communications infrastructure, sufficient investment and maintenance capital, and sufficient expertise and access to maintain the equipment. Financing is also important. Unlike telecenters that rely on a variety of funding sources, cyber cafes are most often "for-profit" "fee for service" establishments and must sustain a profit. People using cyber cafes for access must therefore have sufficient discretionary funds and as a result cyber cafes are appear often to be not available to the "desperately poor".

Though cyber cafes can offer such access, they must meet their financial requirements. To meet these needs, cyber cafes have become remarkably similar in design following a tight range of business models (Ivala, 1999). This "one size fits all" requirement seems to artificially limit the potential with which cyber cafes can help reduce the digital divide – whether in the most disadvantaged rural settings or, as is more likely in the shorter term, to somewhat less disadvantaged urban settings.

2 RESEARCH PROBLEM AND OBJECTIVES

We explore how the concept of cyber cafes might expand their role of providing ICT access by de-emphasizing the "cyber" part and re-emphasizing and reenergizing the "café" part in locally valid ways. This paper attempts to expand the de facto description of cyber cafes by recasting their purpose from that of "access centre" to that of locally situated "third place", i.e., a public, social gathering place (Oldenburg, 1999; Laurier et al., 2001). Places of public congregation, such as bars, coffee shops, news stands, etc., already have the advantage of being attractive and frequented. Recasting what 'access' means in these contexts shifts not only the kinds of access that might be possible and perhaps increases the numbers of people who might gain access, but it can also stimulate the development of alternative technologies and business models that are possibly less capital intensive as well as being less expensive to operate over the long term.

3 RESEARCH SETTING, DESIGN AND METHODS

As part of the People and Practices Group at Intel Corporation, we use ethnographic techniques to gain insight into the lives of people, places and systems of interaction in various places around the world. The work in this paper falls under the auspices of a multi-year project intended specifically to identify potential applications, needs, desires, appropriate technologies and business models relevant to a large corporation for bridging the digital divide. A few years ago, we began the study of the current and potential role of public places and technology in the daily lives of people in various cities and towns around the world; we first focused on traditional "cafes" or coffee shops. More recently, we've been applying these same ethnographic techniques of interviewing, observation and participant observation in the study of cyber cafes in cities worldwide. Based on our work in both domains, we analyse existing patterns and synthesize potentially new patterns as a means of creating technology and business models relevant for crossing the digital divide. In this paper, we take our cues from both cyber cafes and traditional cafes in Brazil, Ecuador, Korea, Peru, Spain and the United States described in the following section.

3.1 Sites Examined

3.1.1 *Cabinas Publicas* (Peru)

The *cabinas publicas* of Peru, in addition to being ubiquitous, small, self-financed businesses (with approximately a one year return on capital), are marked in large part by the variety of services they offer, the variety of services used and the relative sophistication of their user population as a whole. A *cabina* typically has 10 to 20 machines, with relatively few exceeding 20. The largest we saw was about 40 and it was the largest anyone heard about. Services include internet phone service ("voice over IP") to a wide range of countries, email, chat, pornography, business services and media downloading. Individual users also store personal content in password protected, shared and zipped folders and return to that particular *cabina* to access their material. In sum, they are a resource providing support for individuals' entertainment, business and communication needs. They are also marked in that they fall largely within the "informal economy" of Peru, which offers advantages that enable them to exist.[2] As a result, they remain

---

[2] Informal economy refers to a wide range of characteristics all of which involve a dissociation (or lack of association) with certain relevant aspects of the extant national economy, e.g., taxes.

small, family owned businesses, disconnected from one another and mostly from the "formal" economy. They are typically housed in the minimum amount of necessary "available space" with little designed ambience. *Cabinas* were initiated by a group (*Red Cientifica de Peru*, RCP) with a goal to develop the country and install a network for internet access; they taught the general public how to set-up and run *cabinas* through free classes. There are an estimated 2500-3500 cabinas throughout Peru (population 27 million), with approximately half in Lima and the remainder distributed over other large cities (Holmes, 2001; Proenza, 2001).

3.1.2 *PC Baangs* (Korea)

Although Korean *PC Baangs* offer services similar to those of *cabinas*, the PC *Baangs* are situated differently. For one, it's not unreasonable for a PC *Baang* to have 50 to 100 PCs. For many users, they are already social destinations in and of themselves for couples and groups, and the trend seems to be more in that direction. Whereas many people use the *cabinas* as individuals, PC *Baangs* are at their best when groups of kids (e.g. school age) or teens or young adults out for the evening come in and play networked games with each other. It's akin to a computer "playground". The spaces themselves are often purposefully designed to be evocative and to resonate with the clients, e.g. use of structure, lighting, colour, layout, etc., to connote a space ship or something similar. They are places one goes for diversion, after school, before they go out drinking or afterwards to sober up. They are, in many ways, an extension of the home. They are funded through use-fees, but may be owned by small business people or even by large corporations and they may be franchises in that they may have formal arrangements with game companies to reduce license fees, etc. They have far more and more formal associations with large business and government and are thus, explicitly a part of the formal economy. There are an estimated 26,000 through out Korea (population 47 million), in most all cities with the exception of very rural or remote (difficult to access) areas (Herz, 2002).

3.1.3 *Cafés de Internet* (Ecuador)

The situation in Ecuador is a little like Peru, albeit with relatively fewer services and with less and less up to date equipment. The *cafes de internet* in Ecuador are almost completely ad hoc arrangements of equipment, space and use; there was no RCP equivalent in Ecuador. As a result, they are smaller (3 to 10 machines) and less well equipped (various vintages and machines comprised of the working parts from several machines). They also appear to be used by a wide variety of people, including older people, but more narrowly for email communication (and some chatting) and music downloads. At the time, there were about 50 of these small cyber cafes in Cuenca (population 350,000, with no guesses for the number throughout Ecuador, population 11 million).

3.1.4 Public Access (Rio de Janeiro, Brazil)

There were relatively few public access points in Rio de Janeiro (population, 12 million) that were at all obvious to us, and even fewer cyber cafes. Those we did see were incorporated into other gathering places, e.g. just off the lobby of a theatre or as parts of gaming centres (video game arcades) in upscale, popular malls. Access of this nature was primarily for gaming and for passing time while doing something else, for example, while you're in the mall with your parents, or waiting to be let into the cinema, that is, while being social. In Brazil, being inside is not valued or experienced in the same way as in many other places in the world. For example, cafes and juice bars are popular and prevalent, where people stand to chat and drink fresh fruit juices, coffee or other drinks. On the other hand, CDI (Centre for the Democratization of Information) has crossed into the digital divide, arranging with local neighbourhood associations to insert small computer classrooms making use of donated and used equipment. The result is extremely low capital costs. While they serve an education, training and civic function, most of these classrooms are not connected to the internet and are not generally publicly accessible other than during class times.

3.1.5 *Bars* (Spain)

Confusingly, the Spanish word "bar" does not mean what the English word "bar" means. There are numerous types of "bars" in Spain. In Barcelona alone, there are at least eight different types of bars, including "after hours" bars open from about 1am to 6 or 7am, themed and high-styled "design bars", "yuppie bars" with long time regular business people, journalists, etc., tourist bars which are more like discos and "all day" bars, which are the staple of neighbourhood public life in Andalusia, Catalonia, the Basque Country, Asturias, Galicia and everywhere in between. Over the course of the day and into the night, the bar undergoes a transformation in its nature including what's served, the amounts served, the atmosphere, the clientele and the types of social interaction. These bars are heavily integrated into the daily lives of people (including all children), who frequent them (though not necessarily the same one) two, three or more times a day – for example, in the morning for a "*café con leche*", at midday for a "snack" or *tapa*" and in the evening for *cañas* or draft beers, the portions of the latter determined by the barkeep. Other features make these bars very comfortable places, for example, the client is primarily responsible for maintaining his or her tab and there are rarely any queues in which one must wait one's turn. This type of bar is less of a destination, per se, and more of an extension of one's living room, "a third place" where people who know each other, including families, see each other on a daily basis. The types of interactions will also change depending not only on time of day, but with particular events, such as soccer (*fútbol*) on TV.

3.1.6 Coffee Time (USA)

Coffee Time is a locally owned coffee shop in Portland, Oregon, USA. It is, in many ways, unique, and one could argue, the antithesis of a "franchise" shop. For example, at Coffee Time, there is a large population of younger, "edgier" people, whose attire of black garb and chains might be somewhat off-putting to the same 65 year old retired housewife who visits the shop in Pella Iowa. On the other hand, the poetry readings on certain weekday nights attract a more "artsy" crowd. The furniture is also unique; old brown leather couches and other "used" furniture dominate the lounging rooms. The walls, lighting, flooring, layout, signage, and all the other accoutrements and products, e.g. vegetarian (in fact, 'vegan') chocolate chip cookies, also differ from the franchise shops. Moreover, the shop encourages people to spend time there in ways that gather no fee. Coffee Time keeps a stack of board games including chess (for which they solicit additional pieces from their customers) among other items like an internet terminal, some reading materials and a host of customer placed postings. Coffee Time also offers outdoor heating under the awning, something one might only expect in more upscale locales. Coffee Time, like other local shops we've studied, is, in many ways, co-constructed by the owner and the clientele in a sort of informal interaction, unlike the franchise shops which are corporately constructed and consciously replicated. Local shops, like Coffee Time, are more a place to be and less a place simply to retrieve a cup of coffee.

4 DISCUSSION

4.1 Third Places

Traditional cafes have long been recognized as "social spaces" for conversation and general conviviality within communities (Habermas, 1991). Traditional cafes offer far more than coffee and assorted drinks; they also provide a place that permits the emergence of sociality. They offer a place where people can gather, talk, know each other, express community and "hang-out".

Cyber cafes seem to emphasise the "cyber" in terms of "functionality", "efficiency" and "effectiveness" even though they borrow the term "café". One way to think differently about cyber cafes is to consider them fulfilling a role more akin to traditional cafes or, "third places" (Oldenburg, 1999), i.e. as a part of public life. Put another way, we consider the characteristics of traditional cafes and imagine maintaining and/or expanding these characteristics while providing ICT access. One outcome might be a different category of "cyber café" that better serves different populations. Another outcome could be various models for introducing cyber cafes in a town or village and evolving them appropriately.

One pattern worth mentioning here is that whereas the locally owned bars and cafes appear to demonstrate continuous adaptive response in conjunction with their clients, cyber cafes appear less malleable. For example, in Spain and Portland the sites seem to have adapted along a wider variety of "dimensions of daily life" insinuating themselves more deeply into the lives of their patrons – and vice versa – offering everything from games to vegan cookies, morning coffee and afternoon beer, a place to sit and talk, to watching *fútbol* on TV, only some of which actually requires the patron's money. In contrast, the cyber cafes regardless of location appear to offer very similar capabilities, network gaming and internet access. While there are some variations, e.g. internet telephony in Peru, they are remarkably similar. It's not necessarily bad; it's just stifling.

Traditional cafes also appear to have perfected the art of adapting their shops to local conditions by offering not only their explicit products but by providing a space for the emergence of sociality, something they cannot "sell", e.g. watching a sporting event on TV. Thus, at one level, they are the same – they all sell coffee (albeit some better than others) and they all offer small snacks or pastries, etc. However, at the level of actually creating and running the business, they are each quite different. And it's important that they're different because they each attract different clientele depending on the whole set of conditions and characteristics. It's exactly these variations that seem less prevalent among cyber cafes. It's exactly these variations among local, social establishments that are of prime relevance to a more differentiated incorporation and provision of ICT access.

4.2 Examples Emphasising "Café"

In Rio de Janeiro, Brazil, we noted that many people spend significant amounts of time out of doors rather than inside. There were few cyber cafes of any sort, and with one exception in a mall near a wealthy area, publicly accessible internet connected computers were not being used – even in the midst of large numbers of people. However, juice bars and snack bars – which are exactly that, bars where people eat/drink standing up right off the sidewalk – were frequently occupied. The question of course is: What would a "juice bar offering ICT access" be like? That is, in what way can computing and internet technologies be introduced into this existing third place such that it enhances the "café-nature" of the juice bar, which establishing an element of "cyber"?

It certainly wouldn't be a room full of computers, one per person. Perhaps it might be a form of "local area broadcast" with one screen showing content of particular interest to the regular patrons of that juice bar. Or perhaps it might be small flat (but bright) touch screens attached to the bar with local newspapers or magazines that people can pick up, scan for an interesting publication and read an article or, in Brazil particularly, watch snippets of web cast TV news as it unfolds over the course of the day. Of course, these suggest different support technologies as well. For example, small screens on the bar counters might receive information wirelessly from a backroom "server" using the appropriate wireless protocol. Perhaps people coming in can also receive information to their own personal devices as well, e.g. to a personal digital assistant or to a cell phone.

For contrast, in Seoul, almost all gathering places were indoors, off the street. It was very common to enter the side door of a nondescript building, climb a narrow flight of stairs to a fourth floor coffee shop or *PC Baang*. In their early instantiations, they offered simply internet access but quickly became gaming centres in neighbourhoods in almost every city in Korea. In addition to the highest broadband penetration rates in the world (~60% of Korean households), there are an estimated 26,000 *PC Baangs* in Seoul alone (population 10 million). These cyber cafes have over time, increasingly becoming venues for groups of people to come and "hang out". Many cyber café's offer conscious styling and atmosphere, e.g. "space age", comfortable modern chairs and a variety of arrangements designed to permit face-to-face and verbal interaction simultaneously with game play.

Another *PC Baang* we visited had "love seats" in front of several computers for couples to sit together and play games, chat or read email. This same one had a movie viewing area as well. Their most popular times are when school lets out and evenings when groups of people come by. One *PC Baang*, which has recently seen business decline, reinforced this notion of cyber café as "venue" by declaring that the most successful *PC Baangs* were ones located in areas near other venues, such as singing rooms or bars, i.e. places where people go when they go "out".

Thinking about the simple concept of a "love seat", where couples can sit together, hints at different sorts of technological interventions. As a simple example, playing games that require two people with two separate simultaneous inputs to the computer; these were not available. Or perhaps, smaller screens with clever input devices on table tops such that a couple can sit more comfortably with a coffee and a dessert (popular in Seoul and Brazil) while engaging in a lightweight diversion. Alternately, kids in a small village might be better able to play a game together with multiple inputs and one larger screen. A few adults might gather together at various times in the day to learn about a new farming technique of particular interest to local towns or villages (the video for which was downloaded asynchronously overnight). Additionally, the screen might also serve special events, such as "movie night" or more generally as a mini-cinema at appropriate times. Or, we imagined a projected image of a crossword puzzle on the wall of a café in Portland, Oregon, with a few specially tailored input devices that would permit people to causally make a contribution to the puzzle without disrupting the social nature of the place. Of course, these sorts of usages suggest different instantiations of basic technologies, which is, of course, the point, to imagine how technologies have to be recombined, improved or even invented.

By shifting the conversation from "business centre" to "third place" we change the nature of what "access" might mean. Each of these examples, the "juice bar cyber café", the "mini movie cyber café" and the "cross word cyber café", variously incorporates computing and access to the internet in a way that might conform to local sensibilities. None of these sites requires 10 or 20 PCs, network connectivity, broadband or even synchronous connectivity (perhaps taking advantage of store and forward technologies or night time downloads). On the other hand, the foundation laid in each just as well might over time offer additional capabilities as appropriate.

4.3 Practical Considerations

In actual practice, there are business and technology issues to consider. Perhaps the critical business factors from the proprietor's perspective are those of capital and operating costs, while out-of-pocket costs and associated benefits might be considered from the customer's view. As we've noted previously, capital costs are typically covered in two ways. For telecenters, capital is often raised as "donations" from NGOs, foundations and governments and for cyber cafes as "private funds", e.g. personal loans from family and friends. Operating costs in telecenters are often met through a variety of revenue sources, whereas in cyber cafes, customer use fees are the primary and often the only source of revenue. One possible advantage in thinking about ICTs and "third places" is that an appropriate level of ICT inclusion might represent a relatively small proportion of capital and operating costs while producing a disproportionately high rate of return.

Of course, the opposite is also a possibility. It's very difficult to assess the financial impact for either customers or proprietors without detailed knowledge of the specific case. In one case study, we were able to evaluate the finances of a typical *cabina publica*. One result of particular interest was the effect of internet telephony on the profitability of the *cabina*. The cost of an internet call was significantly less than the cost of a normal phone call from one's home, making internet telephony a fairly popular service. From the proprietor's perspective, the incremental capital requirement to add internet telephony to the already-existing *cabina* was small. Yet, the revenues extracted from telephony services were substantial in terms of incremental net income; the internet telephony service nearly doubled the profits of the *cabina*. As a stand-alone service, the internet telephony would not make sense from a return on capital perspective, but it made perfect sense as an "add-on" service to an existing asset base. While the bulk of the revenues in the *cabina* came from computer rentals for internet access, internet telephony made the difference between being a good overall business or not. This example clearly demonstrated the importance of adding relevant services and local content to the *cabina* business.

The argument in this paper is to consider incorporating access such as internet telephony in extant third-places, as an incremental or "add-on" service. The financial considerations are of course critically important. It is very difficult, however, to get a view on finances in a general way as one key element of third-places is their variability. Thus, financial considerations remain an open issue that must be resolved locally, not unlike the sort of access one might consider.

Inextricably bound with financial considerations, are technology considerations. On one level, it's easy to suggest reducing the price of technology as a solution. Another possibility is appropriately adapting the technology to suit the environment.

Again, these issues also remain open and require further research. Yet, various ideas suggest themselves as potential directions. For example, building far more modular systems such that basic components can be swapped in or out as required without the need to change whole systems is one approach. For example, *cabinas publicas* have widely adopted internet telephony as a major portion of their offerings. Internet telephony serves a real need in Peru. Perhaps it's also useful elsewhere. One might consider selling a system that offers only internet telephony, but which can also accommodate added functionality over time in a modular, building-block fashion.

A related approach would be to sell and build systems based on an initially diversified set of components, e.g. offering solar power, electrical power or hand-generated power, e.g. pedal power[3]. Street markets in some developing nations are already disassembling systems to salvage components for replacement in other systems, e.g. replacing cell phone screens if yours breaks by using an old screen from another phone.

In the *cabinas*, at any one time it is an individual using an individual computer as if it were, for that moment, personally owned. The "upgrade path" is to replace an old PC with a new one, or ten old PCs with ten new ones, which is burdensome. An alternative would be to offer technologies that rely on shared computing power, per se. For example, disassociating the processing from the storage, the monitors and keyboards can permit a different form of sharing, potentially reducing capital costs, while maintaining shared access.

Consider the "juice bar cyber café" example. There is presumably one source of compute power while there are various means of interaction, e.g. multiple displays on the counter. Over time, one might imagine evolving the cyber capabilities of the juice bar, an evolution that might well involve replacing parts of the system rather than the entire system (which is also harmonious with a notion of less waste). Upgrading the compute power, or upgrading the input and display devices could be independent events. An open question is how technologies can be designed more modularly to permit this form of mixing and matching of various parts to support a wider variety of applications that could be developed as localization intensifies, much akin to the localization of traditional cafes while better supporting sociality.

5 CONCLUSION

As currently conceived with the emphasis on "cyber", cyber cafes have limited potential to further extend access to computing and the internet given their infrastructure, business and usage requirements. Shifting the emphasis from "cyber" to "café" opens a space of possibilities suggesting various technical and business innovations that better conform to realities and desires of people's lives that have the possibility of extending the reach of information and communication technologies – even if in a limited extent. The current paper represents less a radical approach and more an evolutionary approach. More radical approaches, more research and more understanding are required to sensibly reach the deepest parts of the divide.

---

[3] cf., Jhai Foundation IT Village Project: http://www.jhai.org/jhai_remoteIT.html

One key benefit of thinking about smaller interventions in extant third-places is that more narrowly defined applications, locally considered, appear to be more harmonious with other conditions of these businesses such as thinner margins, lower revenues, lower per customer revenues, variable infrastructure, etc., that might not apply to a cyber café right in the *Plaka* of Athens. This conception of ICT introduction seems to better permit the ongoing co-construction of third-places in which the patrons and the owner symbiotically (even if implicitly) evolve the establishment over time. For example, it appears that one reason for the diversity among traditional cafes around the world is that they have evolved slowly, depending on customer interests (politics, sports, lotteries, etc.), current technologies (TV, radio, video games, etc.) and owner capital. Because they are third-places in which the customers have a stake, a co-evolutionary process seems to ameliorate the risks involved with the owner's capital expenditure.

It is the owners who take the financial risk. And to do so with new technologies is to increase the risk for obvious reasons of costs, complexity and uncertain benefits. Franchises offer an advantage in this regard by seeking similar environments in which to establish their shops such that changes to one shop have a greater probability of being relevant to other shops. Local establishments do not have this advantage, and thus the risks must be lowered and business and technical considerations must be considered in addition to the actual benefits of the ICTs. Considering third-places as a possibility shifts the context of the conversation in a way that reconsiders the applicability of ICTs as they are currently instantiated

Obviously, such a shift in emphasis as outlined here is only one part of an overall solution space. More work is required from all points of view – business, technology and that of the customer. Yet even this simple tactic implies fairly severe departures from current cyber café technologies and business models. Daily conditions among the digitally disadvantaged are quite different from those among the digitally enabled; it only stands to reason that reducing the imbalance that is the digital divide while providing economic opportunity coupled with sustainability requires us to consider alternatives different from extant common practice.

ACKNOWLEDGEMENTS

This work was funded by Intel Corporation.

REFERENCES

Benjamin, P. "Telecenters in South Africa", *The Journal of Development Communication*, (12:2), 2001. Available at: http://ip.cals.cornell.edu/commdev/jdc-1.cfm

Castells, M. *The Internet Galaxy: Reflections on the Internet Business & Society*, New York: Oxford University Press, 2001, p. 248.

Colle, R.D. and Yonggong, L. "ICT-Capacity Building for Development and Poverty Alleviation", *Third Asian Conference for Information Technology in Agriculture*, Beijing, October 2001. Available at: http://ip.cals.cornell.edu/commdev/documents/ictpaper-china.doc

Gáspár, M. "Telehouses in Hungary", *The Journal of Development Communication*, (12:2), 2001. Available at: http://ip.cals.cornell.edu/commdev/jdc-1.cfm

Habermas, J. *The Structural Transformation of the Public Sphere: An Inquiry into a Category of Bourgeois Society*, Translation by T. Burger with F. Lawrence, Cambridge: MIT Press, 1991.

Hashem, S. "Community Telecenters in Egypt", *The Journal of Development Communication*, (12:2), 2001. Available at: http://ip.cals.cornell.edu/commdev/jdc-1.cfm

Herz, J.C. "The Bandwidth Capital of the World", *Wired*, August 2002, pp 90-97.

Holmes, V. "The Internet, Inequality and Exclusion in Peru: The Social Impact of the Cabinas Publicas", *Masters Dissertation*, Institute of Latin American Studies, University of London, 2001. Available at: http://funredes.org/mistica/castellano/ciberoteca/participantes/docuparti/VictoriaHolmesCabinas.pdf

Ivala, E. "The Internet and Distance Education", Papers Presented at the *1st National NADEOSA Conference*, 11-13 August 1999. Available at: http://www.saide.org.za/nadeosa/conference1999/ivala.htm

Kelly, T, Girardet, G and Ismail, M. "Internet on the Nile: Egypt Case Study", *International Telecommunications Union Report*. Available at: http://www.itu.int/osg/spu/wtpf/wtpf2001/casestudies/egypt1.pdf (Accessed at February 14, 2001).

Keniston, K. "Introduction", in *Bridging the Digital Divide Lessons from India*. Keniston, K. and Kumar, D. (Eds), India: Sage, 2003. Available at: http://www.mit.edu/people/kken/PDF/Intro_Sage_1_.pdf

Kyle, R. "Models of Telecenters in Brazil", *The Journal of Development Communication*, (12:2), 2001. Available at: http://ip.cals.cornell.edu/commdev/jdc-1.cfm

Laurier, E., Whyte, A., and Buckner, K. "An Ethnography of a Neighbourhood Café: Informality, Table Arrangements and Background Noise", *Journal of Mundane Behaviour*, (2:2), 2001. Available at http://mundanebehavior.org/

Liff, S., S., Fred and Watts, P. "Cybercafes and Telecottages: Increasing Public Access to Computers and the Internet", *Survey Report, Virtual Society? Programme, Economic and Social Research Council,* United Kingdom, 1998. Available at: http://virtualsociety.sbs.ox.ac.uk/text/reports/access.htm . (Accessed February 14, 2003).

Milanovic, B. and Yitzhaki, S. "Decomposing World Income Distribution: Does the World Have a Middle Class", *Papers 2562, World Bank - Country Economics Department,* 2001, Available at: http://econ.worldbank.org/files/1423_wps2562.pdf (Accessed at February 14, 2003).

Oldenburg, R. *The Great Good Place: Cafes, Coffee Shops, Bookstores, Bars, Hair Salons, and Other Hangouts at the Heart of a Community*, New York: Marlowe & Co, 1999.

Pedrelli, M. "Developing Countries and the ICT Revolution", *European Parliament, Directorate for General Research Report,* 2001. Available at: http://www.europarl.eu.int/stoa/publi/pdf/00-14-01_en.pdf

Proenza, F. "Telecenter Sustainability: Myths and Opportunities", *The Journal of Development Communication*, (12:2), 2001. Available at: http://ip.cals.cornell.edu/commdev/jdc-1.cfm

Proenza, F., Bastidas-Buch, R., and Monter, G. "Telecenters for Socioeconomic and Rural Development in Latin America and the Caribbean: Investment Opportunities and Design Recommendations with Special Reference to Central America", *Food and Agriculture Organization of the United Nations, Inter American Development Bank, International Telecommunications Union, (FAO/ITU/IADB)*, May 2001. Available at: http://www.iadb.org/ict4dev/telecenters/

Short, G. "Lessons Learned in Pioneering Telecenters in Australia", *The Journal of Development Communication*, (12:2), 2001. Available at: http://ip.cals.cornell.edu/commdev/jdc-1.cfm

Thamizoli P. and Balasubramanian K. "Information Management and Knowledge Empowerment: MSSRF Telecenters in South India", *The Journal of Development Communication*, (12:2), 2001. Available at: http://ip.cals.cornell.edu/commdev/jdc-1.cfm

About the Authors

Tony Salvador is a Design Ethnographer at Intel Corporation, Portland, Oregon, USA. He was born in New Jersey, and achieved his PhD in Experimental Psychology from Tufts University. Tony can be reached by e-mail at tony.salvador@intel.com.

John Sherry is Anthropologist at Intel Corporation, Portland, Oregon, USA. He was born in Portland, and achieved his PhD in Anthropology at the University of Arizona. His research interests include technology and economic development. John can be reached by e-mail at john.sherry@intel.com.

Alvaro Urrutia is Strategic Financial Analyst at Intel Corporation. He was born in Lima, Peru, and achieved his MBA from University of Michigan. His research interests include competitive intelligence analysis. He can be reached via email at: alvaro.e.urrutia@intel.com.

21

CAN E-MARKETPLACES BRIDGE THE DIGITAL DIVIDE?

C. Standing[1], I. Sims[1], R. Stockdale[1] and A. Wassenaar[2]

[1]School of MIS, Edith Cowan University, AU; [2]Department of Business Information Systems, University of Twente, NL.

Abstract: In Australia, small and medium enterprises (SMEs), particularly rural SMEs, are perceived as being on the wrong side of the digital divide. Government at local and state levels have taken a leading role in the development of electronic marketplaces with an aim of improving the lot for SMEs. Many government departments now either own or sponsor electronic marketplaces. Government aims in creating e-marketplaces are often motivated by regional economic development issues. Whilst government entities may think e-marketplaces are an effective channel for implementing government policy a number of complications can arise from this model, not least is being seen as stifling free trade. Despite the community development motivation a major argument for e-marketplace development being put forward is the economic one and this has contributed to a narrow view of the e-marketplace concept and one, which for the time being at least, is likely to restrict its impact. Government sponsored e-marketplaces should consider the value of on-line business networks to share knowledge and potentially increase levels of innovation.

Keywords: SMEs, electronic marketplaces, government, digital divide.

1 INTRODUCTION

Within Australia the term "digital divide" is used to describe the gap between the level of sophistication in IT and e-business adoption and usage in rural compared with urban areas and small and medium enterprises (SMEs) compared with large companies. The severity of the issues are expressed in a range of government sponsored reports and reflected by a range of government funding opportunities to address the problems (Curtin, 2001; DoIT, 2001). In Western Australia, State government and city councils have got involved in developing electronic marketplaces with a key aim of encouraging the development of the SME sector and narrowing the digital divide between SMEs and their larger counterparts. Regional commissions in other areas of Western Australia are now beginning to look at the potential of electronic marketplaces to address the digital divide in their own localities.

This paper examines the potential of electronic marketplaces to address the "SME problem" and in particular the "rural SME problem" in Australia. The first section explains the nature of the problems that face the SME sector, which are exacerbated in rural communities. The second section of the paper presents various definitions of e-marketplaces followed by an examination of marketplace ownership structures. Two government sponsored electronic marketplaces are examined to draw out lessons that can be learned for regional commissions thinking of developing their own e-marketplaces. The implications of the findings are discussed in relation to how effective these strategies are for lessening the digital divide within Australia.

2 THE AUSTRALIAN DIGITAL DIVIDE

The digital divide within Australian society exists on a number of levels. Two related forms of the divide exist in relation to the IT/IS sophistication of SMEs compared with large companies and also the IT/IS sophistication of rural Australia compared with the major metropolitan centres (Curtin, 2001; DoIT, 2001; NOIE, 2002). The two are clearly interlinked and are compounded in rural Australia since most rural businesses are SMEs. This section of the paper presents information on the IT/IS sophistication in the SME and rural sectors.

Ongoing research indicates that even where SMEs have some awareness and use of e-commerce there still remain problems (Bode and Burn, 2001; Tetteh and Burn, 2001). Research reports indicate that lack of access to advice on Internet strategy is seen as a major barrier by SMEs (van Akkeren and Cavaye, 1999) and only approximately one third of Web enabled SMEs have any form of an Internet strategy (Kinnes, 2001; Stokes, 2000). SME websites are primarily information sites for customers and only 20% are capable of taking an order online (Korchak and Rodman, 2001). There is little awareness that e-commerce offers companies "unprecedented access to information on IT" (Swatman, 2000).

Less than a third of SMEs use the Web for procurement and there has been low penetration of e-marketplaces. Although e-marketplaces are being increasingly used by large organisations, which have been quick to realise their potential, SMEs have been slow to take up their adoption as a mechanism for buying and selling. Some argue that SMEs are disadvantaged in tendering for large projects, especially government ones, because of their difficulties in areas such as the financial and legal requirements of contracts, software compatibility and their inability to partner and thereby provide as competitive a service as larger companies (Davidson, 2002). Poor performance in winning large government contracts has exacerbated the SME digital divide.

Many of the problems relating to the failure of SMEs to address the importance of electronic marketplaces lies in a lack of understanding of the advantages and how they can benefit from them. SMEs understanding of the global marketplace is 'not good enough' (Erbschloe, 1999) and they lack sufficient awareness of the nature of the Internet and how it interacts with other methods of trading. Smaller companies do not see themselves as part of a large supply chain and they underestimate how the Internet can benefit them by sharing information, buying from suppliers with no paper system, electronic fulfilment, tracking, and efficiencies in cost and time (Jack, 2001). If they do not understand their ability to function within the larger supply chain they will lose out to large firms in electronic markets (Korchak and Rodman, 2001). The developing world markets brought about by e-commerce, and the increased ability to trade globally facilitated by electronic markets, adds to pressure on the SMEs by increasing the number of firms with the ability to trade in each region (Said, 2000).

The problems associated with the SME sector in Australia are exacerbated in rural areas (Curtin, 2001), where markets, expertise, general business support, IT support and specialist consultancy services are limited. Poor telecommunications infrastructure is also cited as a major barrier to e-commerce adoption. Although there is no denying these problems many opportunities are lost because of a lack of awareness and expertise in e-business.

The extensive economic and social benefits of effective e-marketplace participation for rural SMEs include reduced costs (reportedly by a 'factor of five or ten or more' (Lucking-Reiley and Spulber, 2000), improved customer service, reduced communication costs, accelerated flow of news and information and improved market information. These benefits stem from participation in e-marketplaces with a transaction and value-added information focus. However, other types of network/marketplace participation can produce substantial benefits for organisations since the knowledge and expertise that is shared within the network can stimulate new levels of business innovation. These business benefits are supplemented by definable social benefits where the interconnectedness that is a feature of what Raisch (2001) terms Value Trust Networks (VTNs) can contribute to stability and self-esteem in rural communities (Lewis, 2001).

Success in extending the markets of rural industries will bring major benefits both to the businesses and to the community. Conversely, failure to compete in value trust networks will result in regional isolation as the increased ability to interact globally adds to the pressure of competition from other regions.

3 DEFINING ELECTRONIC MARKETPLACES

Before examining government sponsored e-marketplace initiatives in Western Australia we briefly analyse various definitions and e-marketplace concepts. The proliferation of electronic marketplaces in the last five years has been rapid and extensive, although an anticipated period of consolidation is taking place (Forrester Research, 2000). The rapid emergence of this topic has led to a diverse range of definitions highlighting differing perspectives such as the role of the stakeholders (Federal Trade Commission, 2000) or the interactivity of business communities (Brunn et al., 2002). However, Bakos' definition of an electronic marketplace as 'an interorganisational information system that allows the participating buyers and sellers in some market to exchange information about prices and product offerings' retains simplicity but manages to encompass the essence of marketplace activity (Bakos, 1997). It is important to distinguish between a market and a marketplace. A market covers the supply and demand for a product or service but a marketplace is a bounded entity that provides specific mechanisms for the exchange, hence any market could have one or a number of marketplaces associated with it.

The level of e-marketplace activity has been evolving from early matchmaking models to more complex interactive and interconnected marketspaces, which can be termed Value Trust Networks (VTN). Raisch (2001) describes four phases of VTN evolution beginning with a transaction focus at level one that evolves into a value-added marketplace offering transaction support services. The increase in information value-add contributes enhanced industry knowledge and interorganisational collaboration and has the potential for e-marketplaces to develop into a third phase. This will see services enhanced by capturing and utilising rich information flows and creating knowledge exchanges. The ability to integrate the transaction exchange, the value-add services and the knowledge services moves the evolution of e-marketplaces into Raisch's fourth phase.

VTNs offer the promise of a new business platform of integrated and interconnected business communities that will spawn a new era of business innovation (Raisch, 2001). Although Raisch proposes this as an evolutionary model no research has been conducted to verify this as far as the authors are aware. Therefore, it is safer to assume that the four levels represent different functionalities within a complex e-marketplace/network arrangement without being evolutionary in nature.

The roles of electronic marketplace participants are not necessarily mutually exclusive since for example, a buyer could also be a seller and vice versa and a market maker (owner) could also be a buyer and/or a seller. These overlapping roles have the potential to create role ambiguity and even conflict of interests.

It is assumed that an e-marketplace governance structure is either biased or neutral (Kaplan and Sawhney, 2000) and that private e-marketplaces will be, at least to some extent, biased in favour of the owner, and public e-marketplaces will either be biased in favour of buyers or sellers or be neutral. It is suggested that it is in the interests of e-marketplaces to be neutral, neither favouring buyers nor sellers (Sculley and Woods, 2001). The benefits of a neutral marketplace are seen as a perception of fairness, which impacts on increased trust between trading participants. In addition, fewer channel conflict issues are expected to arise due to increased transparency and better exploitation of market and supply chain efficiencies (Brunn et al., 2002).

4 EVALUATION OF BENEFITS CONCEPTS

Benefits can be broadly classified according to whether they produce economic, network, service, or community advantages. Market makers may have one or a variety of motives in creating and maintaining an electronic marketplace. Each potential motive is discussed below.

4.1 Economic Motive

Initial incentives for the development of an interorganisational information system are economic and involve three potential benefits for participants; cost reductions, productivity improvements and product/market strategy (Barrett and Konsynski, 1982). The economic motive for engaging in e-marketplaces is bound up with transaction cost economics. Simply, the costs of a business fall into two categories: production costs and transaction costs. Production costs are concerned with the process of transforming inputs into outputs. Transaction costs are the costs associated with finding someone with whom to do business, reaching an agreement about the price and other aspects of the exchange, and ensuring that the terms of the agreement are fulfilled (McTaggart et al., 1996). The early pioneer in this area is Ronald Coase who contends that it is impossible to understand the workings of the economic system without taking into account transaction costs (Coase, 1937).

A key work on transaction cost economics (Williamson, 1979) identifies the critical characteristics of a transaction and links these to the institutional governance structure of transactions. The significant characteristics of a transaction are uncertainty, frequency of exchange and the extent to which investments are specific to certain transactions. According to Williamson, non-specific transactions are efficiently organised by markets, while recurrent specific transactions are more efficiently governed internally. Porter (2001) argues the main economic benefits of e-marketplace participation for buyers are low transactions costs and sometimes the ability to pool markets, while for sellers the benefits are lower selling costs, lower transaction costs and access to wider markets. Although there are other motives beyond the economic for owning an e-marketplace they each have economic implications.

4.2 Network View

The network view of electronic marketplaces focuses on the relationships and communication infrastructure of groups of organizations, which are bound together in some way. Interorganisational alliances are a form of network with social, political and economic implications. Here, the focus is on the socio-political arrangement. Oliver (1990) proposes six generalisable determinants of interorganisational relationships:

- Necessity — to fulfil legal or regulatory requirements
- Asymmetry — potential to exert power over other organisations
- Reciprocity — desire to cooperate, collaborate and coordinate
- Efficiency — internally focused efficiencies
- Stability — in response to environmental uncertainty
- Legitimacy — related to reputation, image, prestige, or congruence with prevailing norms in the environment

4.3 Service Motive

The service motive is concerned with providing a better service to customers, which may include such things as continuity of supply, convenience and speed of processing and greater choice for buyers. The service motive is closely aligned to the economic but is kept separate as this may not always be the case. Higher service typically comes at a cost but in theory an organization could choose to deliver higher levels of service despite this extra cost. There are five dimensions by which consumers evaluate service quality (Bebko, 2000; Berry and Parasuraman, 1991):

i. Tangibles. The appearance of physical facilities, equipment, personnel and communications materials.
ii. Reliability. The ability to perform the promised service dependably and accurately.
iii. Responsiveness. Providing a prompt service and desire to help customers.
iv. Assurance. The knowledge and courtesy of employees and their ability to convey trust and confidence.
v. Empathy. The caring individualized attention the firm provides its customers.

In relation to e-marketplaces service quality relates to such things as the Web site and e-marketplace software, personnel, marketing literature and supporting documentation, the reliability of the system and help provided.

4.4 Community Motive

Some e-marketplaces are created with a community emphasis. In other words a major objective of the electronic market is to play a role in the development of a community. This is usually done through stimulating economic activity working on the premise that if local/regional business flourishes then so will the communities they are part of. The market maker, usually local or state government, provides encouragement to adopt e-marketplace trading and in doing so raise the level of general e-business knowledge, skills and technologies within the business community.

4.5 Hybrid Arrangement

Of course, a market maker may have a set of objectives to achieve in the construction and management of the electronic marketplace. For example, the community model may be seen as being for the common good of the society or business community but may still need to be economically viable.

Issues such as trust between participants, information systems architectures, revenue models and transaction mechanisms are all features, which can be used to support the primary motive.

5 CASE STUDY APPROACH

Two examples of e-marketplace development have been selected to illustrate the range of government motivations behind developing e-marketplaces in Western Australia. Although examined as extensive case studies, we present the cases here as vignettes (Barter and Renold, 1999) to examine the range of aims, objectives and perceived benefits resulting from e-marketplace creation. Vignettes can take a number of forms. In this paper we use them as concrete examples, which allow the situational context to be explored and influential issues to be identified (Finch, 1987). The original case studies involved the collection of information from official documents and reports, as well as through face-to-face discussions, email correspondence, and attendance at meetings with the e-marketplace sponsors and developers, and the official Web sites of the organisations.

5.1 Vignette 1: Government Electronic Marketplace (GEM)

The Western Australian Government currently spends approximately $A5 billion on goods and services and estimates an average transaction cost for simple purchases of $100 (DoIT, 2002).

Early in 2000 the WA government agency responsible for management of government purchasing, the Department of Contract and Management Services (CAMS), embarked on the development of a major project known as the Government Electronic Marketplace (GEM) (DoIT, 2002). In July 2001, as a result of a major government reorganisation the Department of Industry and Technology (DoIT) assumed responsibility for GEM. GEM is Australia's first comprehensive online government buying service and provides an array of services that cover the range of government buying:

- Purchasing of low value commodities
- Public tendering for high value goods and services
- Contract planning, formation and ongoing management (coming soon)

GEM aims to streamline traditional business partnerships between the public and private sectors and significantly enhance the quality, timeliness and cost-effectiveness of services to the community. The published objectives and benefits of the system, listed on the DoIT web site, (DoIT, 2002) are:

- Saving taxpayers money through the introduction of more efficient procurement practices.

- Increasing the accountability and transparency of government purchasing
- Increasing the levels of compliance with State Supply Commission procurement and purchasing policy (including buy local and common usage contract policies)
- Demonstrating leadership in the implementation of the Australian Procurement and Construction Council (APCC) guidelines and standards for electronic procurement
- Assisting West Australian industry to enter the world of e-commerce in a safe and secure government environment.
- According to the GEM Web site (http://www.gem.com.au):

"Gem gives suppliers access to an enormous market of buyers - initially in government, but ultimately including private schools and hospitals, public benevolent institutions, and third party purchasers such as facilities managers who are looking after government buildings...... Suppliers can rest assured that GEM supports the government's stringent purchasing policies, such as the Buy Local Policy".

The establishment of GEM is not just a tool for implementing market efficiencies, but also for implementing a variety of policies. In the case of GEM, the government owns most of the buyers, some of the sellers and operates the market. To further complicate matters, it also owns the policy-making body that sets the rules for open and effective competition for all government purchasers. This separate body is known as the State Supply Commission.

5.2 Vignette 2: Regional Electronic Marketplace (REM)

The twin cities of Joondalup and Wanneroo in WA have developed what they term a regional electronic marketplace (REM), and a significant number of companies have registered to be participants. The REM will be operational in December of 2002. The e-marketplace aims to provide e-procurement and marketing solutions for business, local government and education organisations within the North West corridor of the Perth Metropolitan area. This corridor includes a mixture of suburban and rural communities. The major drivers for the projects are to increase e-commerce adoption, stimulate greater interaction between businesses in the locality, and produce savings and efficiencies for buyers and sellers, all within a local region. A consortium is funding the initial development of REM. This includes North Metro Community Association Incorporated (NMCOA) On-line Joondalup and Wanneroo Councils, Edith Cowan University, Joondalup Business Association, Wanneroo Business Association and several other local businesses. NMCOA is a not-for-profit incorporated body with most of the sponsors as its founding members.

It is intended that the SME sector will access the REM without high entry cost or EDI compliance barriers; access will be available via a range of communication facilities including Internet, Fax or WAP enabled mobile telephones. The e-marketplace project incorporates three functions:

i. Business to consumer
ii. The e-marketplace will provide local businesses with the opportunity to sell their goods and services to people both inside and outside of the region.
iii. Business to Government
iv. The REM will provide local businesses, who can meet the supply requirements, the opportunity to automatically receive electronic offers to tender or quote for goods and services required by both the City of Joondalup and the City of Wanneroo.
v. Business to business
vi. The same e-marketplace will allow companies to trade with each other electronically.

The key motivations for development of the regional e-marketplace are:

- Increase e-commerce adoption
- The training associated with the e-marketplace will take the form of seminars and laboratory style hands-on sessions. In general, the project aims to improve awareness related to the benefits of e-marketplace trading in the region, which is also part of a larger plan to raise e-commerce adoption and knowledge so that companies can become globally competitive
- Improve business efficiency in the locality
- It is anticipated that e-marketplace participation will reduce costs for local businesses and make them more efficient
- Increase trade within the locality
- It is expected that trade within the region will increase as businesses trade more with one another rather than with businesses outside of the region.
- Expansion into new markets
- The City Councils expect that when companies become comfortable with e-marketplace trading they will be more likely to venture into other e-marketplaces and as a result access other state, interstate and international markets.
- Development of the region generally
- It is hoped that the REM will play a role in developing the Northern suburbs as an attractive proposition for new businesses. The Two Cities e-marketplace Web site states: "This facility will encourage

> the growth and retention of jobs within our region by encouraging a more effective "buy local" attitude (http://www.2Cities.com.au).

The 2cities e-marketplace is part of a broader community portal for the Joondalup/Wanneroo cities. This includes functionality supporting aspects of e-government and information about community groups and facilities for them to interact.

6 DISCUSSION

The digital divide between rural and urban Australia and the difficulties faced by many SMEs compared with large multi-nationals are two, often related, issues which government at Federal, State and local levels are well aware of. The Western Australian State Government is particularly keen to address the plight of rural SMEs as communities are dependent on how parties at various levels succeed. Electronic marketplace participation is seen by some government agencies as a way of stimulating SME business. The two e-marketplace examples discussed in the paper cover rural and rural/urban fringe areas and have a primary aim of encouraging small business development and growth. In this section of the paper, we discuss the lessons that can be learned from these two WA Government sponsored e-marketplaces for examining how WA Regional Commissions and other States can effectively employ e-marketplace business models.

The two electronic marketplaces examined have different ownership structures. In GEM the State Government is the owner but also the buyer. In the 2Cities Regional E-marketplace the city councils of Joondalup and Wanneroo are two of a group of owners. However, they are taking the leading role in setting the direction and in managing the e-marketplace. The lesson for other State governments and city councils is to examine a variety of e-marketplace ownerships models. These may involve using the purchasing power of the government (at whichever level) as a catalyst for e-marketplace participation. Alternatively shared ownership arrangements have the advantage of decreasing the development costs and increasing the demand for goods. In addition, shared ownership structures may be seen as more neutral and hence fairer compared with a solely owned marketplace. A potential problem associated with shared ownership is that it may be more difficult to use the e-marketplace to implement government policy such as a buy local strategy. It will be interesting to observe if in the 2Cities REM government objectives are deflected by other members of the group, which may change the emphasis to obtaining the lowest prices rather than regional development. In other words, the main objective could change to an economic rather than a community focus.

There may be an argument for regional Governments and city councils to serve as information providers and concentrate on disseminating information on e-marketplace participation rather than developing or managing such models. This way SMEs could be exposed to e-marketplaces out of the region with potentially greater returns. The local emphasis to both e-marketplaces would seem to be at odds with the generally held views that SMEs should take part in a global business arena. The underlying assumption is that once exposed to e-marketplace trading SMEs will be more likely to take the next step of trading internationally. However, this view of evolutionary participation in e-marketplaces from local to global has not been thoroughly researched and in fact may not be valid.

The view of electronic marketplaces in both cases is one that is transaction focussed (economic) with a minor importance attached to value added information. It is interesting that the Western Australian e-marketplaces although being seen as way of building the local community do not highlight or include provision for a strong network model through electronic collaboration and knowledge sharing, in addition to providing access to consultants and digital service providers. Although, the e-marketplace owners may wish this to happen there is little mentioned that would facilitate this. This may appear surprising considering the community emphasis.

A value trust network perspective would seem very relevant in fostering SME development and growth. An emphasis on creating value trust networks should be considered by government at state, regional and local levels since the knowledge and expertise that is shared within the network has the potential to create significant levels of innovation. The lesson for government agencies wishing to develop e-marketplaces is to take a broader view of their purpose and to explore ways to create business networks in addition to transaction (selling and procurement) mechanisms.

The main emphasis of the two e-marketplace developments is on the economic motive so that costs can be reduced and potentially markets expanded. This is to be achieved by the reduction in search time for suppliers and a faster transaction process and hence provide a reduction in transaction costs. Porter (2001) warns that buyers may yet turn away from marketplaces to building relationships with fewer suppliers and focus on reducing costs through efficiency gains. Government e-marketplace owners should consider providing advice to the community as to where e-marketplaces can be of value and where extranet style arrangements have advantages.

The Joondalup and Wanneroo regional e-marketplace expected that the financial savings accruing to participants would create regional economic stimulus. Just how this will be achieved and sustained is not highlighted in the published reports, documents and Web sites, other than through a 'buy local' policy. Government owned e-marketplaces should not assume regional development, and for that matter community development, is an automatic result of e-marketplace participation by SMEs. Indeed, the benefits for the community and region need to be properly evaluated.

The governments appear to see the electronic marketplaces as vehicles to implement policy. This is apparent in GEM and the 2Cities REM. Western Australian Government at State and regional levels are currently promoting buy local policies. It could be argued that this is in conflict with an open market policy. The 2Cities REM, for example will only allow companies to sell in the regional e-marketplace if they are within the council boundaries.

Government agencies developing e-marketplaces should not only state their aims and objectives to suppliers but also how they plan to make it a fair marketplace. Neither of the e-marketplaces addresses this issue. The government acting as both buyer and owner may appear suspicious to some suppliers. The government in such circumstances forms a formidable power structure, hence they need to be sensitive to this perception.

7 CONCLUSIONS

Western Australian government sponsored e-marketplaces have the potential to narrow the digital divide both for SMEs rural and urban areas. The enormity of the challenge should not be underestimated. However, regional commissions in Western Australia examining the potential of e-marketplaces should learn from the experiences of the first two governments sponsored e-marketplaces in that State. To fully harness the benefits of online business a broader vision for e-marketplaces should be considered. This should include the value of information generated, knowledge sharing and the facilitating the development of on-line business networks or clusters. A result of this would not only be a lowering of costs and access to new markets but better access to expertise and potentially higher levels of innovation.

REFERENCES

Bakos, J. Y. "Reducing Buyer Search Costs: Implications for Electronic Marketplaces", *Management Science*, (43:12), 1997, pp. 1676-1692.

Barrett, S., and Konsynski, B. R. "Inter-Organization Information Sharing Systems", *MIS Quarterly*, Special Issue (6:5), 1982, pp. 93 - 105.

Barter, C., and Renold, E. "The Use of Vignettes in Qualitative Research", *Social Research Update*, Issue 25, 1999, Department of Sociology, University of Surrey. Available at: http://www.surrey.ac.uk/sru/SRU25.html.

Bebko, C. "Service Intangibility and its Impact on Consumer Expectations of Service Quality", *Journal of Services Marketing*, (14:1), 2000, pp. 9-26.

Berry, L., and Parasuraman, A. *Marketing Services: Competing Through Quality*, New York: Free Press, 1991.

Bode, S., and Burn, J. M. "Website Design Consultants, Australian SMEs and Electronic Commerce Success Factors", *International Journal of Business Studies*, (9:1), 2001, pp. 73-85.

Brunn, P., Jensen, M., and Skovgaard, J. "E-Marketplaces: Crafting a Winning Strategy", *European Management Journal*, (20:3), 2002, pp. 286-298.

Coase, R. H. "The Nature of the Firm" in *The Economics of Transaction Costs*, O. E. Williamson & S. E. Masten (Eds.), Cheltenham: Edward Elgar Publishing Ltd., 1937, pp. 4-22.

Curtin, J. "A Digital Divide in Rural and Regional Australia?" *Current Issues Brief 1 2001-2002*, Department of the Parliamentary Library, Information and Research Services, 2001.

Davidson, P. "Giving Our SMEs a Fair Go!" *Information Age*, Oct/Nov 2002, pp. 36-47.

DoIT. "E-Commerce: A Primary Objective": *Department of Industry and Trade, Western Australian Government*, Available at: http://www.indtech.wa.gove.au/doit/publications/pireport.pdf.

DoIT. "GEM", 2002, Available at: http://www.gem.wa.gov.au (Accessed at: September 24, 2002).

Erbschloe, M. "Working with SMEs in the Supply Chain", *Digital Systems Report*, (21:4), 1999, pp. 22-23.

Federal Trade Commission. "Entering the 21st Century: Competition Policy in the World of B2B Electronic Marketplaces: The Federal Trade Commission B2B Public Workshop", 2000.

Finch, J. "The Vignette Technique in Survey Research", *Sociology*, (21), 1987, pp. 105-114.

Forrester Research. "The E-Marketplace Shakeout", *Forrester Research*, 2000, Available at: http://www.forrester.com/ER/Baseline (Accessed at: April 19, 2001).

Jack, B. "Information Services for Profit: Small Business Embracing Net Procurement", *M2 Communications Ltd*, 2001. Available at: http://www.presswire.net (Accessed at: November 7, 2001).

Kaplan, S., and Sawhney, M. "E-Hubs: The New B2B Marketplaces", *Harvard Business Review*, May-June 2000.

Kinnes, K. "Digital Union: Digital Union Launches Free E-Market Seminar Series to Help SMEs Reap the Benefits of E-Procurement", *M2 Presswire*, 2001. Available at http://www.digitalunion.com/ezm.htm (Accessed at: November 7, 2001).

Korchak, R., and Rodman, R. "E-Business Adoption Among US Small Manufacturers and the Role of Manufacturing Extension", *Economic Development Review*, (17:3), 2001, pp.20-25.

Lewis, V. "E-Commerce Ready", *America's Network*, (105:13), 2001, pp. 49-54.

Lucking-Reiley, D., and Spulber, D. F. "Business-to-Business Electronic Commerce", *Journal of Economic Perspectives*, (15:1), 2000, pp. 55-68.

McTaggart, D., Findlay, C., and Parkin, M. *Economics*, Sydney. Addison-Wesley, 1996.

NOIE. "Digital Divide", *Canberra: National Office for the Information Economy*, 2002. Available online at: http://www.noie.gov.au/publications/noie/access/connecting_communities/digitaldivide.doc (Accessed at: October 20, 2002).

Oliver, C. "Determinants of Interorganisational Relationships: Integration and Future Directions", *Academy of Management Review*, (15:2), 1990, pp. 241-265.

Porter, M. E. "Strategy and the Internet", *Harvard Business Review*, March 2001, pp. 63-78.

Raisch, W. D. *The eMarketplace. Strategies for Success in B2B E-Commerce*, New York: McGraw Hill, 2001.

Said, A. J. "Helping Small Firms Trade Effectively with the Internet", *International Trade Forum*, (3), 2000, pp. 16-19.

Sculley, A. B., and Woods, W. A. *B2B Exchanges. The Killer Application in the Business-to-Business Internet Revolution*, New York: HarperCollins, 2001.

Stokes, A. "Small and Medium-Sized Enterprises and the Environment: Business Imperatives", *International Small Business Journal*, (19:1), 2000, pp. 100-102.

Swatman, P. "Internet for SMEs: A New Skill Road?", *International Trade Forum*, (3), 2000, pp. 22-24.

Tetteh, E., and Burn, J. M. "Global Strategies for SMe-Business: Applying the S-M-A-L-L Framework", *Journal of Logistics and Information Management*, (14:1/2), 2001, pp. 171-180.

van Akkeren, J., and Cavaye, A. L. M. "Factors Affecting Entry-Level Internet Technology Adoption by Small Business in Australia - Evidence From Three Cases", *Journal of Systems and Information Technology*, (3:2), 1999, pp. 33-48.

Williamson, O. E. "Transaction Cost Economics: The Governance of Contractual Relations", *Journal of Law and Economics*, (22:2), 1979, pp. 33-261.

About the Authors

Craig Standing is Professor of Strategic Information Management at Edith Cowan University, Western Australia. He was born in the UK and achieved his PhD at the University of Western Australia. His research interests include the impact of e-marketplaces on organisations. He can be reached by e-mail at c.standing@ecu.edu.au.

Ian Sims is a doctoral student in the School of MIS at Edith Cowan University. He worked for many years as a procurement manager and has taught supply chain management at university level. He can be reached by e-mail at i.simms@ecu.edu.au.

Rosemary Stockdale is a doctoral student in the School of MIS at Edith Cowan University. She is researching the use of e-marketplaces in a large mining multi-national organization. He can be reached by e-mail at r.stockdale@ecu.edu.au.

Arjen Wassenaar has been researching inter-organisational information systems for a number of years. He has published widely on the topic in IS journals and conferences. He is part of an international team studying e-marketplaces. He can be reached by e-mail at d.a.wassenaar@sms. utwente.nl.

22

THE FOOTPRINT OF REGULATION

How Information Systems are Affecting the Sources of Control in a Global Economy

Prodromos Tsiavos, Ian Hosein and Edgar A. Whitley
Department of Information Systems, London School of Economics and Political Science, UK

Abstract: Whilst the issue of jurisdiction – the question of how far control extends – has always been controversial, the introduction of information and communications technologies only exacerbates the problem. The first generation of scholars to consider this question tended to see technology creating a separate space – cyberspace – with its own legal boundaries. A second generation of scholars, however, has argued that there is nothing new with cyberspace and that conflicts over boundaries have always existed in the law; as a result, they argue that technology is not as remarkable a factor as the first generation believe. By considering the case of copyright and peer–to–peer technologies together with the regulatory environments surrounding their development, use and control, this paper proposes a further refinement in the dialectics of control through technology that refines our notions of jurisdiction in an era of globalisation enabled by new technologies.

Keywords: Jurisdiction, regulation, peer-to-peer.

1 INTRODUCTION

The bursting of economic bubbles and the emphasis upon global security has given rise to a new form of scepticism regarding technology and society. It is common nowadays to dismiss previous technologically - optimistic claims regarding the information society. Whereas earlier it was claimed that "no one knows who or where you are on the internet" (Turkle, 1996), or "governments are powerless to regulate global networks" (Angell, 2000) such views are now considered by many to be unrealistic, or worse, deterministic. Indeed, it is now claimed that regulating data flows is no different than regulating other activities.

Deciding which of these views holds has important implications for understanding the developing nature of globalization and the ways in which information and communications technologies may be shaping the globalization debate. The future is likely to be very different if governments have (or believe that they have) the sovereignty to act than if they don't. This principle of sovereignty is often summarised as the "government's exclusive power within its borders and virtually nowhere else" (Leeds, 1998: pp. 6-7).

The jurisdiction of regulation and control, regardless of what is being regulated, has a particular domain of application. For instance, when referring to the regulation of copyrighted material by the UK Copyright Act 1988, its domain would be the geographical area and the natural and legal persons on which it may be enforced. Video rental records privacy laws in the United States specify a specific geographical area. The domain is sometimes even more limited than geography; the jurisdiction of U.S. privacy protection regulatory regimes include video rental records, but does not include electronic commerce conduct. The domain of regulation constitutes the **footprint** of each regulatory form and often is found to have the same meaning as the term jurisdiction. There is a series of diverse definitions regarding what is and what is not jurisdiction, but for the time being this paper will refer to the term jurisdiction as having the same notion as the application domain of a set of regulations. We note that many of the examples cited in this paper draw on cases in Western legal systems, this is primarily because other than censorship, these issues have not yet arisen in many non-Western contexts. Similarly, the developing nature of these issues in relation to new technologies means that no transnational structures exist, as yet, to address them.

Jurisdiction has been traditionally linked with the concept of the state since the latter has been the institutional mechanism supporting and cultivating the main regulatory mechanisms of the modern society. The globalization debate, (see for example, Avgerou (2002); Beck (2000), Walsham (2000)), however, has highlighted that the arbitrary physical boundaries of countries are becoming increasingly irrelevant to the global world.

One useful articulation of this argument is given by Ulrich Beck who argues that we have entered "a second modernity". He defines the first modernity as being based on the assumption that "*we live and act in the self–enclosed spaces of national states and their respective national societies*" (2000 p. 20 emphasis in original) and suggests that this assumption no longer holds in an era of globalism.

That is, our actions are no longer restricted to the self–enclosed spaces where they take place. Rather our new global connections mean that things as diverse as "money, technologies, commodities, information and toxins 'cross' frontiers as if they did not exist. Even things, people and ideas that governments would like to keep out (for example, drugs, illegal immigrants or criticisms of human rights abuses) find their way into new territories" (p. 20). In the global economy of the second modernity it does not make sense to speak only of 'national' effects; any actions can only sensibly be considered in terms of their global effects, even if the immediate effects might appear to be fairly localised.

As a result, "nothing which happens on our planet is only a limited local event; all inventions, victories and catastrophes affect the whole world, and we must reorient and reorganize our lives and actions, our organizations and institutions, along a 'local–global' axis" (pp. 11-12).

This globalization debate, therefore suggests that regulation as a source of control is not limited to the State. Coupled with this, since the 1960s, the Chicago School of Regulation and Economics has been arguing that norms and markets are also sources of regulatory control. Each has different effects on jurisdiction. Markets are increasingly globalized, bringing with them a form of regulation that is not necessarily state-sponsored; the same may be said for norms.

This broader conceptualisation of regulation and its impacts on jurisdiction in a global environment are considered further below. First, however, the paper presents the two major ways in which the regulation of cyberspace has been considered.

2 TECHNOLOGY AND CONTROL OF CYBERSPACE: TWO VIEWS

The international regulatory environment has been of interest to a number of scholars as they studied the control of "cyberspace" or "the Net". The views and findings of these scholars may be divided into two generations.

2.1 First Generation: Staking Out a New Place

The first generation of scholars argued that under traditional notions of sovereignty and jurisdiction, governments relied on borders to enable their power, give effects to their rules, create legitimacy to their enaction, and notice to those who were regulated. Jurisdiction was therefore essential to regulation; but over computer networks, this was challenged.

Johnson and Post (1996) argued that:

> *(M)any of the jurisdictional and substantive quandaries raised by border–crossing electronic communications could be resolved by one simple principle: conceiving of Cyberspace as a distinct "place" for purposes of legal analysis by recognizing a legally significant border between Cyberspace and the "real world".*

In effect, they were making both descriptive and normative arguments (Goldsmith, 1998a). First, their claim was that applying regulation limited by geographic–borders to an a-geographic-border environment would be senseless. This claim for a new space for cyberspace was famously developed by John Perry Barlow in his "Declaration of the independence of cyberspace" (Barlow, 1996) which asked "On behalf of the future" to be left alone by those of the past as they "have no sovereignty where we gather" (Barlow, 1996).

Second, they claimed that regulation by one jurisdiction would have spillovers immediately upon another because of the a-geographic-border nature of cyberspace. If one country decided that the internet was to be in their jurisdiction, then the rules dictated by this country would affect the entire internet. For example German hate speech laws could have a chilling effect on the willingness of service providers worldwide to host controversial content.

2.2 Second Generation: Reclaiming the Space

The second generation of scholars, however, regard this view as promoted by "regulation–sceptics" who exaggerate the problems of regulating cyberspace and ignore how similar it is to older forms of regulation and infrastructure (Goldsmith, 1998b: p.1130).

The critique is that cyberspace transactions are not all that different from transnational transactions, in that they both involve people in "real space" in different territorial jurisdictions causing "real-world" actions and effects.

Most "cyberspace" issues, therefore, have real–space analogies. According to Goldsmith, data havens may be created in the same way that tax havens may in the real world. Conflicts of law will occur in cyberspace the same way as they occur regularly in real-space; and the response to the real-space problem in the past has been international harmonization strategies. The feasibility of regulation will thus increase as the practices become more common in various jurisdictions, and once knowledge of laws in other jurisdictions grow and experiences are shared (Perrit, 1996).

The second generation argues that the changes in transportation and communications technologies of modernity have made multi-jurisdictional activity more common. International law permits States to "apply its law to extraterritorial behavior with substantial local effects"; that is "in modern times a transaction can legitimately be regulated by the jurisdiction where the transaction occurs, the jurisdictions where significant effects of the transaction are felt, and the jurisdictions where the parties burdened by the regulation are from" (Goldsmith, 1998a: p.1207). That this is happening with the internet and global data flows, to these thinkers, is just more of the same.

As a result countries have been establishing regulation of data flows. The European Union finalised a harmonizing directive on data protection in 1995 that included two articles regulating transborder data flows (European Union, 1995). Countries as diverse as Australia (Australian Broadcasting Authority, 1999), China (BBC news, 2002a), and Saudi Arabia (Lee, 2001) have implemented censorship regulations to control the kind of information that was sent, received, or both (cf. (Rapporteurs sans Frontieres, 1999)); despite claims of infeasibilities and inaccuracies (cf. (Clarke, 1999; Dogcow, 1999; Electronic Privacy Information Center, 1997)). In 2000, a French court decided that Yahoo! was obliged to prevent French users from accessing material on Yahoo!'s sites that were illegal according to French law. This was done despite the fact that the servers and services were in the United States (Akdeniz, 2001); and technological arguments about the infeasibility of implementing the ruling and legal claims about the jurisdiction of the French courts.

The changing nature of technology also affects the ability to regulate a technological infrastructure. However, the regulation literature notes that *every* new technology can be seen to disrupt existing regulatory regimes, not just computing technologies. For example, Peltzman includes "changes in the 'politics' and changes in the 'economics' of the regulated industries" as factors that alter regulatory regimes, adding that political change also comes from "changes in the underlying organization and information technologies" (Peltzman, 1989: p.108).

While this second generation argues that technological change is not that big a deal, this paper contends that ignoring the constitution of technology may lead to problems in understanding regulation and jurisdiction.

There may be value in investigating the details of how technology changes regulation; or how technology enforces regulations, and other such interactions between technology and regulation. Then it is possible to have a deeper appreciation of how technology and jurisdiction interact.

3 A THIRD GENERATION UNDERSTANDING

Whilst the first generation argument perhaps placed too much emphasis on technological issues, the second generation places too much emphasis on how the environment is similar to previous practices, technologies, infrastructures and laws. A third generation argument has recently evolved that seeks to combine an appreciation of the role of technology in changing the regulatory habitat (Hood, 1994) whilst also allowing for the ongoing evolution of established legal practices.

Drawing on the Chicago School of Regulation which argues that markets and norms, together with laws play a role in controlling action, Laurence Lessig has also used ideas from Bentham and Foucault to include architecture (Lessig, 1998) and more particularly code / technology (Lessig, 1999) as other modalities of regulation. In his proposal for a New Chicago School, Lessig warns that it is necessary to look to how code or technological architectures are also capable of regulating human action. His model incorporates all these elements as modes of regulation.

The third generation understanding therefore acknowledges both the new practices enabled by technology and the role of existing laws (and markets and norms) in regulating behaviour. In the next section the paper considers how a third generation understanding can be used to analyse peer–to–peer networks and copyright before coming back, in the concluding discussion to review the impacts of this understanding for our views of effects of technology on regulation and the globalisation more generally.

4 CASE: COPYRIGHT, NAPSTER AND PEER-TO-PEER SYSTEMS

"Copyright means many things to many people"

The aphorism, which Sterling uses as an introductory phrase for his "World Copyright Law" (1998), reflects in a very eloquent way the idiosyncratic – if not protean – nature of copyright. Copyright affects people in a multitude of different ways and thus it becomes a myriad of different things for them.

In the past few years, MP3 files, services like Napster and peer-to-peer networks more generally were seen by many as the latest and perhaps the most significant challenges to the copyright *status quo*. The small file sizes of MP3 files that allowed near CD quality music to be searched for and downloaded quickly from all over the internet were leading to an order of magnitude increase in the problems of the copyright management industry.

Throughout its history copyright has tended to be used to protect the interests of the content disseminators rather than those of the content creators. This does not mean that the interests of the creators or the consumers of intellectual creations were not protected; it is just that the mapping of the interests was much more in favour of the disseminators and other intermediaries than anyone else. The only case where such a situation is reversed is that of the authors' collecting societies, especially those of continental Europe and Germany in particular (Sterling, 1998), although again the collecting societies are intermediaries. This is due to both the content and structure of copyright and author's right laws, and the way these laws are constructed. In the terms of content, almost all the rights granted to a creator can be—and are—transferred to intermediaries that vary from a disseminator, such as a music label, to a collecting society or a software house. In terms of structure, copyright laws have been built based on the assumption of a hierarchical system of distribution where the points of original dissemination are controlled.

In the digital version of copyright laws, especially as it has been manifested in the DMCA or the European Copyright directive, the control becomes even more stringent, with copyright being transformed into an access–right. In terms of the way copyright laws are created, it seems that content disseminators or intermediaries of all kinds have a much stronger lobbying presence than consumers or creators. This becomes increasingly apparent after the mid 1990s, when there has been a realisation of the potential of digital dissemination.

4.1 Napster and Peer-to-Peer File Sharing

In May 1999 Shawn Fanning, an undergraduate student at Northeastern University, created an application called Napster. The idea behind Fanning's software was to enable end–users to share the MP3 files stored in their computers, using a centralised indexing service to locate the files. Two years after its launch, Napster had experienced an exponential growth to reach an audience of over fifty million users. Napster's popularity resulted in a lengthy legal battle between the music industry and Fanning's newly founded company on issues of copyright infringement (Alderman, 2001).

Most of the peer-to-peer services have been entangled in legal disputes with the media industries. Napster was sued in 1999 and a shortly thereafter Scour faced a very similar fate (Alderman, 2001). Both companies had to suspend services (from July 2001 and December 2000 respectively). In the midst of a series of legal developments and extensive media hype, the Recording Industry Association of America (RIAA) and Motion Picture Association of America (MPAA) have also gone after a number of other peer–to–peer services based on the most advanced peer–to–peer technology available at the time: that provided by FastTrack. Morpheus, KaZaA, Xolox and Grokster have been the targets of the media industry both in the U.S. and in The Netherlands. Xolox was shut down in the process only to open some months later when the case against FastTrack in The Netherlands was resolved in favour of the technology company. Morhpeus and KaZaA have survived to become two of the most widespread file–sharing applications.

5 ANALYSIS

In order to understand these developments, this paper will use the third generation understanding and, in particular, Lessig's four modalities of regulation: markets, laws, norms and technology. These will be used to show the different ways in which the audio-visual industry has sought to address the threat to copyright from peer-to-peer networks.

Markets. The main strategy of the content industry up to now for dealing with the copyright problems offered by Napster and peer-to-peer networks has been to identify bottlenecks of control and then charge them with contributory, vicarious or wholesale infringement of copyright, thus directly impacting the 'market' for such files. This is a tactic that makes sense, especially in terms of copyright law, since the latter is structured in such a way that it encourages the monitoring of the content users by the content distributors through liability clauses. Copyright law is structured on the assumption that the distribution of copyrighted material happens in a more or less centralised and therefore more or less controllable way: if the bottlenecks can be controlled then the end users can be controlled. RIAA and MPAA have followed a variety of methods for controlling bottlenecks, from high profile lawsuits against peer-to-peer services to recommendations to colleges and universities, where much file sharing is happening, accompanied by warnings that if no measures are taken in due time, legal action would follow. Prominent examples include that of Carnegie Mellon University in Pittsburgh, where in late 1999 seventy-one students were disciplined for the illegal use of MP3 files on the University's intranet (Wired News, 1999b).

Laws. Another approach for the copyright holders is to sue directly the infringers, i.e. the end users. This approach has some inherent problems since it is problematic for a company to start suing its own actual and potential customers. This was manifested in the Levy case. Geffrey Gerard Levy was a student at the University of Oregon and was prosecuted under the No Electronic Theft (NET) Act v1997 for illegally distributing MP3 files as well as pirated software and clips from theatrical movies (Wired News,

1999a). He was reported by network administrators at the University of Oregon. Despite its success in legal terms, the Levy case has proven to be a public relations disaster for the audio–visual industry and it does not appear to have deterred any users from file–sharing practices. The content industry has repeatedly leaked to the press its intentions to go after the file–swappers themselves. It has also been the case that internet service providers have addressed e–mails to their customers asking them to stop sharing certain files under the penalty of having their accounts terminated.

Norms. The litigation instigated by the audio-visual industry have been supplemented by a series of awareness initiatives, but these have tended to have limited success. For instance, in October 2001 the Disney Channel aired an episode of the Proud Family, a cartoon series aimed at the pre–teen audience, where the heroine Penny Proud realises the dark side of file-trading after she has been threatened with arrest by the police, been deprived of her computer and found that her local store had gone out of business (Wired News, 2001). It would, however, be quite naive to believe that an industry that has been based on the production of content that emphasizes anti–conformism and rebellion could manage to pass the message of compliance as easily as some of the top executives would like it to happen. "Homer's gaffe" (whereby the UK website for the Simpsons briefly gave advice on how users could circumvent the technological measures that limit where the DVDs could be played (BBC news, 2002b)) is only the latest in a series of self-conflicting practices that the media industry has followed.

Technology. Other strategies that the audio-visual industry has adopted include the development and evolution of Digital Rights Management (DRM) systems for the protection of their content. Despite the enthusiasm that followed the introduction of DRMs as an alternative way to protect content, it has become increasingly apparent that the very construction of DRMs has been their main problem. Most of the existing DRMs impose restrictions on the way the users experience music either in terms of the time(s) they can listen to a track, or the players they can use to listen to it or in terms of the portability of the files. The existing distribution systems used can match neither the ease of use nor the variety of content that the peer-to-peer services provide their users with. Nevertheless research on DRMs continues and is still seen as one of the possible solutions to the peer-to-peer "problem". Most of the last generation file-sharing networks allow the display of information concerning the files being shared to help users in their searching and this could be seen as a form of DRM forerunner. Files distributed by EMI's 2Ksound music label can be found over the KaZaA network and this has been used as an argument for supporting the distribution of music files with DRMs over peer-to-peer networks.

All these developments happen in the footprint of the U.S. Digital Millennium Copyright Act that contains provisions supporting technical measures of protection and under the debate concerning the proposed Security Systems Standards and Certification (SSSC) Act, which would make it a civil offence to sell or create any kind of computer equipment that "does not include and utilize certified security technologies" approved by the federal government.

5.1 Providing Safe Harbours: Where Technology and Jurisdiction Interact

Despite their multi-level nature, the efforts of the audio-visual industry to combat the file-sharing phenomenon have not been met with particular success. Following a trajectory of increasingly stricter measures and after having exhausted legal, technical and audience–shaping methods, the copyright holders have to find refuge in other kinds of measures. These are called "self-help measures" by their proponents and "hacking" by their opponents.

This solution has been advocated by Congressman H. L. Berman and is based on the idea of allowing a safe–harbour status for copyright holders when trying to protect their content even if in the course of their attempts they are committing acts including "interdiction, decoy, redirection, file-blocking, and spoofing" (Berman, 2002; CEI, 2002).

6 DISCUSSION

The third generation approach supports the presence of multiple modalities of regulation, including norms, markets, and technologies. Law is merely one form of regulation that is subject to national jurisdiction, and sometimes even more limited than that. Norms and markets are increasingly global; while technology may be used across borders. In this sense, any given user is subjected to a number of regulatory jurisdictions at a given time and information and communication technologies only increase this diversity.

There are two main implications for this approach. The first is the mixing of modalities that subject the user to greater control and possibly coercion, with decreased accountability. The second is the notion of regulatory patching that may indeed grant the user some autonomy. These are then combined in the notion of the commodification of regulation (Romano, 1985).

6.1 Mixing Modalities

As there are multiple sources of regulation, each may establish jurisdiction over the user. The sources of regulation also intermix with each other. Laws are created to shape technology. As discussed above, DRM technologies are protected in the U.S. by the DMCA.; if you reverse-engineer the protections within a DRM technology you are in contravention of U.S. law, regardless of whether under traditional copyright laws you are permitted to access the protected data. National laws and policies that dictate the form and structure of information technology are interpreted and embedded within a technological solution, and when this solution is used world-wide, it carries the politics of its creation: in effect, it enforces foreign law and policies. The jurisdiction of the law therefore extends. In one recent case, because the reverse-engineered solution was made available on the internet and was to be discussed at a conference within the U.S., the U.S. Department of Justice established jurisdiction over conduct occurring abroad when a Russian computer programmer visited the U.S. for a conference (Sklyarov, 2002).

As a result, the first generation thinkers may have had a point: information and communications technologies do extend space; but it is not necessarily a space of its own. The interaction amongst the modalities is what is interesting; not the idea that one modality operates in isolation of the others. Whilst the second generation thinkers would argue that being regulated by multiple jurisdictions at one time is what they have been arguing all along, they fail to appreciate exactly how the technology takes on a regulatory form, how the other modalities may shape its regulatory tendencies, and how the modalities together may extend (or decrease) each others jurisdictions.

The challenge for a user then becomes a situation of trying to understand their jurisdiction. While users are normally subject to national law while operating within a country, they may be subject to foreign laws: technology-use policies of DRM systems, acceptable–use policies of service providers, and even the regulations embedded within peer-to-peer services. The legal and ethical problem is then to identify the political system that can be held accountable for the existence of the regulation, and who to appeal to for changes to it.

6.2 Regulatory Patching

A second implication can be considered the inverse of the regulatory polyphony that exists under the mixing of modalities. This inverse situation can be called "regulatory patching". It refers to the situation where the subjects "build" the regulatory 'ecology' that they wish to be subjected to. Unlike the traditional regulatory model where there is a single non–modifying regulation for all subjects, traditionally the state, there may be multiple regulatory modalities acting upon the user. As a result some form of regulatory selection may occur.

For instance, a person may choose to use MP3.com with its streaming technology in order to access independent label artists and KaZaA to download mainstream music. That means that the person is subjected to the KaZaA regulatory regime that allows the sharing of music amongst the participants of what is referred to by the KaZaA company as the "KaZaA community", but at the same time it installs spyware on the participants computers. Spyware is a kind of software that transmits data about web usage and the kind of files that users store on their hard drive.

Users may not want to be subjected to such kind of regulation. Therefore, they have the option of altering the regulatory regime and thus shifting the boundaries of jurisdiction with the addition of software patches that allow the bypassing of spyware during the execution of the original KaZaA program – with the use of KaZaA lite – or the removal of spyware components once the use of the program has terminated, e.g. with Ad-Aware.

In addition, in order to regulate the phenomenon of free–riding, where participants of a peer-to-peer network only download files and do not post files for others to download, KaZaA has implemented the participation level specification in their software. The participation level relates to the computing and content resources, in terms of computing power, bandwidth and number of files that an individual user is ready to devote to the network. The higher the participation level the more the access to the content that the user has. Thus, in order to download files you must provide files for upload. This is a regulatory mechanism enforced in order to encourage participation of the users. However, the user can always download the patch of KaZaA lite 2.0 which creates a fake participation level for users and thus allows them to continue to download files without really sharing them. In such a case there is not a common regulatory mechanism for everyone but each user has – at least in principle – the option to construct their own regulatory scheme. This can be considered as 'regulatory patching'.

There are also alternative technologies that allow users to change their internet protocol addresses, in effect changing their network–apparent geographical location. Many of these services actually exist in the U.S., however, and as it pertains to intellectual property laws, locating one's self virtually in the U.S. may actually be more hazardous; but if the users are concerned about other regulations such as free speech protections in their home country, then the shift to the U.S. jurisdiction might be understandable.

To some extent, there are parallels with real-world legal issues. A classic example is that of off-shore companies that choose to be located in jurisdictions that have a more relaxed taxation regulations. While this practice of regulatory arbitrage (i.e. of companies in particular, choosing to be subjected to one particular set of laws and not to another because of the better regulatory options that one jurisdiction offers over another) is not uncommon, what is special about this case is the way in which the arbitrage can be created through the technology on an individual basis. Users may, even in the absence of knowing what regulations they are necessarily subjected to, choose to alter the controls through regulatory-patching.

6.3 The Commodification of Regulation

The practice of "regulatory patching" proves the fundamental change regulation has been subjected to. It is not a new theoretical understanding of what regulation does or how regulation operates in a global setting; it is how individuals actually interact with regulatory structures (including technology) that seems to become substantially different to all pre-existing regulatory notions. The findings presented in this paper are just based on policy makers' interventions or theoretical assumptions regarding the nature of interventions. Sources like these were certainly helpful for understanding the trajectory of regulatory evolution, however, it is the practice of those subjected to regulatory regimes that leads this new regulatory debate. This constitutes a paradigmatic shift from traditional regulatory studies. Even in cases where formulations such as self-regulation were studied, the sources of regulatory intervention were not coinciding in their entirety with the subject of regulation. When, for instance, the self regulation of the banking sector is studied, the consumer who is essentially affected and regulated by such a mechanism is either not examined at all or is approached as a passive receptor of the regulatory effort. On the contrary, in this paper, the information systems that have been studied allow regulation to move from a "turn key" agreement, where the only choice for the regulated is that of which regulation to choose but to a "do it yourself" construction, where the end–users "construct" their own regulatory ecology.

REFERENCES

Akdeniz, Y. "Case Analysis of League Against Racism and Antisemitism (LICRA), French Union of Jewish Students, V Yahoo! Inc. (USA), Yahoo France, Tribunale De Grande

Instance De Paris, Interim Court Order, 20 November 2000", *Electronic Business Law Reports*, (1:3), 2001, pp. 110-120.

Alderman, J. *Sonic Boom: Napster, P2P and the Future of Music*, London: Fourth Estate, 2001.

Angell, I. O. *The New Barbarian Manifesto: How to Survive the Information Age*, London: Kogan Page, 2000.

Australian Broadcasting Authority. "Broadcasting, Co-Regulation and The Public Good", 1999, *NR* 101/1999.

Avgerou, C. *Information Systems and Global Diversity*, Oxford: Oxford University Press, 2002.

Barlow, J. P. "Declaration of the Independence of Cyberspace", 1996. Available at: http://www.eff.org/~barlow/Declaration-Final.html. (Accessed at: February13, 2003)

BBC news, "China Internet Firms 'Self-Censoring'". 2002a. Available at: http://news.bbc.co.uk/1/hi/world/asia-pacific/2098530.stm. (Accessed at: February13, 2003)

BBC news, "Homer Says Hack your DVD Player", 2002b. Available at: http://news.bbc.co.uk/hi/english/sci/tech/newsid_2124000/2124345.stm. (Accessed at: February13, 2003)

Beck, U. *What is Globalization?*, Translated by P. Camiller. Cambridge. Polity Press, 2000.

Berman, H. L. "Regarding Solutions to Peer-to-Peer Piracy". 2002. Available at: http://www.house.gov/berman/p2p062502.html. (Accessed at: February13, 2003)

CEI. "IP and Technological Self-Help: Barbed Wire for the Digital Age". 2002. Available at: http://cei.org/gencon/016,03114.cfm. (Accessed at: February13, 2003)

Clarke, R. "Subject: ABA Demonstrates its Ignorance to the World", Forwarded to the Politech Mailing List, message titled *FC: More on Australian official demanding Net-regulation - demonstrating ignorance to the world*, November 3, 1999 10:42:30 -0800 1999.

Dogcow. *Evading the Broadcasting Services Amendment* (Online Services) Act 1999. 2600 Australia, 1999.

Electronic Privacy Information Center. "Faulty Filters: How Content Filters Block Access to Kid-Friendly Information on the Internet", 1997.

European Union, "Directive 95/46/EC Of the European Parliament and the Council of 24 October 1995 on the Protection of Individuals with Regard to the Processing of Personal Data and on the Free Movement of Such Data", *Official Journal* L281, 1995.

Goldsmith, J. L. "Against Cyberanarchy", *University of Chicago Law Review*, (65), 1998a, pp. 1199-1250.

Goldsmith, J. L. "Symposium on the Internet and Legal Theory: Regulation of the Internet: Three Persistent Fallacies", *Chicago-Kent Law Review*, (73), 1998b, pp. 1119-1131.

Hood, C. *Explaining Economic Policy Reversals*, Buckingham, England: Open University Press, 1994.

Johnson, D. R., and Post, D. G. "Law and Borders – The Rise of Law in Cyberspace", *Stanford Law Review*, (48:1367), 1996.

Lee, J. S. "Companies Compete to Provide Saudi Internet Veil", *The New York Times*, November 19, 2001.

Leeds, J. W. "United States International Law Enforcement Cooperation: A Case Study in Thailand", *Journal of International Law and Practice*, (7:1), 1998, pp. 1-14.

Lessig, L. "The New Chicago School", *Journal of Legal Studies*, (27), June 1998, pp. 661-691.

Lessig, L. *Code: And other Laws of Cyberspace*, New York: Basic Books, 1999.

Peltzman, S. "The Economic Theory of Regulation after a Decade of Deregulation", *Brookings Papers on Microeconomics*, 1989.

Perrit, H. H., Jr. "Jurisdiction in Cyberspace", *Villanova Law Review*, (41), 1996, pp. 1-64.

Rapporteurs sans Frontieres. "Report: The Twenty Enemies of the Internet", 1999.

Romano, R. "Law as Product: Some Piece of the Incorporation Puzzle", *Journal of Law Economics and Organization*, (1:2), 1985, pp. 225-283.

Sklyarov. "Free(d) Dmitry Sklyarov!", 2002. Available at: http://www.freesklyarov.org. (Accessed at: February13, 2003).

Sterling, J. A. L. *World Copyright Law: Protection of Authors' Works, Performances, Phonograms, Films, Video, Broadcasts and Published Editions in National, International and Regional Law*. London: Sweet and Maxwell, 1998.

Turkle, S. *Life on the Screen: Identity in the Age of the Internet*, London: Weidenfeld & Nicholson, 1996.

Walsham, G. "Globalization and IT: Agenda for Research". A paper delivered at the *Organizational and Social Perspectives on Information Technology*, Aalborg, Denmark, June 10-12, 2000.

Wired News, "DOJ Cracks Down on MP3 Pirate", Wired, 1999a. Available at: http://www.wired.com/news/print/0,1294,21391,00.html. (Accessed at: February13, 2003).

Wired News, "The Student Jukebox Sting" Wired, 1999b. Available at: http://www.wired.com/news/print/0,1294,32444,00.html. (Accessed at: February13, 2003).

Wired News, "Disney's Peer-to-Peer Pressure" Wired, 2001. Available at http://www.wired.com/news/business/0,1367,47806,00.html. (Accessed at: February13, 2003).

About the Authors

Edgar Whitley is a Reader in Information Systems at the London School of Economics and Political Science. He has a BSc (Econ) Computing and a PhD in Information Systems, both from the LSE. He has taught undergraduate, postgraduate students and managers in the UK and abroad. Edgar was one of the organisers of the First European Conference on Information Systems and is actively involved in the coordination of future ECIS conferences. He has recently been appointed Joint Editor in Chief for Information Technology & People and was one of the programme chairs for the IFIP 8.2 conference on global and organisational discourse and information technology held in Barcelona in December 2002. He can be contacted at e.a.whitley@lse.ac.uk and his homepage is http://is.lse.ac.uk/staff/whitley.

Prodromos Tsiavos is a Teaching Assistant in Information Systems Development and Design at the London School of Economics. He was born in Athens, Greece. He is a barrister in Athens Barrister Association and holds an LLB (Athens Law School), an LLM in Telecommunications and Information Technology Law (Queen Mary and Westfield College) and an MSc in Analysis Design and Management of Information Systems (the London School of Economics). He is currently completing his doctoral thesis on Copyright regulation and open source development of peer-to-peer networks in the department of Information Systems at the London School of

Economics. He is funded by the Greek Scholarships Foundation for conducting research in Information Technology Law. His research interests include e-regulation, technical measures for Copyright protection, peer-to-peer networks and open source software development. He can be reached by e-mail at p.tsiavos@lse.ac.uk.

Ian (Gus) Hosein is a Visiting Fellow in the Department of Information Systems at the LSE, where he is also pursuing an elusive PhD. His prior research has been in technology policy and regulation, and his current research is on international cooperation in criminal matters and anti-terrorism policies; as a result making himself unpopular with both IS students and governments. He can be reached by e-mail at I.Hosein@lse.ac.uk. More information can be found at http://is.lse.ac.uk/staff/hosein.

Part 5:

Cultural, Philosophical and Policy Issues

23

THE LINK BETWEEN ICT AND ECONOMIC GROWTH IN THE DISCOURSE OF DEVELOPMENT

Chrisanthi Avgerou
London School of Economics, UK.

Abstract: In this paper, I examine the validity of the relationship between ICT and economic development that has been constructed in the discourse of some influential international development organizations. I argue that the tool-and-effect association suggested in such discourse is dubious and misleading. It is based on narrow economic theory and ignores both the controversies that surround it and empirical evidence of alternative development policies. I point out that the policy analyses and recommendations of major development organizations influence the interventions of information systems professionals in developing countries with misguided perceptions and prescriptions that stifle the undertaking of situated efforts to put ICT to effective use.

Keywords: Economic development, economic growth, economic theory, economic policy, ICT and development.

1 INTRODUCTION

A striking feature of the world at the beginning of the 21st century is the gross inequalities between the socio-economic conditions of different communities. The most visible of these relate to the world development problem of inequalities among nations. Contemporary discourses on development consistently identify ICT as a requirement for economic growth and the improvement of social conditions. Strictly speaking, this is not a new discourse; ever since the advent of computers, government policy advisors and international development agencies have pointed to the opportunities the technology opens for development. More recently, however, the link between ICT and development has been articulated in the alarming terms of the 'digital divide'. There is concern that developing countries are deprived of the opportunities for economic growth and life improvement generally enjoyed by advanced economies because of the scarcity of ICT, particularly limited Internet connectivity.

The lack of ICT is understood to be an important factor contributing to the widening of the gap between 'developed' and 'developing' countries, as shown by world socio-economic indicators published in the annual reports of international development agencies, such as those from the World Bank and the United Nations Development Programme (UNDP) discussed later in this paper. Many high profile initiatives have been undertaken to remedy this problem. They typically aim to create awareness on the benefits of ICT, raise investment, and promote policy measures for the deployment of telecommunications infrastructures and the diffusion of ICT applications in all societal sectors. Notable examples of these projects include the Digital Opportunity Task Force of the eight major industrial nations, G-8 (Dot force initiative, http://www.dotforce.org), the World Summit for the Information Society of the United Nations and the International Telecommunications Union (WSIS initiative, http://www.itu.int/wsis) and the World IT Forum of the International Federation of Information Processing (WITFOR programme, http://www.witfor.lt).

For many information systems scholars and professionals, such a general association of ICT with socio-economic effects is of questionable validity. It is well understood in information systems studies that the actual 'effects' of ICT in the place where it is used cannot be identified in terms of the potential of the new technologies as manifested in the laboratory or as realized in other social settings. ICT innovation is a process that takes place within the formative conditions of a particular social and organizational context (Suchman, 1987; Ciborra and Lanzara, 1994; Avgerou, 2002a). With specific reference to the question of ICT and development, the literature on information systems in developing countries includes a substantial amount of empirical evidence, mainly case studies, that reveals the situated manner in which information systems projects take shape within communities striving to improve their life conditions – see, for example, the publications of past IFIP 9.4 conferences (Bhatnagar and Bjorn-Andersen, 1990; Bhatnagar and Odedra, 1992; Odedra-Straub, 1996; Avgerou and Walsham, 2000; Sahay, 2000; Krishna and Madon, 2002). At the organizational level of analysis, information systems researchers and professionals are well aware of the tension between the situated nature of the course of change and general, apparently rational, theoretical propositions on the way ICT impacts – or should impact – on organizational performance.

Nevertheless, the discourse of international development agencies on the role of ICT merits attention in information systems research because it constitutes part of the institutional context of the micro-level processes involved in the formation of information systems (Avgerou, 2002b). This discourse influences the legitimacy of professional interventions towards specific objectives and sensitizes 'users' to a particular view of the way ICT may affect their lives. The current emphasis on the digital divide as the major contemporary problem facing developing countries also determines the way the meaning of ICT-based information systems is understood in universalist terms. It conveys specific views on why Internet connectivity is important and what it should achieve for even the remotest communities of the world. For example, interventions to develop community ICT services in poor regions bear implicit promises for economic benefits through participation in the global market and for rationalized citizens/government interactions. Moreover, there is a tendency to see such ICT centres as sustainable businesses in their own right (Best and Maclay, 2002). In other words, a universalist discourse on ICT and development constructs and spreads in developing countries specific development visions of new, technology-mediated modern lives.

In this paper, I examine the relationship between ICT and economic development in four recent influential publications: UNDP's 2001 Human Development Report, *Making New Technologies Work for Human Development* (United Nations Development Programme, 2001); the 2002 World Development Report, *Building Institutions for Markets*, of the World Bank (2002); and two publications by the Center for International Development at Harvard University, *The Global Information Technology Report: Readiness for the Networked World* (Kirkman et al., 2002), and *The Global Competitiveness Report 2001-2002* (Porter et al., 2002a).

All these publications propose ICT as an instrument for economic and social gains within a market regime, and they elaborate on the conditions under which ICT plays this kind of developmental role. The central issue in the discourse in these texts concerns the socio-economic conditions that are favorable for the mutual re-enforcement of ICT innovation and an effective market. To examine the logic underlying the suggested conditions, I then look briefly at the theories that inform these documents and the controversies surrounding them. I conclude by arguing that the tool-and-effect link between ICT and economic development exemplified in these publications is dubious. My contention is that such a link is based on a narrow economic perspective of human action which ignores recent socio-economic theory of development and is not informed by the evidence on processes of development that has emerged from the few countries which achieved substantial economic growth in the last decades of the 20th century.

2 EXAMPLES OF THE DISCOURSE ON ICT AND DEVELOPMENT

UNDP's 2001 Human Development Report (United Nations Development Programme, 2001) is a good example of the association international organizations make between ICT and development, not least because this series of UNDP reports takes a broad view of development as a socio-economic condition that goes beyond a narrow consideration of economic growth. The 2001 UNDP report seeks to present a clear association between technology and desirable development effects, giving special attention to ICT – particularly the Internet. Indicatively, it quotes a World Bank study (Preker et al., 1999) that showed 'technical progress accounted for 40-50% of mortality reductions between 1960 and 1990 – making technology a more important source of gains than higher incomes or higher education levels among women' (ibid.: p.29). The UNDP report asserts that, '(c)ross-country studies suggest that technological change accounts for a large portion of differences in growth rates' (ibid.).

More importantly, this report attempts to qualify how technology, especially ICT, is 'enabling' development effects. The association between technology and human development is presented as follows. Technological innovation enhances human capabilities – such as a healthy life, knowledge, creativity, and participation in the social, economic, and political life of a community – and impacts on economic growth through productivity gains. At the same time, human capabilities are an important means for achieving technological innovation. Therefore, technology innovation and development are 'mutually reinforcing, creating a virtuous circle' (ibid.: p.28).

This 'virtuous circle' model is a significant step towards tracing the dynamic relationship of technology innovation and development, which goes beyond the static association of ICT diffusion and growth rates. Nevertheless, the UNDP report chooses to elaborate mainly on the argument that ICT innovation achieves developmental goals, thus retaining and cultivating a view of an instrumental relationship between the means of 'technological advance' and the desirable effect of 'human development'. Although the authors of the report explicitly recognize that technology may well be 'a reward of development', rather than being a tool for development, they are keen to dismiss this interpretation and assure readers that 'technology is a tool for, not just a reward of, growth and development' (ibid.: p.27).

This indicates a need to take a closer look at the reasoning that sustains the tool-and-effect association of ICT and development, in order to help answer the question: How is ICT understood to contribute to the process of economic development? For this, relevant explanatory insights are provided by the two recent publications of the Center for International Development at Harvard University mentioned in the previous section.

The report on the 'networked world' begins with the statement that 'the Internet and other ICTs have fundamentally changed the way the world works' (Kirkman et al., 2002). It then sets out to analyse, understand, and measure the link between ICT and development, with a particular focus on the issues of developing countries. The economic reasoning of its premise on the role of ICT in the development process is that the technology enhances the functioning of the markets because it provides information to producers and consumers in order to help them make efficient choices (Eggleston et al., 2002).

Kirkman et al. (ibid.) propose a framework of factors contributing to a country's capacity to exploit the opportunities offered by ICT. On the basis of this framework, their report derives the 'Networked Readiness Index' and ranks 75 countries on it. This index is composed of (a) measurements of the diffusion of the Internet and other ICT components and (b) assessments of a number of factors considered to be preconditions for high quality use of the Internet and its further proliferation and application. Important factors to that end identified by the authors of the report include: infrastructures for network access; the level of competition in the economy, particularly in the telecommunications and ICT sectors; social conditions, such as education level and the incorporation of ICT in education; and the extent to which ICT has been incorporated into business and government activities.

In this analysis, the market is the mechanism through which ICT is associated with economic development. In effect, this report identifies the virtuous circle process more specifically in terms of ICT and an effective market regime. But the relationship between ICT and market-driven development is presented in a self-referential way. The existing capacity of ICT in the socio-economic fabric is considered a condition of 'readiness' for further ICT development through network-based activities. The diffusion of ICT in all sectors of the economy and society, together with the liberalization of the telecommunications sector, are set up as desirable policy targets in their own right. And it is assumed that market mechanisms are required to achieve the developmental potential of the technology. For example, 'quality of learning' is taken to be the extent to which ICT is incorporated into education, and the privatization of telecommunications is identified as the main criterion for assessing network policy. However, at present there is little evidence that ICT contributes to better educational systems, even in industrialized societies, while there are studies showing that market mechanisms cannot be relied upon to provide telecommunications access for poor communities in remote areas of developing countries (Bhatnagar, 2003).

Porter et al. (2002b) in their *Global Competitiveness Report 2001-2002* provide an analysis that differentiates the role of ICT for development in different socio-economic conditions within the global market. They present economic development as a process that moves successively through three general states based on national income levels. At the low-income level, economic growth is determined mainly by the mobilization of land, primary commodities and unskilled labour. At middle-income level, national economies get integrated into the international production system and economic growth is increasingly achieved by adopting foreign technologies in local production. Economies at the high-income level achieve global competitiveness through rapid technology innovation and high rates of learning, especially science-based learning.

In discussing these three states of development, Porter et al. (ibid.) suggest that technology innovation has little significance in low-income economies, for which the main challenge is to get the basic market factors of land, labour, and capital to work properly. The harnessing of 'global technologies' acquires greater importance as countries move from low to middle income. The institutional characteristics of the knowledge-based economies at the high-income level include continuous training and upgrading of the workforce, high labour mobility across enterprises, and a dynamic combination of fierce competition and cooperation among enterprises. Governments play a crucial role in the higher education, R&D, and market regulation that supports start-ups and high-tech enterprises, while business firms become less hierarchical and form flexible buyer-supplier networks.

From this report's perspective, the role of ICT therefore varies according to the extent to which a country's market economy has developed the capacity to enter the global market and to sustain competitive advantage. The problem of pursuing the virtuous circle of ICT innovation and development in the global competitive market surfaces again here. The analysis of the role of ICT in terms of competitiveness does not explain how progression on the ranking scale occurs. The linearity of the model and the notion of competitiveness suggest that the more successful economies in the global market are more capable of technology innovation to enhance their economic gain and, thereby, to disadvantage those less techno-economically capable of doing so.

Indeed, the authors of the report note that the hardest transition along their three-stage model of development is from a technology-importing, efficiency-based 'middle' level economy to the innovation-based, high-income knowledge economy. They point out that the challenge for policy lies in the process of adaptation to new institutional conditions at the transition points of the model. This observation suggests the need to shift the study of the way ICT is associated with development from tracing the purely economic reasoning of the market to consideration of the social conditions that sustain it. What are the institutional conditions of the mutually reinforcing processes of ICT innovation and a growth-fostering competitive market?

The 2002 World Development Report of the World Bank elaborates on institutional conditions considered conducive to competitive economic action (World Bank, 2002). It explains a particular view of how institutions support markets: they channel information about market conditions, goods, and participants; they define property rights and enforcement mechanisms; and they increase competition in markets. Thus, state government is seen to be important for the regulation of property rights and their enforcement.

This report highlights the importance of the historical context of an economy and advises that new institutions should be built by complementing existing ones. Nevertheless, it provides clear direction towards 'good governance'. This is understood in a somewhat cyclical manner as the state provision of: institutions for the creation, protection, and enforcement of property rights; a regulatory regime to promote market competition; and sound macroeconomic policies that create a stable environment for market activity. In short, to the extent that social change is understood to be implicated in the dynamic intertwining of technology innovation and an effective market economy, the current discourse on development seeks to emulate the institutions of the few societies that have achieved the mobilization of ICT innovation to sustain economic growth through competitiveness in global markets.

However, this leads to the question of whether it is possible for developing countries around the world to emulate the human and institutional conditions of the few techno-economically advanced societies, and whether such emulation is an effective development strategy. I examine these questions in the following section by addressing the economic theory that underpins the views of market-driven development.

3 CONTROVERSIES IN ECONOMIC THEORIES OF DEVELOPMENT

The central concern preoccupying the four publications examined in the previous section focuses on how ICT can become an instrument for development in the context of the global market economy. This reflects the prevailing views in this domain and takes us to the core issue of development theory since the mid 20th century: economic growth. However, these views are not uncontroversial. To understand better the limitations of these propositions on ICT and development, and the reasoning that seeks to justify them, I now examine the theoretical underpinnings of economic development and their related policy debates.

In a nutshell, the prevailing perception of the development problem focuses on how to establish efficient markets in societies that, for various historical reasons, have low production capacity, ineffective allocation of existing productive resources, and inadequate trade mechanisms. In neo-classical economic theory, economic growth rests on two fundamental assumptions: the rational behaviour of economic agents – individuals and business organizations – and the capacity of market competition to eliminate inefficient producers and to create equilibria of production and consumption at optimal conditions of full employment and the lowest consumer prices. From this view, development therefore consists of efforts to transform the socio-economic regimes that exist in different countries into such free markets. To achieve this, development policies are expected to include: abolishing protection of national industries from foreign competition and eliminating trade barriers, in particular barriers to the flow of capital; privatization of organizations governed by political control, such as telecommunications; and exploitation of natural resources, mainly oil and minerals. Governments following such policies assume the minimal regulatory role of overseeing the legal framework of property rights that is required for the free market.

Two serious problems arise when putting this theory into practice. First, the experience of applying these principles in Western economies has led to an understanding that the free market tends to run into crises, known as 'market failures'. Imbalances of production, prices, and consumption give rise to combinations of undesirable conditions, such as inflation and unemployment. The second problem is that although, arguably, the theory is the basis of economic performance of the few rich countries of the world, little is known about how it can be fostered in the countries that are currently poor.

The main challenge to the neo-classical economic perspective of development stems from the theory of new institutionalist economics (NIE). This assumes that the rational individual choice on which neo-classical theory is built is unrealistic because individuals rarely possess complete information about the market. As a result, economic transactions require the search for relevant information, with such a search entailing costs. In addition, economies involve negotiations and the establishment of contracts for minimizing risks; they also need mechanisms to enforce such contracts. Moreover, the rationality of individuals' decisions is biased by their 'mental models' (North, 1995), their culturally-formed values about the world. Collectively, these factors can lead to inefficient economies.

With these observations, NIE broadens the analysis of rational behaviour across cultural and political dimensions. It shifts attention to the significance of institutions, that is, the formal rules and informal conventions that govern the behaviour of economic actors – whether individuals or firms. Institutions are important because they limit the scope of search of economic choices, thereby reducing transaction costs. They also reduce uncertainty by providing enforcement mechanisms. While market competition remains the core mechanism for increasing efficiency, NIE shows economic and non-economic institutions play a key role in shaping the economy. Indeed, the economic history of today's advanced economies shows the importance of government organizations in preventing and correcting market failure. A network of organizations standing watch over the market dynamics are ready to intervene, primarily in financial and capital markets, not only with regulatory capacity but also as economic actors in their own right. Moreover, state organizations play a significant role in shaping the political rules and social norms that drive, or at least influence, the decisions of economic actors.

In effect, NIE regards non-market institutions as mechanisms for overcoming the costs and risks of the market, thus serving the economic interests of rational individuals and firms. It provides both the grounds for an interventionist state and legitimacy for the involvement of multiple civic actors in the development effort, which are seen as necessary for providing support in the transition to an efficiently functioning, largely free-market economy. Moreover, NIE recognizes the diversity of mental models that are found in different cultures and implicated in economic behaviour and institutional arrangements. Policy possibilities depend on the institutional history of an economy. In the jargon of NIE, economic change is 'path dependent', in that it is constrained by the historically developed institutions of a society (Toye, 1995).

Policies of development in almost all countries demonstrate the influence of varying combinations of neo-classical and NIE theories. Many policy analysts, though, are critical of the prevalence of neo-classical ideas in the policy recommendations of the most powerful international development agencies. For instance, Stiglitz voices strong criticisms of the policies of the International Monetary Fund (IMF) and US Treasury – what he calls the new 'Washington Consensus' about the 'right' policies for developing countries – because they are based on the orthodoxy of the neo-classical economic prescription of economic restructuring towards a global free market regime with unrestricted trade and capital flows (Stiglitz, 2002). This kind of policy, he argues, does not target the poor who are only marginal in the underdeveloped markets; the expectation is that the incorporation of developing countries into the global free market will have 'trickle-down' beneficial effects for their poor populations. Stiglitz's critique emphasizes the importance of building institutions for the free market along locally-meaningful routes, such as the World Bank's advice to establish a competitive banking system and a legal system able to enforce contracts (World Bank, 2002).

Stiglitz (ibid.) goes even further than suggesting the importance of developing institutions for the market. He is critical of policies that aim to reduce the role of government in developing countries as he argues that strong and effective government, rather than less government, is required for the transition to effective markets. He looks at the course of development during the 1990s of different countries, some of which followed the development prescriptions of the IMF while others resisted these pressures by implementing policies and taking action they considered locally appropriate. He cites China, Malaysia and Poland among the developing countries that have done better in raising their populations above poverty, avoiding crises, and coping with international economic crises. He argues that their governments have succeeded by seeking to change towards free-market regimes through a careful sequencing of the opening of the economy to the building of necessary institutions and by the pursuit of 'homegrown' policies sensitive to the specific needs and concerns of the country.

Other critics of neo-classical economic theory highlight the significance of politics in the development process (Bates, 1995; Leftwich, 2000). The basic premise here is that the assumption of rational choice for the economic actor is complicated by a diversity of interests, preferences, values, and ideas. Thus, economic behaviour – basically the use, production, and distribution of resources – involves activities of conflict, cooperation, and negotiation. This consideration of politics alters significantly the understanding of the market-led development process in at least two ways. First, it highlights the centrality of state government in shaping the course of development. Second, it reinforces the contextual, historically contingent nature of a society's development process.

Evidence for the validity of such an understanding of the development process comes from the history of the so-called 'developmental states' clustered in East and South East Asia: South Korea, Taiwan, Singapore, China, Indonesia, Thailand, and Malaysia (Wade, 1990). These are the very few countries that managed to achieve average annual GNP per capita growth rates in excess of 4% between 1965 and 1997. Governments in all these countries had a much more prominent interventionist role in the economy than suggested by neo-classical economic theory and most versions of the NIE. For instance, the economic growth of Taiwan, South

Korea, and Japan in the 1980s shows that the governments of these countries used state power heavily to manage the market (Wade, ibid.). This included raising surplus for investment, ensuring that a high proportion of this was invested in productive capacity within the country and in industries that would yield higher wages in the future, and exposing investment projects to international competition.

Studies of technology innovation in the developmental states have also recognized the significance of interventionist government policies in fostering high-tech industries, as well as in steering technology innovation across industries (Freeman, 1987; Hobday, 1995; Archibugi and Michie, 1997). However, it is important to note that studies of this region have not resulted in alternative universal theories of development, or general models of techno-economic action. In looking at the way Korea, Taiwan, Singapore and Hong Kong achieved innovation-driven growth, Hobday (1995: p.4) notes that '(h)ighlighting the plurality of government policies and development models within the region, the study shows that there was no single path to development, nor any single model or lesson for other developing countries'. The main conclusion towards which such studies converge is the broad observation that the political institutions of the developmental states diverge from those associated with successful economies in western democracies.

The position outlined above can be summarized as suggesting economic development is a situated, context-specific process that is entangled with indigenous politics and historically-formed institutions. This echoes Granovetter's (1985) thesis of economic action embedded in social structures; his critique of economics as an under-socialized conception of rational action is particularly relevant here. In this light, it is not surprising that ICT does not serve all societies equally well as a tool for development. Therefore, although universalist models of rational economic behaviour may offer useful working hypotheses, the social relations within which economic action is embedded may drastically change the scope of desirable, feasible – and therefore rational – action.

4 CONCLUSION

The static picture of ICT and development measures presented in the tables of development indicators assembled by international development agencies makes a strong association between ICT and development: the more successful economies have more technologies and are better prepared for using them to their competitive advantage. This paper's brief discussion of four such publications has shown that this association tends to be interpreted as indicating that ICT is an instrument for development.

However, if we consider the dynamics of ICT and development, that is, if we consider ICT innovation and development as processes rather than as states exemplified by existing societies, the close correlation between ICT innovation capabilities and success in the market tells a different story. It shows that a few economies have historically developed an institutional setting that sustains the mutual re-enforcing of competent free-market economic activity and ICT innovation, but that such a process has not been set in motion in developing countries. Yet, developing countries are now advised to simultaneously: acquire the ICTs that served the advanced market economies well; emulate their institutions; and engage in innovation-driven free market competition. This is an unrealistic expectation because, as the critics of neo-classical economic theory and policy have pointed out, economic and institutional change is a path-dependent, historically-contingent process. Thus, ICT continues to be a factor responsible for the widening of the huge difference between the rich and the poor societies measured along the multiple linear scales of progress in the global market economy.

This argument does not suggest that ICT is inappropriate for developing countries, but it does indicate the misguided nature of the universalist visions of economic and institutional development that currently accompany efforts to promote the diffusion of the technology. These visions frustrate efforts to make sense of locally meaningful ways of accommodating ICTs in socio-economic activities. They prescribe what ICT is used for and restrict the scope for the improvisation that is necessary for making technology a trusted actor amidst the negotiations which bring about effective courses of action for change in industry or government.

Information systems professionals in developing countries have for several decades been called on to support the transfer of business practice that has been considered to be effective in the successfully competitive economies, such as business process re-engineering, integrated enterprise information infrastructures, or customer relationship management systems. More recently, they have been channeling their professional skills into e-government projects, which has involved them in intervening in the explicitly political setting of government administration. There is a widespread expectation that government can be transformed into a network of rationalized institutions, as seen desirable from an acontextual view of economic development.

It is important that information systems professionals in developing countries should be aware that this view is controversial in economic theory and policy, and that there is hardly any evidence to date that it delivers its promised results of entering a virtuous techno-economic circle. As emulation of western organizational practices in developing countries has rarely succeeded, the pleas in information systems literature for situated action appropriate to formative contexts have taken on a particular poignancy.

REFERENCES

Archibugi, D. and Michie, J. "Technological Globalisation and National Systems of Innovation: An Introduction", in *Technology, Globalisation and Economic Performance*, D. Archibugi and J. Michie, Cambridge: Cambridge University Press, 1997, pp. 1-23.

Avgerou, C. *Information Systems and Global Diversity,* Oxford: Oxford University Press, 2002a.

Avgerou, C. "IT as an Institutional Actor in Developing Countries", in *Information & Communication Technologies and Development: New Opportunities, Perspectives & Challenges*, S. Krishna, and S. Madon (eds.), proceedings of 7th International Working Conference of IFIP WG 9.4, Bangalore: Indian Institute of Management Bangalore, 29-31 May, 2002b.

Avgerou, C. and Walsham G. (eds.) *Information Technology in Context: Studies from the Perspective of Developing Countries*, London: Ashgate, 2000.

Bates, R. H. "Social Dilemmas and Rational Individuals: An Assessment of the New Institutionalism", in *The New Institutional Economics and Third World Development*, J. Harriss, J. Hunter, and C. M. Lewis, London: Routledge, 1995, pp. 27-48.

Best, M. L. and Maclay, C. M. "Community Internet access in rural areas: solving the economic sustainability puzzle", in *The Global Information Technology Report: Readiness for the Networked World*, G. S. Kirkman, P. K. Cornelius, J. D. Sachs, and K. Schwab, Oxford: Oxford University Press, 2002, pp. 76-89.

Bhatnagar, S. "Development and Telecom Access: Cases from South Asia", in *Information Systems and the Economics of Innovation*, C. Avgerou and R. L. La Rovere (eds.), Cheltenham, UK: Edward Elgar, 2003.

Bhatnagar, S. C. and Bjorn-Andersen, N. (eds.) *Information Technology in Developing Countries*, Amsterdam: North-Holland, 1990.

Bhatnagar, S. C. and Odedra, M. (eds.) *Social Implications of Computers in Developing Countries*, New Delhi: Tata McGraw-Hill, 1992.

Ciborra, C. and Lanzara, G. F. "Formative Contexts and Information Technology: Understanding the Dynamics of Innovation in Organizations", *Accounting, Management and Information Technology*, (4:2), 1994, pp. 61-86.

Eggleston, K., R. Jensen, and Zeckhauser, R. "Information and Communication Technologies, Markets, and Economic Development", in *The Global Information Technology Report: Readiness for the Networked World*, G. S. Kirkman, P. K. Cornelius, J. D. Sachs and K. Schwab, New York: Oxford University Press, 2002, pp. 62-75.

Freeman, C. "Technology Policy and Economic Performance", *Cambridge Journal of Economics*, (18), 463-514, 1987.

Granovetter, M. "Economic Action and Social Structure: The Problem of Embeddedness", *American Journal of Sociology*, (91:3), 1985, pp. 481-510.

Hobday, M. *Innovation in East Asia: The Challenge to Japan*, Cheltenham, UK: Edward Elgar, 1995.

Kirkman, G. S., Cornelius P. K., Sachs J. D., and Schwab K. *The Global Information Technology Report 2001-2002: Readiness for the Networked World,* New York: Oxford University Press, 2002.

Krishna, S. and Madon, S. *Information & Communication Technologies and Development: New Opportunities, Perspectives & Challenges,* Proceedings of 7th International Working Conference of IFIP WG 9.4, Bangalore: Indian Institute of Management Bangalore, 29-31 May, 2002.

Leftwich, A. *States of Development: On the Primacy of Politics in Development,* Cambridge: Polity, 2000.

North, D. C. "The New Institutional Economics and Third World Development", in *The New Institutional Economics and Third World Development,* J. Harriss, J. Hunter, and C. M. Lewis (eds.), London: Routledge, 1995, pp. 17-26.

Odedra-Straub, M. (ed.) *Global Information Technology and Socio-Economic Development,* Nashua, New Hampshire: Ivy League, 1996.

Porter, M. E., Sachs, J. D., Cornelius, P. K., McArthur J. W., and Schwab K. *The Global Competitiveness Report 2001-2002,* New York: Oxford University Press, 2002a.

Porter, M. E., Sachs, J. D., Cornelius, P. K., McArthur J. W., and Schwab K. "Executive Summary: Competitiveness and Stages of Economic Development", in *The Global Competitiveness Report 2001-2002,* M. E. Porter, J. D. Sachs, P. K. Cornelius, J. W. McArthur and K. Schwab, New York: Oxford University Press, 2002b, pp. 16-25.

Preker, A., Bos, E., Wang, J., Peabody, J., and Jamison, D. T. *Measuring Country Performance on Health,* Washington, DC: World Bank, 1999.

Sahay, S. *Information Flows, Local Improvisations and Work Practices,* Proceedings of 6th International Working Conference of IFIP WG 9.4, Cape Town, 24-26 May, 2000.

Stiglitz, J. *Globalization and its Discontents,* London, Penguin, 2002

Suchman, L. *Plans and Situated Action,* Cambridge: Cambridge University Press, 1987.

Toye, J. "The New Institutional Economics and its Implications for Development Theory", in *The New Institutional Economics and Third World Development,* J. Harriss, J. Hunter and C. M. Lewis, London: Routledge, 1995, pp. 49-70.

United Nations Development Programme. *Making New Technologies Work for Human Development,* New York: UNDP, 2001.

Wade, R. *Governing the Market: Economic Theory and the Role of Government in East Asian Industrialization,* Princeton, New Jersey: Princeton University Press, 1990.

World Bank *Building Institutions for Markets,* New York: Oxford University Press, 2002.

About the Author

Chrisanthi Avgerou (C.Avgerou@lse.ac.uk) is Senior Lecturer in Information Systems at the London School of Economics. She holds a degree in Mathematics from Athens University, an MSc in Computer Studies from Loughborough University and a Ph.D. in Information Systems from the London School of Economics. Dr. Avgerou teaches information systems implementation and information systems in developing countries. She chaired the IFIP WG 9.4 group from 1996 to 2003; since 1999 she has been vice-chairperson of the IFIP Technical Committee 9. Her research interests include the relationship of information technology to organisational change, and the role of IT in socio-economic development.

24

GOOD GOVERNANCE AND DEVELOPMENT AID

Risks and Challenges of E-Government in Jordan

Claudio Ciborra[1], Diego D. Navarra[2]

[1]Information Systems Department – LSE, UK; IULM – Milan; Institute of Informatics – University of Oslo, Norway; [2]Information Systems Department – LSE, UK

Abstract: Reforms and innovations in the governmental and bureaucratic apparatus of the nation-state are an important pre-requisite for development. This paper provides suggestive evidence that e-government and the related view of the neo-liberal state implicit in the good governance discourse put forward by the International Development Agencies might not be conducive to rapid late development. The New Institutional Economics as a background theory and Jordan as a case study show that a number of risks arising from external aid interventions are likely to exacerbate, and not resolve the problem of late development. The study is based on the analysis and design of e-government global solutions aimed at the creation of a neo-liberal state. It shows that implementing standardised ICT portfolio to support good governance may prove to be a very difficult task when contextual and other locally relevant variables are taken into account.

Keywords: E-government, good governance, LDCs, institutions, development, risk, Jordan.

1 INTRODUCTION

The recent efforts of G-8 countries to launch e-government projects to support 'good governance' in Late Developing Countries (LDCs) invite us to reflect on what is the role of the state proposed by the 'good governance' argument put forward by the International Development Agencies (such as the World Bank and the International Monetary Fund) and the scope of Information and Communication Technologies (ICTs) to enable the creation of such state. Specifically, to what extent can good governance as advocated by the 'Monterrey Consensus' contribute to development? What are the risks and challenges arising from the application of e-government in a late developing as opposed to an advanced state?

The alignment of International Development Agencies (IDAs), governments, civil society members and non-government organisations produced the 'Monterrey Consensus' in 2002 as a way to address key financial and development issues and to set development priorities for the next 20 years. Governance, which is a wide encompassing process involving the political, administrative and societal spheres of a country, is intended as 'good' when 'good policies' are implemented. In general terms, these policies have been reiterated in the Monterrey Consensus and are broadly aimed at: a) establishing democracy; b) eliminating corruption; and c) expand human capabilities.

A main point of the Monterrey Consensus is the introduction of ICTs in the development discourse and policy recommendations for LDCs, indicating that the creation of an information society, and especially e-government implementation, would lead to good governance by increasing transparency and accountability of governments, reducing the transaction costs involved in service delivery by public bureaucracies (Ciborra, 1993) and enhancing the participation of citizens, businesses and civil society. E-government, in very general terms, is the use of Information and Communication Technologies (with a special new emphasis on the Internet) by government agencies and institutions, and the on-line provision of government services. This represents a new project for LDCs with plenty of risks and challenges still to be fully investigated.

In what follows, the case of Jordan will be discussed as an example of an e-government policy initiative. The case is of interest because of its advanced programs for the creation of the information society, including the launch of a world class software industry and the networking of the nation. It is one of the rare countries in the Middle East with a history of commitment to ICTs related initiatives and to good governance: in the words of Mahmoud Khasawneh, Chief Information Officer at the Jordanian Ministry for Information and Communication Technology, '*We have no choice but to do it*'.

Section 2 will shed some light on the current debate surrounding e-government and good governance in LDCs through a review of the New Institutional Economics literature. Section 3 presents the main contextual and organisational features of the Jordanian e-government initiative. In Section 4 the case study findings are put forward. Conclusions follow.

2 ON GOVERNMENT, GOVERNANCE AND THE ROLE OF ICTs IN LDCs

2.1 The Relationship Between E-Government and Good Governance

E-government implies a drastic change in the way ICTs are used in government and a sharp departure from the 'traditional' way of working of government itself. While previous ICTs related efforts in the public administration were mostly based on the utilisation of technology for the automation of procedures, the achievement of cost savings and to streamline bottom-line performance, the goals of e-government are more ambitious. Namely, e-government means the use of the Internet for the provision of government services and the application of ICTs by government agencies in order to increase access to information and operational transparency, to improve service quality and delivery, and to raise social welfare. In general, the three main objectives of e-government are: a) restructure administrative functions and processes; b) overcome barriers to coordinate and cooperate within the public administration; and c) monitor government performance.

A crucial aspect is the way in which the relationship between state and citizens is expected to change. The shift to e-government requires an entirely new mindset in government and public administration; the latter becoming the interface between *customers* and *provider of services*, and not any longer between *citizens* and *state*. Hence, a cooperative interaction between citizens and public administration is crucial for successful implementation together with an overall organisational re-design of government's departments and agencies. Such a re-design should adhere to the tenets of the methodologies variously inspired to the Business Process Reengineering movement of the 1990s (see Ciborra, 2002).

To be sure, e-government is not applied in isolation: it is supposed to deeply affect the way in which a country is governed. Thus, according to Okot-Uma (2001: p.5) 'EGovernance seeks to realise *processes* and *structures* for harnessing the potentialities of information and communication technologies [...] for the purpose of enhancing *Good Governance*' (emphasis in the original). Governance is 'the way society collectively solves its problems and meets its needs [...]. In a framework of good governance, government services across administrative levels co-ordinate their activities in order to enhance the global effectiveness of policies and minimise conflicting action' (OECD, 2001: p.13). Accordingly, 'E-government initiatives should be measured by the degree to which they contribute to good governance' (UN, 2002: p.4).

The process whereby conflicting action is minimised involves the creation of well-functioning institutions to smooth the operations of the market and allow free relationships of exchange to prosper. In particular, given the presumed role of ICTs in moving public bureaucracies toward a market model, one can conclude that e-government implementation tries to create a *neo-liberal* (or minimal) state.

2.2 Good Governance, Development Aid and Policies

The concept of good governance is better understood by looking at the underlying theory and the policies the government should implement. The development aid perspective is the one here privileged. Three main stages can be identified within the international aid allocation regime, with the multi-lateral IDAs setting widely followed guidelines for all donors: in the 1980s there were the Structural Adjustment Programs (SAPs); 1989 brought the 'Washington Consensus', while in the 1990s the emphasis shifted to 'Good Governance'. Table 1 below contains the main policy recommendations of the three approaches advised and supervised by the World Bank (WB) and the International Monetary Fund (IMF). The first two stages present obvious similarities in their stressing of privatisation, deregulation, and devaluation of the currency to contribute to macroeconomic balance: these policies have not been entirely abandoned in the 1990s.

Table 1. The evolution of the official guidelines for development assistance

From 1980s: SAPs	1989:Washington Consensus	1990s: Good Governance
• Privatisation • Tax reform • Financial reform • Trade liberalisation • Devaluation of the exchange rate • Reduction of the fiscal deficit • Deregulation of markets and agricultural price reform	• Privatisation • Tax reform • Public expenditure in health, education, etc… • Trade liberalisation • Competitive exchange rate • Interest rate deregulation • Deregulation of markets • Fiscal discipline • Secure property rights	• Establishment of a foundation of law • Maintenance of a non-distortionary policy environment • Investment in basic social services and infrastructure • Protection of the vulnerable • Protection of the environment
Source: Weinhold (2001)	Source: Di John (2001)	Source: WB (1997a: 41-60)

Crucially, good governance requires the state to be involved in the creation, protection and enforcement of property rights, to provide a sound macroeconomic regime (thus including the previous recommendations) and to create institutions that limit state capacity for arbitrary action in order to improve its ability to support broad based markets (WB, 2002: pp. 99-101). The circulation of information through the free press and a vibrant civil society are also relevant, because the availability of information increases transparency, reduces corruption and promotes democratic governance (WB, 2002: pp.102-109).

2.3 New Institutional Economics Principles and Implications

The importance of institutions in the economy goes back to the pioneering work of Coase (1937; 1960), who introduced the concepts of information and transaction costs to complement neoclassical economic analysis. Central to the research agenda of the New Institutional Economics (NIE) are the emphasis on property rights, the transaction costs of measurement and enforcement, and the role of incomplete information (North et al, 1996: p.1).

Institutions are humanly devised constraints that shape human interaction, define property rights and determine the costs of enforcement of rules and transacting in society. These costs, in turn, affect the costs of exchange and production and provide the incentive structure of the economy. By allocating property rights (or rents[1]), institutions provide opportunities, and are thus able to alter the behaviour of the agents in the economy. If the opportunities (or incentives) put forward by institutions are growth enhancing, the agents would respond by undertaking productive investments, otherwise they would engage in rent-seeking, leading the economy to stagnation (North, 1990: pp.1-10). According to North (1981) the lack of appropriate incentives and a weak property rights structure are responsible for state failure, and, one may add, underdevelopment.

In the NIE the next best alternative is the neo-liberal model, of the minimal, efficient state. But Khan (1995; 2000; 2002) stresses that it is not only a matter of transaction costs and static efficiency. As a late developing state is typically in the process of *transformation*, there are high *contestation* and *political transition* costs caused by the bargaining process between the state-led reforms and other agents in the economy. Wade (1990) and Khan (2002) indicate that there is not enough empirical evidence to prove that the

[1] Rents are incomes that individuals can earn that are higher than in their next-best opportunity and so rents exist if those in the next-best activities are prevented from getting access to particular resources or opportunities (Khan, 2002: p.7).

minimal (or liberal) state is conducive to development, while states that have intervened aggressively and engaged in rent-management, according to the theoretical model of the developmental state have historically performed better. One has to manage effectively not only transaction costs but also transition costs (Ciborra, 1990). We refer to Figure 1 to illustrate the current debate.

Late development is a complex political process, with fast changing political and institutional structures and rapid capital accumulation. The outcome of policies and interventions can be successful (i.e. growth enhancing rents) or fail (i.e. growth reducing rents). The range of policies outcomes is represented by the horizontal axis in Figure 1. The 'appropriateness' of particular policies will be case specific and will depend on how these influence the consequent allocation of rents by state institutions. For example, enforcement will be easier if the policies promoted by the state will not go against politically organised, powerful or wealthy groups of LDCs. If the government is not able to win support from these groups the likely result will be fragmentation of interests leading to high rent-seeking costs. In Figure 1, such costs are expressed on the vertical axis as 'enforcement costs', measuring the aggregate rent-seeking costs (including lobbying bribery and corruption) of the policies promoted by the government. Growth will be highest when these aggregate costs are kept at a minimum coupled with effective enforcement by state institutions (as expressed in the bottom-right quadrant of Figure 1).

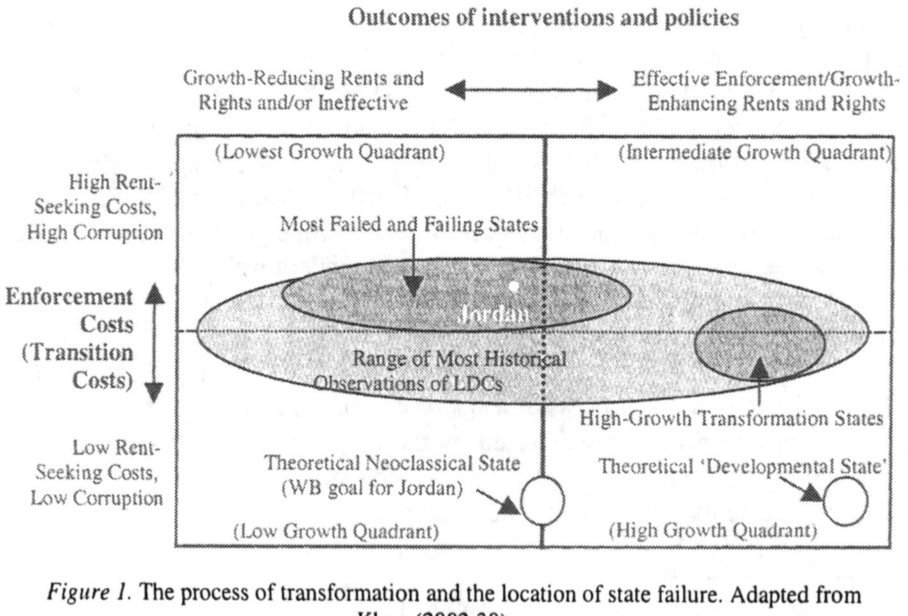

Figure 1. The process of transformation and the location of state failure. Adapted from Khan (2002:30)

This depends not only on the ability of the state to choose specific policies, and allocate rents to productive groups accordingly, but also on the provision of incentives for long-term investment. The government needs to be credible over the long term, meaning that expropriation of the rents gained by the eventual investments of productive groups, will not take place. If the government is not able to initialise such productive processes, the risk would be to create a situation of high rent-seeking costs and growth-reducing rents, where most failed and failing states are located (see the top-left quadrant of Figure 1).

On the other hand, credibility, low costs of enforcement and appropriate rents allocation, seem to have been the main strength of the development of South- East Asian economies in the 1960s, although those governments were not following the neoclassical state model. Particularly in Japan and Taiwan, the government played a key role for the allocation of rents and the overall industrial policy direction of their countries. Thus, South-East Asian economies are high-growth transformation states located between the top and bottom-right quadrant of Figure 1, while the theoretical neoclassical state (the WB 'ideal') lies between the bottom-right and bottom-left quadrant.

The Jordanian state belongs to the category of 'most failed and failing states' (see UNDP, 2002). According to the G-8 vision (DOT Force, 2002) e-government represents the promise for a rapid transition towards the theoretical neoclassical (neo-liberal or minimal) state. But what are the likely effects of e-government implementation in the Jordanian institutional and political context? What will be the likely costs of its enforcement? And which are actual, as opposed to the expected, outcomes?

3 THE CASE STUDY[2]

3.1 Jordan: Recent History and Key Statistics

Jordan is landlocked between Iraq, Israel, Saudi Arabia, Syria and the West Bank; has a population slightly above 6,5 million and a land surface of about 91,000 sq. Km, mostly desert. The country is governed by a constitutional monarchy headed by King Abdullah II. Since 1989 regular democratic elections have taken place and the government has implemented major reforms programs aimed at further increasing democratic participation, improving the provision of health care and education, and modernising the private sector as well as government agencies.

Literacy is very high in Jordan, with a formidable rate of 90% (UN Statistics Division, 2002). However, despite the dismantlement of the telecommunications monopoly started in 1997, together with market deregulation and privatisation, there are still less than 10 landlines and 2 internet users for 100 people. Hence Jordan ranks only 49 on the e-readiness report compiled by the Centre for International Development at Harvard University (see CID, 2002). Part of the explanation is that only 11% of Jordanian families have an income higher than US $ 1070 per month (MoICT, 2000: pp.2-20), which makes personal computers and Internet connections virtually unaffordable to the greatest majority of the population.

With the assistance of the IMF and the WB, Jordan has tried to reduce the role of the state in the economy and boost the private sector's role and activities. However, progress has been inconsistent since the process began in the late 1980's. Despite Jordan being a member of the WTO since 2000, and benefiting from preferential trade agreements with a number of countries (including the US, the European Union and Japan) government services

---

[2] The methodology used was aimed at gathering qualitative material to describe both the planned and the vision of e-government in Jordan and collect elements of the current state of implementation. A case study research methodology was chosen because of the complexity of the observation, the inter-locking dependencies of multiple actors and organisations, and the paucity of empirical data available. The case study observation in Jordan lasted from the 24th of June to the 6th of July 2002. Primary sources of data have been collected from interviews with key people responsible for the major ongoing projects. The interviewees (about 20) included professionals and officials working in the government, the private sector, external consultants, non-governmental and donor organisations. During the interviews, which lasted between 45 and 60 minutes, open-ended questions were asked aimed at gathering an understanding of visions, strategies, models and methods being used; expected and actual organisational impacts; the influence of cultural factors in adopting standard solutions; especially the major risks and challenges facing e-government initiatives and specific projects.

remain the largest single contributor to GDP, accounting for nearly 17.6% in year 2000 (Central Bank of Jordan Annual Report, 2000).

3.2 E-Government in Jordan: Institutional Aspects

The implementation of e-government in Jordan is at an early, but crucial stage. The governmental entity in charge of implementing e-government is the former Ministry of Post and Communication, now Ministry of Information and Communication Technology (MoICT). Having completed the design stage, the MoICT is in the delicate phase of co-ordinating with all other (often reluctant) government Ministries and departments to create a shared vision on e-government. According to Al-Jaghoub and Westrup (forthcoming) 'Jordan is an example of a nation state trying to develop using ICT in an increasingly globalised world', by promoting an ICT industry and following the competitive market-based model of the neo-liberal state.

The King is the single most important initiator of all ICT led projects. These are then filtered by the Ministry of Planning, supported and monitored by the MoICT, and then reviewed and implemented by other Ministries (especially the Ministry of Industry and Trade), the private sector and other non-governmental organisations (such as AMIR). The REACH (Regulatory Framework, Estate, Advancement Programs, Capital, Human Resources) initiative, for instance, represents the spirit of a public-private partnership approach. REACH, is supported by the Information Technology Association of Jordan (int@j), a non-profit private sector initiative that resides under the patronage of the MoICT. 'Connecting Jordanians' and the International Computer Driving Licence are other two parallel programs, the former under the patronage of the Jordanian government, while the latter is sponsored by the United Nations. Table 2 contains an initial list of the key actors involved in the above mentioned initiatives and the sources of funding (as of July 2002).

Jordan's e-government strategy is aimed at using new technologies to facilitate inter and intra-agency communication and cooperation, as well as provide information and services to its citizens more effectively. The program relies on four foundations: introduction of e-services, infrastructure development, education and training, and legal change. A number of Fast Track projects have been launched in 2001. They include motoring services (at Department of Driving Licenses and Motor Vehicles), taxation (income and sales) services, and land registry. Next will be the Government-to-Business and Government-to-Customer portals and a Government Personnel Directory. A new network infrastructure is envisaged to enable government introduce knowledge management, empower and connect government staff.

Table 2. External actors, sources of funding and aid projects

EXTERNAL ACTORS	
Consulting firms	Deloitte & Touche, EDS
Vendors	Helwett-Packard, Cisco Systems, Microsoft, Intel, Oracle
Multi-lateral donor agencies	USAid, UN Development Programme
Non-Governmental Organisations	AMIR

DONORS	PROJECTS
Spanish Government	Hardware
Japanese Government	Technical assistance
British Council	System testing and quality assurance
World Bank	Hardware, technical assistance, learning and resources centres
Islamic Bank	Teacher training
National Plan	Basic schools
Cisco Systems	Teaching assistance

In the Jordanian case, controlling the essential components of the e-government apparatus involves a certain degree of centralisation. At the same time the organisations promoting it demand co-ordination and a certain degree of local autonomy. But as a steering committee in charge of the monitoring, co-ordination and evaluation of the projects linked to the e-government initiative had not yet been established (at least at the time of the present study) many crucial projects were receiving the go-ahead without the appropriate conditions for competitive tendering.

4 RISKS AND CHALLENGES

Note that the creation of new institutions and organisations aimed at promoting the appropriate conditions for a neo-liberal form of state, which uses decentralisation and competition as a way to provide better services for the citizens/customers, is ultimately crafted under the directives of the current (non liberal) government. This leads to a number of inconsistencies. For example, Internet Cafes in Jordan are still under strict government regulation. This might be a first clue to suggest that 'e-Jordan' is not approaching the model of the neoclassical state wished for by the WB and other IDAs. More specifically, what are the costs of allocating the new rents of e-government and what is the likely level of effectiveness of the current projects? What could be the side effects when issues specific to the Jordanian context are taken into account?

4.1 A Mindset Change?

The Minister of the MoICT has many times emphasised that a mindset change is needed for progress. Unfortunately, it is no guarantee that implementing e-government models and work methods will solve this issue.

Once the new system will start working, citizens will be required to pay for services. Presently, however, the value of these services is extremely difficult to estimate let alone understand the willingness and ability of the population to pay for them. In fact, there is no standard methodology to calculate the costs and benefits associated with providing such services. Last but not least, a senior member of government states that '*Investors and aid agencies demand reforms, not the people*'. This statement is of great importance as it emphasises that while democratic governance is the objective, it is not appreciated by the citizens. Indeed, many decisions are taken at a level where there is no popular control, electoral or otherwise.

4.2 Co-ordinating the Public-Private Partnerships

There are a number of tensions in the dialectic between the political economy argument put forward by the IDAs and the reality of LDCs. For instance, e-government requires strong leadership to be successful, while at the same time it tries to promote decentralised and democratic organisational structures. The presence of a variety of consulting firms and donor agencies (see Table 2) requires a gigantic steering effort for their coordination. Unfortunately, neither the donors nor the government may be able to learn from past experience. A senior government official noticed that in the 1990s the greatest problem of a large ICT venture in Jordan was the lack of a single authority. Hence, it was difficult to define roles and track responsibilities and the multi-million US$ project became almost a data graveyard.

In a nutshell, the theories, models and methodologies underlying e-government implementation seem not be followed in Jordan's context. While these focus on a standard set of applications and call for a set of sequentially defined stages which are meant to be adhered to during implementation, according to a senior member of government: '*It is difficult to find who is in charge*', and thus to promote real enforcement.

4.3 Placing Jordan on the Map

A number of administrative and contextual features help placing Jordan's trajectory among the quadrants of Figure 2. Along the horizontal axis what matters is the feasibility and the most likely outcome of the policies needed for e-government, given the specific structural characteristics of the Jordanian state. The overall approach is aligned with the current models of e-government in Western countries (Deloitte Research, 2001). The extent to which this conduct will be growth-enhancing, if compared with the exigencies of late developers as opposed to a developed state, is uncertain at this stage. On the other hand, IDAs and donors do not seem to be able to bring any alternative know-how to the one of the leading consultancies and vendors.

The Fast Track applications, the portals and the internal network, once implemented, are supposed to bring about gains in efficiency of service delivery and reduce internal co-ordination costs. However, the gains may be limited because certain services will not be fully privatised (e.g. driving licences), nor will be fully marketable. For example, a 'simple' product like the driving licence is in reality a security document, the issue of which requires several authorities of various Ministries. Hence, efficiency gains will result only if the Jordan state reforms its inner workings and political purpose. Unfortunately these wide-ranging and cross-sectoral reforms do not feature in the Jordan e-government agenda. As a result, one would expect only limited, even if positive, growth enhancing rents being promoted by e-government: hence the location of e-Jordan on the right –half of the horizontal axis, but not far from the central axis.

Moving now to the vertical axis, other cultural and context specific issues need to be taken into account. The first point to note is the absence of feedback mechanisms, which only a previously instituted democracy, a critical press and competitive recruitment of public sector officials could bring about, thus resulting in a serious problem of credibility of the government with respect to the Jordanian productive groups. Also, the presence of Palestinian immigrants holding many of the private sector activities in the country is a factor that should not be undervalued, since the reforms proposed by e-government are skewed in favour of the private sector. This is a signal of the presence of fragmented interest groups, meaning that public sector Jordanian workers might hinder the process with a certain degree of resistance, halting effective coordination of projects and policies.

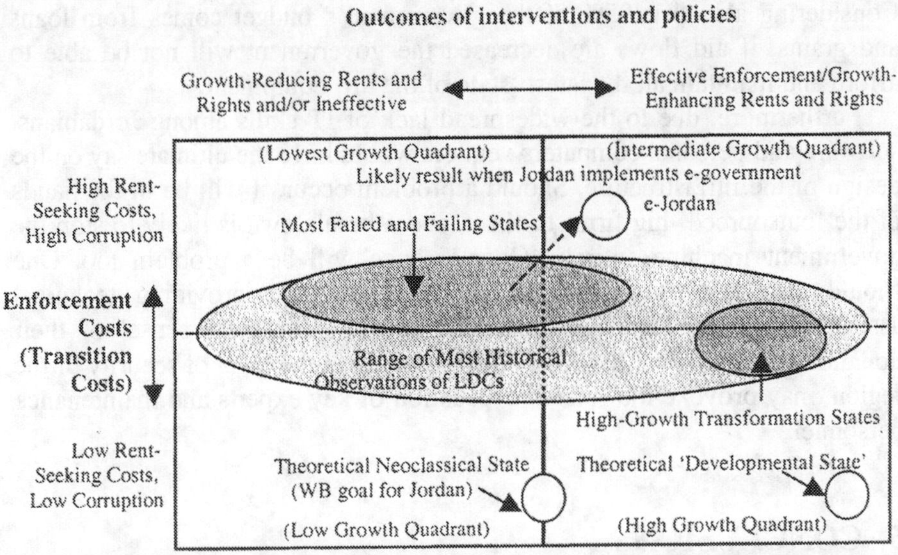

Figure 2. The location of the Jordanian state as a result of e-government implementation

Furthermore, respondents reported cases where some Ministries and private sector lobbies have often come together to write the new legislation required for the implementation of e-government. Whereas this would be a practice approved in a 'developmental' state model, it is what e-government is supposed to remove. In the case of Jordan, this could signal a reluctance to fully implement the changes required for e-government applications to take off with a citizen-centred focus. In sum, the present analysis suggests that e-government related methodologies may not reduce the costs of transition, hinting instead at the fact that transaction costs and other challenges might even rise in the immediate short-term.

As a result, while e-government aims at moving failed and failing nations towards the theoretical neoclassical state, it can be inferred that even if the Jordanian institutions were able to implement their interventions, the costs of enforcement would still remain high, resulting at best in a situation of intermediate growth as shown by placing Jordan in the upper right quadrant of Figure 2.

4.4 Technical Knowledge Dependence and Other Concerns

At the moment, while different platforms are 'donated' to different Ministries in an incremental fashion, the risk is not only to create the basis for future problems of incompatibility, but also to implement solutions that are not fit for government agencies. The latter would be locked-in by high replacement costs, which would only benefit the suppliers of such systems. Considering also that 95% of the government's budget comes from loans and grants, if aid flows are decreased the government will not be able to afford and maintain all the latest 'state of the art' equipment.

Furthermore, due to the widespread lack of IT skills among Jordanians, software and personal computers vendors would have the ultimate say on the design of the infrastructure. Should a problem occur, it will be in the hands of the 'outsourced' big firms to fix it, and a breakdown is likely to stop the government machine entirely. Obsolescence will be a problem too. One should also keep in mind that the multinationals providing technical assistance and products need after all to make 'business sense' of their donations. On the other hand, any downgrading in the level of security of the region, may provoke the sudden evacuation of key experts and maintenance personnel.

5 CONCLUSIONS

The present study of the Jordan's e-government initiative suggests that the current view on how to achieve good governance embedded in the 'Monterrey Consensus' is not frictionless. Also, when taking into account the specific context in which LDCs' governments operate, potentially damaging side effects could arise. Perhaps, the greatest challenges to the implementation of e-government are the low competence of donors in allocating and managing the funds, the increased complexity of the state apparatus and the push for democratic reforms without real popular participation from either the people in LDCs or from the developed nations.

The implementation of e-government is bound to encounter the same old problems (such as redundancy, incompatibility, unclear requirements, difficulties in control) related to the creation of major ICT infrastructures (Ciborra et al., 2000). Moreover, even if implemented, there are still serious doubts that it would be the best solution for Jordan (as well as for other LDCs), due to concerns of *ex post* opportunism by vendors and external interventions exacerbating the structural constraints of LDCs (Wade, 2002).

The case study indicates that:

1) Even if public-private partnerships can help rebuild state capacity in developed nations, this may not be substitute for effective governance in LDCs;
2) LDCs willing to implement e-government should have a strong efficient state *ex ante*, as e-government may not be conducive to a strengthened state capacity, *per se*;

3) Better chances of e-government success may reside in using it as a learning experiment for local adaptation, rather than as an implicit regulatory instrument.

ACKNOWLEDGEMENT

We thank all the people who spared some of their precious time to help us during our study in Jordan. Special thanks for their kindness and hospitality are for Dema and Mahmoud of the MoICT and to Basma.

REFERENCES

Al-Jaghoub, S. and Westrup C. "Jordan and ICT Led Development: Towards a Competition State?", (forthcoming in *Information Technology and People*).

Central Bank of Jordan, "Central Bank Thirty Seventh Annual Report 2000", *Research Department, Central Bank of Jordan*, 2001.

Ciborra, C. "X-Efficiency, Transaction Costs and Organizational Change", in *On X-Efficiency Theory*, Weiermair. P. (Ed.), Ann Arbor: University of Michigan Press, 1990.

Ciborra, C. *Teams, Markets and Systems*, Cambridge.: Cambridge University Press, 1993.

Ciborra C. (ed), *From Control to Drift: The Dynamics of Corporate Information Infrastructures*, Oxford: Oxford University Press, 2000.

Ciborra, C. *The Labyrinths of Information: Challenging the Wisdom of Systems*, Oxford: Oxford University Press, 2002.

CID, "The Global Information Technology Report 2001-2002: Readiness for the Networked World", *Center for International Development, Harvard University*, 2002.

Coase, R.H. "The Nature of the Firm", *Economica*, (4), 1937, pp. 386-405.

Coase, R.H. "The Problem of Social Cost", *Journal of Law and Economics*, (3:1), 1960, pp. 1-44.

Deloitte Researce, *E-Government's Next Generation: Transforming the Government Enterprise Through Customer Service*, New York: Deloitte Consulting (Global), 2001.

Di John, *From Bretton Woods to the Post-Washington Consensus*, unpublished, DV 400 Public Folders, LSE, 2001.

DOT Force, *Digital Opportunities for All, Report Card, Digital Opportunities Task Force*, 2002.

Economist Intelligence Unit, *Country Profile: Jordan*, March. London: The Economist Intelligence Unit Limited, 2002.

Khan, M. and Jomo, K. S. *Rents, Rent-Seeking and Economic Development: Theory and Evidence in Asia*, New York: Cambridge University Press, 2000.

Khan, M. "State Failure in Weak States: A Critique of New Institutionalist Explanations", in *The New Institutional Economics and Third World Development*, Harris et al. (eds.), London: Routledge, 1995.

Khan, M. "State Failure in Developing Countries and Strategies of Institutional Reform", draft Paper for the *Annual Bank Conference on Development Economics*, 2002.

MoICT, *Launching E-Government in Jordan: Readiness and Approach*, 2000.

Nahapiet, J. and Ghoshal, G. "Social Capital, Intellectual Capital and the Organisational Advantage" in *The Strategic Management of Intellectual Capital and Organisational Knowledge*, Choo, C. W. & Bontis, N. (eds.), Oxford: Oxford University Press, 2002.

North, D.C. *Structure and Change in Economic History*, New York: W.W. Norton, 1981.
North, D.C. *Institutions, Institutional Change and Economic Performance*, Cambridge: Cambridge University Press, 1990.
North, D.C., Alston, L.J. and Eggertsson, T. (eds.). *Empirical Studies in Institutional Change*, Cambridge: Cambridge University Press, 1996.
OECD, "Local Partnerships For Better Governance", *OECD Publications*, 2001.
Okot-Uma, R. "Electronic Governance: Reinventing Good Governance", London. *Commonwealth Secretariat*, 2001.
UN Statistics Division, *On-Line Database*. Available at: http://unstats.un.org/unsd /default.htm. (Accessed at: March 12, 2002).
UN, *Final Communiqué of the International Conference on E-government for Development*, Conference Proceedings, 2002.
UNDP, *Human Development Report 2002: Deepening Democracy in a Fragmented World*, New York.: Oxford University Press, 2002.
Wade, R.H. *Governing the Market*, Princeton. Princeton: University Press, 1990.
Wade, R.H. "Bridging the Digital Divide: New Route to Development or New Form of Dependency?", *Global Governance*, (8:4) (forthcoming), 2002.
Weinhold, D. "Stabilisation and Structural Adjustment: A Very General Overview", unpublished, *DV 409 Public Folders*, LSE, 2001.
World Bank, "World Development Report 2002: Building Institutions for Markets", Washington: *World Bank*, 2002.
World Bank, "The World Bank Annual Report 1997", Washington: *World Bank*, 1997a.

About the Authors

Claudio Ciborra is PwC Professor of Information Systems and Risk at the London School of Economics and Political Science. His research has focussed on issues of technology, organization, strategy and innovation. He also teaches at IULM, Milan and is Visiting Professor at the Oslo University. He has published many articles and books on information systems. His latest is titled "The Labyrinths of Information – Challenging the Wisdom of Systems". He can be reached by e-mail at: c.ciborra@lse.ac.uk.

Diego D. Navarra is a doctorate student in the Information Systems Department at the London School of Economics. He is doing research on the role of the government in late developing countries for the implementation of ICTs policies and the institutional obstacles to effective ICT implementation. He has recently finished his MSc in Development Management at the London School of Economics and can be reached by e-mail at: d.d.navarra@lse.ac.uk.

25

THE EMERGENCE OF ICTs FOR DEVELOPING COUNTRIES

Using Dramatistic Analysis to Identify the Heroes and Villains

Kenneth E. Kendall, Julie E. Kendall, Muhammadou M.O. Kah

School of Business-Camden, Rutgers University, USA

Abstract: We analyze discourse on Information Communications Technology (ICT) forums on the Web to extract relevant messages on the issues and challenges raised, the experiences shared, and the policies framed for developing countries. Using dramatistic analysis, we explore specific language of the discourse from actual discussion texts and examine how these discussions reveal a shared drama that helps shape ICT policy, justification, diffusion, use, and implementation. We identify the hero, villain, main plot and subplot of the ICT policy drama that took place from 1997-2000 in an Internet discussion forum. The conclusions derived from these discourses should be useful both for researchers and policy makers who may use it as a basis for creating a comprehensive set of sustainable information technology strategies and policies that can be appropriately implemented.

Keywords: Developing countries, ICT, dramatistic analysis, information technology, global IT, digital divide.

1 INTRODUCTION

Through a form of discourse analysis called dramatism, we explore the role of discourse in shaping Information Communications Technology (ICT). In this paper we analyze discourse on an ICT policy and development forum on the Web to extract relevant messages on issues and challenges raised; experiences shared; policies framed; justification of ICT for development, and the need for appropriate ICT impact assessment and implementation for national economic development. We analyze specific language of the discourse from actual Web discussion texts and examine how these discussions reveal a shared drama that helps shape ICT policy, justification, diffusion, use and implementation in developing countries.

Our conclusions are based on an assessment of the themes and issues raised during the discourse in the forum and the consensus arrived at after numerous iterations of the discourses using dramatistic analysis. The conclusions derived and the drama we have identified in these discourses should be useful both for researchers, who can use it for developing testable research hypotheses, and for policy makers in developing countries who may use it as a basis for creating a comprehensive set of sustainable information technology strategies and policies that can be appropriately implemented.

We examine discourse of participants in an Information Communications Technology (ICT) listserv in which the interest group discourse narrates the entire drama and the intended audience is composed of those who may eventually show an interest in the technology adoption. In this paper, we use dramatism as a critical method to examine multiple levels of the drama serving to unify the interest group community and its many constituents in a shared rhetorical vision. It is then possible to examine how discourse among Internet discussion groups function to keep the community united, encourage measurement of ICT impacts, and serve to promote the wise adoption of emerging information technologies.

2 ICTS IN DEVELOPING COUNTRIES

Information Communication and Technology is viewed from optimism to pessimism. The optimist associates ICTs with largely positive impacts such as wealth creation and improvements in service quality, while pessimists view it largely by negative impacts such as unemployment and alienation (Heeks, 1998). It is also suggested that majority of ICT based initiatives end in total failure of a system that never works; partial failure in which major goals are unattained; sustainability failure that succeeds initially but eventually fails; or replication failure of a pilot that cannot be reproduced (Heeks and Davis, 1999).

Overall, research appears to indicate that for developing countries benefits from ICTs are more likely to accrue from consumption rather than productivity (Kraemer and Dedrick, 1998). This is particularly true for poorer countries where the ability to use ICTs can dramatically improve, even in the short run, the capability of firms to face competition from developed countries. Conversely, there are arguments forwarded that there are fewer advantages for 'latecomer' countries to develop indigenous ICT production, which requires large capital investments and specialized skills (Pohjola, 1998).

While there is consensus that more effective and efficient use of information and of ICTs can stimulate economic growth and development, concerns remain on how such benefits can be shared among the minority ICT users and the majority non-ICT users. Because of the wide access gap between rich and poor, it is indeed possible that ICTs may reinforce, or widen, existing social and economic inequalities in developing countries.

Within this context, there is a question whether government should play a key role in establishing a competitive, private sector led communication market, and by promoting supportive measures to enhance the capabilities for accessing ICTs (Kah, 1999; 2000; 2001). The former would involve the liberalization of the telecommunications sector, the privatization of public monopolies, and the creation of an independent regulator with the mandate of reaching universal access in basic communication services.

Since there is so much conflicting material written about what needs to be done concerning ICTs and developing countries it behooves us try to sort out these issues. It is therefore our goal to explore the role that discourse plays in shaping ICT policies.

3 APPROACHES TO DISCOURSE ANALYSIS

There have been a number studies that have attempted to classify or categorize discourse analysis. One defining article, however, makes a distinction between interpretative approaches and critical analysis (Mumby and Clair, 1997). Most of the recent work in discourse in information technology has conveniently fit into either of these categories.

The interpretive approach to the development of information systems can be seen most clearly in the works of Klein and Myers (1999) and Walsham (1993). An example of this approach can be found in the work of Ellingsen (2003) who studies the development of knowledge work in hospitals in Norway. Furthermore, Dube and Robey (1999) interpret stories that emerged from a software project and the complexity of negotiating multiple perspectives and Boland and Schultze (1996) suggest that although the narrative may be a convincing tool for persuasion, there is always an alternative narrative waiting to step in as the dominant narrative. A criticism of the interpretive approach is that it is "not strongly associated with a model of organizational change" (Heracleous and Barett, 2001).

Recently the study of discourse using a critical analysis approach (CDA) was based on the work of Fairclough (1989; 1999) who "locates social structures within a dialectic relationship with social activities" (Thompson, 2003). CDA makes it possible to examine interrelationships that go unnoticed, such as the use of language and the exercise of power. Thompson

(2003) notes that "In fusing power with semiotic activity (a wider category than straight linguistics), CDA views texts as examples of wider discourses, thus blending the approaches to discourse of social theorists ... with linguists". One example of CDA can be found in the work of Oliver and Oliver (2003) who follow Fairclough's approach when they study ERP adoption. One problem with the CDA approach is that researchers tend to conceptualize discourse as power/knowledge relations.

A third approach, described by Heracleous and Barett (2001) is functional analysis. In this approach, researchers see discourse as a tool, used to communicate, persuade, or facilitate in order to achieve certain outcomes. In the functional approach, discourse is not just a way to exchange information, it is seen as a way to achieve a social construction of reality and to influence that reality. An example of functional discourse analysis successfully applied can be found in a study of how banks conceptualize and treat their customers (de Graff, 2001). Early studies in this functional approach can be found in the analysis of organizational metaphors (Kendall and Kendall, 1981; 1984).

It is this third approach, functional discourse analysis that includes the methodology used in this paper. We use the dramatistic approach, and classify words into God and Devil (good and evil) terms. The methodology assumes that words are chosen to represent reality and to achieve certain goals. It does not assume a power/knowledge relationship. Consequently, it is useful for the researchers to adopt this approach rather than an interpretive or critical approach.

4 THE DRAMATISTIC PERSPECTIVE

Dramatism is an established social science method. Dramatism is a way to understand, interpret, predict and even change modern social interactions. There is ample research that suggests the usefulness of adopting a dramatistic perspective as a legitimate and meaningful research method (Combs and Mansfield, 1976; Mangham and Overington, 1987).

4.1 Dramatism as Theatre

The scholar that is most often identified as the father of dramatism is Kenneth Burke (1969). Since he first developed a framework he referred to as the dramatistic pentad (act, scene, agent, agency, and purpose), many studies in communication have accomplished successful rhetorical critiques using his method. Mangham (1978) extended the works of Burke and others to map out a practical application of dramatism for use in organizational interventions.

Mangham and Overington (1983) acknowledge their debt to Burke and extend Burke's dramatic perspective, arguing that the dramatistic metaphor is a means to demystify the drama and return to consciousness elements that may have been ignored or repressed. Mangham and Overington's drama metaphor was helpful in establishing another method used by Kendall and Kendall (1981; 1984) to analyze the action of executives and employees in organizations using a methodology called STROBE. In another dramatistic manifestation, systems analysts are encouraged to describe the activities of managers using a playscript (Kendall and Kendall, 2002).

4.2 Dramatism and Fantasy Themes

A somewhat different school of thought was developed by Bormann (1972). His particular use of dramatism, (which he also calls "fantasy theme analysis") is rooted in the study of messages, created through social interaction of small groups, which then "chain out" into larger society.

After many years of observation, Bormann noticed small group members dramatized events, casting heroes and villains, placing blame and praise, finding a sanctioning agent for their actions. Bormann theorized that dramatizing helped group members create a rhetorical vision, built of what he described as "fantasy themes".

The fantasy theme form of dramatism developed by Bormann differs from the "organization as theatre" metaphor developed by Mangham and Overington. Bormann made a distinct break from other rhetorical critics by suggesting that dramas (which manifest meaning, emotion, and motive; heroes, villains, sanctioning agents and so on) were present in the discourse itself as it is written. Alternatively, Mangham and Overington seem to suggest the utility of seeing the whole of the organization as a metaphor, including the space, props, costumes of key actors, and their scripts.

Kendall (1993) adapted Bormann's dramatistic approach to demonstrate how dramatism can be used for discovering and interpreting corporate dramas inherent in the language of the boiler plates of the Dow Jones Industrials. She concluded that by attempting to understand corporate dramas researchers can see how companies create a shared rhetorical vision to unify their shareholders with management and employees, label actions as good or evil, and influence the public by putting forward a positive corporate self image. This approach does not require additional physical or interactionist elements of the dramaturgical metaphor as suggested by Mangham and Overington (1983) or as introduced by Kendall and Kendall (1981; 1984). Therefore, we choose and adhere to the this framework for this study.

5 METHODOLOGY

We looked at all text messages dating from July 24, 1998 to January 17, 2002 extracted from one of the Bellanet list forums, called the inet-Impact list forum, that grew from 30 to 117 members. The inet-Impact is a public, unmoderated forum for discussion of how best to measure the development impact of the Internet and other Information and Communication Technologies (ICTs). This discussion started on July 24, 1998 as a follow-up to a session called "Measuring the Development Impact of the Internet and Evaluating and Implementing ICT Strategies for the Information Age", which took place during the first Global Knowledge Conference in 1997 in Toronto, Canada. This Listserv, LEAP IMPACT (inet-Impact-L) aims to improve the institutional performance of monitoring and evaluation practice related to information services, information products and information projects. It is a community of practice open to all individuals/organizations interested in the evaluation of information. LEAP IMPACT is a joint initiative of several development agencies including IICD, Bellanet, KIT, FAKT and GTZ. The site is hosted by Bellanet, www.bellanet.org, an international nonprofit initiative governed by a steering committee representing their donor institutions.

5.1 Applying the Dramatism Methodology

Because of the advantages presented above, data were analyzed using Bormann's dramatistic perspective as adapted by Kendall (1993). Generally this process entails looking for dramas present in each message, then attempting to generate structures which are found to recur. Dramatism is useful in order to understand why some approaches to technology policy succeed while others fail, and why some external entities help support new technologies, while others are perceived to impede progress.

In order to analyze a discourse data set using a dramatism perspective, we first searched for key words that represented good and evil and appeared frequently in the online discussions. We call these words God and Devil terms. (When we provide quotes, we quote directly. We did not correct misspellings that are common in online discussions.)

The God and Devil terms were chosen after reading the discussions. Then the discourse data set was searched and the terms were highlighted and colour-coded.

We then were able to read the discussions and pay particular attention to those issues and technologies that appear around the God and Devil terms. We then focused on these issues to determine the presence of any patterns. By doing this, we were able to identify who the heroes and villains of the drama were.

God terms are words or phrases used in a drama which characterize highly revered values and attitudes towards actors and objects. They are unquestioned assumptions of "good" within the world of ICT policy for the developing world. These God terms include:

Access (ibility)	Goals
Collabor (ation, tive etc.)	Growth
Community	Independen (ce, t etc.)
Connectivity	Infrastructure
Demo (cratic, cracy etc.)	Initiate (Initiative)
Diffusion	Innovat (e, ion)
Dissemination	Open (ness)
Efficien (t, cy,c ies, etc.)	Plan
Empower (ment)	Prior (ities, or y)
Encourage (ment)	Productiv (e, ity)
Facilitate	Progress
Free	Support
Global	

Devil terms are words or phrases that have very negative connotations. In a discussion group these words or phrases promote cohesiveness by targeting enemies and all evil actions or objects which are assumed to hamper the otherwise successful study and diffusion of ICTS and ICT policies for developing countries. These Devil terms include:

Bias (ed)	Failure
Bureaucra (cy, cies, cratic)	Hatred
Constraints	Hype
Controlled	Inefficient
Costs	Land locked
Crime, Criminal	Loss
Destabilization	Problem
Diminish	Repercussion
Disadvantage (d)	Risk
Evil	

As we analyzed the God and Devil terms from the listserv discussion on ICT policy and development and reflected on their use, a clear hero emerged from the rhetoric.

6 THE IT POLICY RESEARCHER AS HERO

The hero is an idealized IT policy researcher who is able to change, influence and shape policy through the results of enlightened impact assessments of the Internet and ICTs in developing countries, which they undertake. Often this means the championing of democracy. For example, one contributor wrote:

On democratization for example, I recently discovered an unusual indicator of Internet impact in Africa ... My conclusion was that the openness of the Internet allows the "good guys" to win out in the long run. But it takes work.

The hero completely comprehends the culture and power dynamics of the developing country under study. The hero is also exquisitely attuned to the desires of the community being served (and studied through impact assessments), realizing that one's expertise carries only so far:

... what is the purpose of the assessment and who is supposed to take advantage of it. For me, impact assessment is a community process for the benefit of the community. It is perhaps more important as a consciousness awakening process than as an input to policy making. This is one more reason why al stakeholders, and not only experts should be involved.

The work of the hero is not constrained by a mere physical manifestation either. The rhetoric created a hero who is cognizant of the virtual world created and sustained by the Internet on an individual, group and community level. One contributor wrote:

...Pipelines and infrastructure are features of that space and important in terms of how they contribute to a built virtual landscape. Access, connectivity, telecom regimes, prices, ext. determine the scope, scale and accessibility of this electronic space. Measuring those and changes in those is useful because it tells us something about scope, range, and accessibility to this new "space". That is useful but there is a lot more to be measured. In my work I define this space as a collaborative work venue AND a social process venue. I also tend to stress equally individual, group, and community use and presence. I use the notion of an electronic persona to characterize presence in the space and ask what things do these (individual/group/community) persona do in this space, what do they bring to it, and what do they take from it.

The idealized IT policy researcher also searches for the best indicators of Internet supply; Internet penetration; its environment support; institutional use of the Internet; its impact on democracy and civil society; and the Internet's impact on private sector development. The hero possesses a keenly developed sense of how to go about the complex tasks of assessment of technology impacts in developing countries.

7 THE AUTOCRATIC RULER AS VILLAIN

The villain is composed of those bodies that would misuse, exploit, or bend the Internet in a developing country for purposes that are anti-democratic; or solely for material gain for the already wealthy.

> *Moreover, there are probably anti-democratic impacts of the introduction of the Internet (use of the Internet by anti-democratic factions, possible deterioration of the standards of validation of statements made in public discourse), which counteract the positive impact identified in the anecdote.*

The villain is also embodied in those who would use the Internet for biased, repressive purposes; those who would use the Internet to spread hatred and factionalism; and those who are isolationist, or propagandistic in their use of the Internet.

> *But I would finish by saying that Internet can also help on improving the those preconditions even though it could deteriorate the situation when used for such destructive ends as destabilisation or propagating hatred.*

Often, but not always, the villain in this discourse is identified with the ruling regime or government; or those who currently possess political power in a developing country. A contributor notes the possible deleterious effect on the citizenry of a developing country when a villain distorts the use of the Internet:

> *I worry too about the negative impacts of the Internet. I suspect African countries will continue to fall behind in mastery of the technology, and as a result will lose competitive advantage to other countries that do master the Internet. I can imagine some countries in which factions will appropriate the Internet for purposes antithetical to the welfare of most citizens.*

The villain has a curious, yet important part to play in that the hero must often interact with them to achieve their own goals of policy influence via comprehensive studies. Therefore, the hero is faced with crafting a relationship with the villain that does not sanction what they do, but somehow affords entry into the country and permission to study the impact of the Internet.

Moreover, there are probably anti-democratic impacts of the introduction of the Internet (use of the Internet by anti-democratic factions, possible deterioration of the standards of validation of statements made in public discourse) which counteract the positive impact identified in the anecdote.

Finally, taking a broader perspective of what it means to work for a villain, the ICT policy researcher is left to ponder what it means to strengthen the villain's power relationships via technology:

I wonder what similar stories the anthropologists will tell in the future about the impact of networking on communities? It seems clear to me that a lot of institutions are based on authority relationships which in turn are based on command of knowledge and understanding. I doubt that many community networking initiatives are neutral with respect to these authority relationships, and I doubt that many networking projects are very thoughtful about which "authorities" are strengthened and which are weakened by the differential access to the network that develops.

8 THE MAIN DRAMA: THE STRUGGLE FOR A DEFINITION OF IMPACTS

The central plot of this discourse revolves around a struggle to define what is meant by the term "impact" of information technology and what indeed are meaningful indicators of Internet usage in developing countries.

...We may need to specify in first place our working definition(s) of impact. Mine is: The changes in people's or organizations' ability to cope with their problems as a result of the use of information. Rather than trying a quite tricky sampling of Web sites, one may just count those which are offering contents bearing on our 3 impact areas (economy, education and democracy); multiple allocations being allowed. The higher dividends the number, the richer the information environment.

Possibly add a quality rating, though this is too sensitive to individual biases. That gives us the input. Remains the outcome ...

The drama reveals tensions among IT policy researchers and consultants who are grappling with whether ICTs are best measured via qualitative or quantitative methods.

...I certainly agree that we want to add qualitative observations, and be careful of too great a realiance on quantitative analysis. On the other hand, I don't want to through out quantitative analysis entirely. We might for example seek intermediate indicators of the impact of the Internet where causality is credable even if the causal linkages are less clear than "substituting email for fax saves money", but for which strong theoretical causal arguments suggest positive indirect impact at the national level.

An extra concern arises when the plot starts moving beyond measuring impact to discuss a new layer, that of the relationship between ICTs and development. They are frustrated with the inadequacy of current approaches to explain these relationships.

What we do need is a notion of what the relationship is between ICTs and development. We have some early models, all device-centered and build on mechanical process models. They include the information pipeline (information highway) and the information infrastructure (production capacity). They underly the various measures of 'penetration', use, connectivity, access, bandwidth, megaflop measures, etc. I would classify your approach as being in this area. ...My critique is that such approaches are: (A) too device focused (counting things) and when they include process they focus on mechanical processes (who uses what); and (B) they measure intermediate stages without looking at the ultimate deliverables.

Further clarification of the relationship between ICTs and the development are warranted:

The purpose of the measurment - and I should have said this earlier - is to "unconfound" the relationship between the the inputs - of which ICTs are just one set - and development deliverables. It is to tease out the identifiable roles played by ICTs and evaluate good and bad uses of ICTs in the development process. Much of what we read is about good and bad process in the introduction of ICTs to this or that. Interesting but only a small piece of the question.

Other strands of the drama broach a concern about the ultimate use to which impact study results are put. Indeed, one of the subplots involves questioning whether resources for studying the impacts of ICTs would be better allocated to other types of development studies, for instance those dealing with educating or feeding a population in an underdeveloped nation. There is often dramatic tension between those who are working within a developing country, and those who are working from outside.

Another elaboration of the central drama involves whether the hero can secure approval to conduct an impact assessment without becoming a tool of the villain, who holds the political power in the country, and who certainly has the right to give life to the impacts assessment, deciding whether the study will go forward, who will be involved, and on what terms.

9 THE SUBPLOT: ACTIONS OF SUPPORTING DONORS

A subplot of the central drama gives depth to what is being played out across the international stage. The hero as the idealized IT policy researcher and the villain as a political power in the ruling country, are not alone on the stage. In fact, the discourse recognizes that there is a producer who underwrites ICT impact assessment studies, the donor organization or funding agency. The subplot develops the theme that perspectives on impacts between the donor and the community being studied may be only partially, and imperfectly shared. There is, of course, another legitimate constituency, the funding institution for a program, where the questions are bottom-line what-should-we-fund. In a more perfect world, that funding institution would be primarily interested in the community's own view of impact. In the real world, the perspectives may well only partly coincide.

In an amplification of this subplot, it is brought forward that those who are involved in the bringing of the Internet to Africa are obligated to examine what the impact of their subsidies has been:

> *...as people of other countries are subsidizing some aspects of the process of introduction of the Internet into Africa, those people have a right and obligation to assess the impact of their efforts.*

In this section we identified the idealized ICT policy researcher as the hero, the autocratic ruler as the villain, and the struggle to meaningfully define the impacts of ICT in developing countries as the main drama played out in the discourse of the forum participants. Subplots included the need for donors and funding agencies to partake in assessments of impacts, along with the realization that their views of development may be only partially shared.

10 CONCLUSION

In this paper we explore the role of discourse in shaping Information Communication Technology (ICT) policy, adoption, diffusion, donor support, and investment in developing countries. We conceptualize discourse here as the communication on the Web jointly mediated and participated in by ICT experts with experience in IT initiatives in developing countries.

In its purest form the rhetorical vision is manifested in an ICT discussion group drama pitting good (the IT Policy Researcher) versus evil (the autocratic ruler). The discourse functions to create a consensus among stakeholders about how to proceed in conducting ICT impact assessments. In doing so, it promotes unity among stakeholders. The hero, villain, main drama and subplots were all discovered through dramatistic analysis.

Most of the participants in the discourse advise national governments or donor agencies and independent ICT researchers. Thus, the multiple perspectives arising from this discourse do eventually affect ICT policies or specific technology adoption, the rate of diffusion, as well as the level and direction of support to a particular country or region in developing countries. This communication results in bringing up pending issues, challenges, and independent positions which are influenced positively or negatively by participants in the discourse.

This has larger implications for the organizational change process in these countries and helps to shape issues. There is a need to understand the language used in the discourse not only as an exchange of information, but as evolving dramas that, when shared, could impact ICT policies, adoption, donor support and direction. The extent to which these discourses could facilitate, hinder and shape Information Communication Technology (ICT) policy, adoption, and influence on donors' level of investment needs to be studied further.

REFERENCES

Boland, R. and Schultze, U. "Narrating Accountability: Cognition and the Production of the Accountable Self", in *Accountability: Power, Ethos, and the Technologies of Managing, R. Munro and R. Mouritsen (eds.).* London. International Thompson Business Press, 1996.

Bormann, E. G. "Fantasy and Rhetorical Vision: the Rhetorical Criticism of Social Reality", *Quarterly Journal of Speech*, (58), 1972, pp. 396–407.

Burke, K. *A Rhetoric of Motives*. Berkeley, California: University of California Press, 1969.

Combs, J. E. and Mansfield, M. W. (eds.) *Drama in Life*. New York: Hastings House, 1976.

de Graff, G. "Discourse Theory and Business Ethics: The Case of Bankers' Conceptualizations of Customers", *Journal of Business Ethics*, (31:4), 2001, pp. 299-319.

Dube, L. and Robey, D. "Software Stories: Three Cultural Perspectives on the Organizational Practices of Software Development", *Accounting, Management, and Information Technologies*, (9), 1999, pp. 223-259.

Ellingsen, G. "Knowledge Work in Hostpitals," in *Global and Organizational Discourse about Information Technology*, E. Wynn, E. Whitley, M. Myers, J. DeGross (eds.) Boston. Kluwer Academic Publishers, 2003, pp. 466-484.

Fairclough, N. *Language and Power*. Harlow, UK: Longman Group, 1989.

Fairclough, N. "Linguistic and Intertextual Analysis with Discourse Analysis", in *The Discource Reader*, A. Jawaorski and N. Coupland (eds.) London. Routledge, 1999.

Heeks, R. *Management Information and Management Information Systems*. School of Oriental and African Studies, University of London, London, 1998.

Heeks, R. and Davis, A. "Different Approaches to Information Age Reform", in R.B. Heeks (ed.) *Reinventing Government in the Information Age*. Routledge, 1999.

Heracleous, L. and Barett, M. "Organizational Change as Discourse: Communicative Actions and Deep Structures in the Context of Information Technology Implementation", *Academy of Management Journal*, (44:4), 2001, pp. 755-778.

Kah, M.M.O. "An Econometric Analysis of Telecommunications, Information Technology, & Economic Growth in Africa", *UMI*, 1999.

Kah, M.M.O. "Strategic Significance of Information Technology: The Gambia, West Africa", *Proceedings of the 2000 Americas Conference on Information Systems*, Long Beach California, 2000.

Kah, M.M.O. "The Futures of Economic and Technological Development in Africa", *Proceedings of Conference and Summit on Africa & Taiwan on a Global Track*, Asia-Pacific Public Affairs Forum (APPAF)-National Chengchi University, Taipei, Taiwan, 2001.

Kendall, J.E "Good and Evil in the Chairmen's Boiler Plate: An Analysis of Corporate Visions of the 1970s", *Organization Studies*, (14:4), 1993, pp. 605-626.

Kendall, K.E. and Kendall, J.E. "Observing Organizational Environments: A Systematic Approach for Information Analysts", *Management Information Systems Quarterly*, (5:1), 1981.

Kendall, K.E. and Kendall, J.E. "Structured Observation of the Decision-Making Environment: A Validity and Reliability Assessment", *Decision Sciences*, (15:1), 1984.

Kendall, K.E. and Kendall, J.E. *Systems Analysis and Design*, 5th edition. Upper Saddle River, NJ. Prentice Hall, 2002.

Klein, H. and Myers, M. "A Set of Principles for Conducting and Evaluating Interpretive Field Studies in Information Systems", *MIS Quarterly*, (23:1), 1999, pp. 67-94.

Kraemer, L.K. and Dedrick, J. "Globalization and Increasing Returns: Implications for the US Computer Industry". *ISR*, (9:4), Dec. 1998 , pp. 303-322.

Mangham, I. L. *Interactions and Interventions in Organizations*. New York: John Wiley & Sons, 1978.

Mangham, I. L. and Overington, M. A. "Dramatism and the Theatrical Metaphor", in *Beyond Method*, G. Morgan (ed.), Beverly Hills, California. Sage Publications, 1983, pp. 133–219.

Mangham, I.L. and Overington, M.A. *Organizations as Theatre: A Social Psychology of Dramatic Appearances*. Chichester, U.K.: Wiley, 1987.

Mumby, D. K., and Clair, R. P. "Organizational Discourse", in *Discourse as Social Interaction*, T. vanDijk (ed.), London. Sage, 1997, pp. 181-205.

Oliver, D. and Oliver, L. "ERP Adoption: Selling the System", in *Global and Organizational Discourse about Information Technology*, E. Wynn, E. Whitley, M. Myers, J. DeGross (eds.) Boston. Kluwer Academic Publishers, 2003, pp. 507-523.

Pohjola, M. "Information Technology and Economic Development: An Introduction to the Research Issues", *Working Paper No.153.*, The United Nations University, WIDER, November 1998.

Thompson, M. "ICT, Power, and Developmental Discourse: A Critical Analysis". in *Global and Organizational Discourse about Information Technology*, E. Wynn, E. Whitley, M. Myers, J. DeGross (eds.) Boston. Kluwer Academic Publishers, 2003, pp. 347-373.

Walsham, G. *Interpreting Information Systems in Organizations*. Chichester: John Wiley, 1993.

About the Authors

Kenneth E. Kendall, Ph. D., is a professor of ecommerce and information technology in the School of Business-Camden, Rutgers University. He is one of the founders of the International Conference on Information Systems (ICIS). Dr. Kendall is the past Chair of IFIP Working Group 8.2 and served as a Vice President for the Decision Sciences Institute. He is an Associate Editor for *Decision Sciences* and the *Information Resources Management Journal*. Dr. Kendall has been named as one of the top 60 most productive MIS researchers in the world, and he was awarded the Silver Core from IFIP. He is a Fellow of the Decision Sciences Institute. Professor Kendall's research focuses on emerging information technologies and developing new tools for systems analysis and design. His e-mail is ken@thekendalls.org and his home page can be accessed at www.thekendalls.org.

Julie E. Kendall, Ph. D., is an associate professor of ecommerce and information technology in the School of Business-Camden, Rutgers University. Dr. Kendall is Chair of IFIP Working Group 8.2. Professor Kendall has published in *MIS Quarterly, Decision Sciences, Information & Management, Organization Studies* and many other journals. She is a Senior Editor for *JITTA*. She recently was elected as Treasurer of the Decision Sciences Institute, and has served as a Vice President for DSI. She and her co-author (and spouse) Ken are currently examining the strategic uses of Web presence and ecommerce for off-Broadway theatres and other nonprofit organizations in the service sector. Her email is julie@thekendalls.org and her home page can be accessed at www.thekendalls.org.

Muhammadou M.O. Kah, Ph. D., is an Assistant Professor of ecommerce and information technology in the School of Business-Camden, Rutgers University. Dr. Kah's current research focuses on how businesses and governments can effectively use information technologies and the Internet in particular to add value and the applications of real options in IT strategic decision making. His research also includes exploring ways in which advances in information communication technologies contribute to economic growth, development and quality of life for developing economies. Dr. Kah's e-mail address is mkah@crab.rutgers.edu and his home page is www.crab.rutgers.edu/~mkah.

26

THE 'EAR' AND 'EYE' DIGITAL DIVIDE

Mike Metcalfe, Carmen Joham
University of South Australia, AU

Abstract: This paper is about using technology to help people who share knowledge orally. The objective is to appreciate the social and technical needs of this preference so as to narrow the divide developing between these people and those who earn their living from written knowledge sharing. Writing is not the preferred method of knowledge sharing for the majority of peoples on earth nor is it appropriate for the majority of problems. A mix of both literacy and orality is believed to be the ideal, so a failure to develop cheap and relevant synchronous and asynchronous oral knowledge sharing technology may down play the importance of orality in the social and economic development of both the developed and developing nations. This paper will argue that IS designers interested in global diversity, equity, innovation and economic development through communication technology need to place more emphasis on orality. The difference between oral and written knowledge sharing will be discussed to explain the need for both synchronous and asynchronous communication technologies. A small study comparing asynchronous oral and written communication is presented as is an attempt to design an Internet based oral conferencing system to link Aboriginal communities. It was found that there was a need for developing cheap community based conferencing facilities and to improve the asynchronous oral communication technologies.

Keywords: Orality, literacy, asynchronous, knowledge sharing.

1 INTRODUCTION

Ashrawi (1999) estimated there were about 600 million telephones in the world, eighty percent of which were located in the wealthiest 25 countries. While presented by Ashrawi as a digital divide issue, it also attests to the popularity of oral communication in the more developed countries.

However, the importance of phones to those not in the 25 countries, which equates to 9/10 of countries of the world (about three billion people), as an economic development technology needs to be appreciated. Access to appropriate technology to even communicate over long distances may be the highest priority for less developed nations. Oral communication technologies, like the phone and the radio, have already proved their worth as technologies that can improve living standards. The lack of reading and writing skills in many developing countries makes the written communication services typical of Internet (email, web pages) nearly useless. The majority of people on earth either can not, have not, do not want to, or do not need to communicate in the written form.

Jones (1995) acknowledges the Internet is a technology invented and developed by what he calls the "writing class" and who Chandler (2002) calls the 'eye people', people who learn and make their living by trading in the written form. Urbanised scientists, lawyers and public servants are obvious examples. However, these people, and the explicit text-based knowledge in which they trade, are a minority use of knowledge sharing and maybe even they do not use written communication for innovation. Over-emphasis of written over oral knowledge sharing may be stifling cultural diversity and tacit knowledge sharing (innovation). Writing cultures appear to *economically*[1] dominate oral ones but this may not be because of writing per se but rather because of the advantages of having the appropriate mix of oral and written, synchronous and asynchronous communication technologies.

This is an important digital divide issue but it is also a social construction of technology issue. Costs aside, the Internet, email, databases and web pages have been socially constructed around text-based and therefore asynchronous communication. The oral functionality of the Internet is underdeveloped. This paper will argue that IS designers need to give more thought to technologies that assist synchronous and asynchronous oral knowledge sharing and move away from what Chandler (2002) calls Graphocentrism; giving writing an assumed privilege over orality. This is not just a developing countries issue, as the popularity of the telephone and radio in the more developed countries suggests. Written knowledge is inextricably linked to issues of objectivism, a universal rather than personal value system and the dominance of legalised relationships. Moreover, written forms of communication are of limited use for complex skills-based knowledge sharing like engineering and farming. The majority of IS services focus on written communication so fail the "ear people" who need more diverse knowledge sharing technologies (Savard, 1998).

---

[1] But, may be not in terms of social development.

2 THE EYE PEOPLE'S KNOWLEDGE SHARING

What are the attributes of writing? Importantly, it is asynchronous and thus fixed over time. Olson and Torrance (1991) define it as having a higher definition but lower activity medium than conversation, a technology that turned city-states into nations. Ong (1982) reports that, of the 10,000 languages used by mankind, only a little over 100 developed meaningful writing. Writing seems to have been developed to allow the recording of asynchronous trade. As a communication system, it includes books, scientific articles, newspapers, dictionaries, diaries, e-mails, web pages, databases, definitions, contracts, reports. It encourages a culture that, if a thought is not in print, then it has not yet matured, is not legal, is only opinion, and may best not be believed. Writing requires years of training; to attain a reasonable standard maybe 4 years, to be very good at it maybe 20 years for the average person. It is a system that is very good for storing a ritualised version of your memory at one point in time. It helps one remember objectified detailed knowledge like telephone numbers and exact legal phrasing. It allows communication between people who have never met - even after the writer is dead. It gives durability to the ideas of writers over non-writers. Writing enables readers to work at a speed convenient to themselves; pieces can be re-read or skipped. Long and complicated arguments can be correctly structured, as editing of previous phrases is possible. It is not a good way to keep secrets and is an unwise medium to record rumour as the act of writing gives more "concreteness" to a thought than may have been intended.

In the anthropology literature (Olson and Torrance, 1991), there has been a lot of discussion about the impact of literacy on thinking, including much backtracking over assumptions that orality equated to a lack of reasoning skills. It has been necessary to separate the influences of written cultures on subsequent educational influences such as the West's absorption of the writings of Plato and Aristotle with their concepts of essences and concept classification. Ong (1982) provides an example using the set: a hammer, a saw, a log and a hatchet. Which is the odd one out? To those of us brought up to thinking in Plato's essence and collective nouns, we may identify the saw, hammer and hatchet as tools, so the log stands out. Ong reports on oral people who are not familiar with this decontextualisation of classifications, so were not comfortable with the question but being polite responded by saying that if he meant which one they could do without, then they would choose the hatchet because the saw did a better job. When it was pointed out that the log was not a "tool", the response was that the other objects would be no use without the log and the log would become a tool once fashioned. The oral person was focused on relative use, i.e. situational thinking. This may be a more useful way for innovators to think.

In their review of the empirics on the impact of writing on thinking, Olson and Torrance (1991) discuss some thing they call the mental skill of 'differentiation' and 'connectiveness' pointing out that hunter gatherers seem to place more emphasis on connectiveness. Differentiation is a skill that involves problem solving by looking more closely at the separate elements that make up the problem, which sounds similar to analysis and the scientific method. Connectivity is a problem solving approach that seeks connections between the problem and the rest of the world to see if that offers solutions. Dewey (1910) uses the word 'synthesis' for something that sounds similar but he puts more emphasis on seeking a hierarchy of broader and broader (zooming out) perspectives that enable a problem to be seen in context. This reminds the reader of the old systems thinking advice that problems need both synthesis and analysis. Olson and Torrance (1991) go on to suggest writing encourages decontextualisation but also encourages differentiation thinking over connectiveness thinking.

Chandler (2002) points out that writing is seen as attractive by some as it encourages the objectification of knowledge. Olson and Torrance (1991) argue strongly that this is because writing decontextualises initially in terms of who is talking to whom which in turn encourages generalisations as the audience is unclear. Indeed, it introduces the concept of audiences as 'strangers'. This decontextualisation of knowledge is problematic outside science research judging by the many articles written in IS and social inquiry (e.g. Walsham, 1999) calling for human activity problems to be more contextualised. For example, it has lead to a "product" image of education based on a perspective of knowledge as a "thing" that can be purchased and handed over. Ong (1982), in his study of orality vs literacy, points out that writing also tends to encourage conservatism. Oral cultures do not have history books, dictionaries and operating manuals. Further, writing encourages standardised learning both in terms of process (because teachers all read the same "how to do it" books) and in terms of content because once an explanation of a phenomenon is written down, there is a tendency not to think about any background reasoning. This conservatism and standardisation of thinking is expected to save time by not "reinventing the wheel" but in other ways it acts against innovation. Ong also points out that text cannot be questioned or change its mind. If "bad" knowledge is written down, it can be corrected only if all the copies are destroyed, else it will "infect" readers for years to come. Writing also encourages the "objectivity" problem solving and decision making, while allowing interpersonal skills to decline with the people centered view of problem solving - something that has plagued IS research (Butler, 2000).

As mentioned, the clever thing about writing is that it is asynchronous. Kock (2001) undertook a study of asynchronous use of group support systems. He found a difference between synchronous and asynchronous problem solving. First, he found that, while synchronous increased the quantity of ideas, asynchronous improved group process efficiency especially with 'group set-up' costs. There was little time wasted organising meetings, in informal social interaction and in introductions when asynchronous methods were used. While there was less volume of ideas generated by the group, the responses were more reflective and considered. Shirani et. al. (1999) have reported similar results. Though there was less number of ideas generated using e-mail over face-to–face meetings, the ideas were more considered, and more in-depth (inferential). They argue that asynchronous communication has the advantage of a deeper analysis of problems. Therefore, they considered a need for the appropriate mix of synchronous and asynchronous communication. In both these studies, the asynchronous communications were in text format.

3 THE EAR PEOPLE'S KNOWLEDGE SHARING

Oral includes aural, spoken or heard - meaning conversation by listening and questioning the knowledge sharer. This is primarily a synchronous and interactive kinaesthetic communication medium even if the listener is only providing body language responses. Ong (1982) sees the shift from oral to literate cultures as a side effect of the enlightenment that starts from Plato noting the human eye was the most used and trusted human sense (don't believe it until you see it) through to empiricism and on to Hollywood. Those who prefer to use 'the ear' tend to rely on phone calls rather than email, discussion rather than books (reports), argument and debate, telling stories, hearing good speakers, listening and watching movies, thinking about riddles and thinking about problems in a group. This form of communication is usually associated with tacit and skills knowledge sharing. Ong (1982), clearly a 'phonocentric', argues that oral communication emphasises different mental skills compared to text. Examples include memory and an ability to adjust the story to the immediate audience.

Orality also aligns with Habermas' (1979) concern for more social integration through immediate communicative interactions (synchronous) and less systems integration that encourages imposed order through distant experts and impersonal media (writing). Further, written communication rather imposes a one-on-one communication style characterised by a person sitting alone with a book. It is not a community sharing, discussing, bonding technology. Oral communication often also means a small group in discussion. Technology that is to address the needs of oral communication needs to be more centred around group conferencing rather than lone PC to lone PC communication. Also secrets need to be handled differently. Most cultures have taboos or secrets that cannot be written down. For example, governments discourage the publishing of information about bomb and drug making and adults try to restrict children's access to pornography. If written, then controlling access is difficult. Secrets are better managed in the oral form but they run counter to the principles of enlightenment (Olson and Torrance, 1991).

People with a preference against learning through writing live not only in developing countries, but also throughout the developed world. Indigenous peoples and the vast majority of the skilled labour classes don't learn about factors that affect their livelihood from reading. The same may be true of everyday business discussions as the recent spate of research recommending more thought be given to dialogue in organisational change attests (Issacs, 1993). Nearly all the innovation and commercial achievement prior to mass literacy in the late 20th century was achieved with writing restricted to scorekeeping (accounting). Lawson (1999) argues for social routines, trust and the opportunity for tacit knowledge sharing as necessary for raising the innovative potential of a region. He points out that the empirical evidence to date is that physical clustering results in more innovation than virtual clustering (use of the Internet to communicate). This is possibly why cities and high-density population areas tend to be more innovative. It is thought experts (persons with core competencies) need to meet, argue, see, feel and discuss new ideas for innovation to occur.

However, if orality were available in an asynchronous form, it might be even more useful. Dictation machines and telephone answering machines are asynchronous oral communication technologies that have proved very popular. Voice over the Internet (VOIP) technologies are not only cheaper, node independent communication but asynchronous. The lack of quality is because the technology is basically asynchronous, the voice file is broken into packages, which are sent separately and reassembled. Voice based emails, where a recorded voice is emailed as a .wav or .mp3 file is a distinctive asynchronous oral technology. However, voiced based emails have not caught on, while email has; (the authors are aware of an exception with a group of people who have disabilities that makes typing an email difficult). Rather, at least in the writing classes, mobile phones are now being used extensively for SMS messages. It is unclear if this is because these people do not want to talk or because of the functionality of asynchronous text communication.

4 A TECHNICAL PERSPECTIVE

While oral people like to talk face to face in groups, there must be occasions when they would like to talk with people they cannot meet. This is the problem communication technology was invented to overcome. Clearly, we have some understanding of the functionality of telephones. Conferencing over short ranges seems best done using CB, which has many of the characteristics of listserve, partly because telephone conferencing is not cheap. A slightly newer technology is voiced based emails (as oral asynchronous communication).

In order to understand its present functionality and usability better, the authors and a colleague[2] devised a small experiment using the Internet and a voice-based web-board, Wimba, with a large first-year university subject. These were "communication" majors studying modes of Internet communication. Initially, the students were required to submit voice based messages. Later, this was relaxed to allow the choice between voice and text. The intention of the experiment was to gain a better feel for some of the problems with the present asynchronous voice based technology for those comfortable with modern communication technologies.

> *I found that typing allowed you to delete mistakes more easily and gather thoughts, while with the voice postings you really had to think and plan and concentrate on what you were going to say. I also think that voice postings are a little embarrassing! (novice user of Wimba).*

Their most common comment was on the embarrassment of talking into a void - there is no one at the other end, no feedback; recording is like talking to yourself. One simple solution to this would be to develop voice-activated heads for the sender that at least give the illusion of listening. Software such as 'psychologist for Mac' couples a voice synthesizer with an Eliza type Rogerian analysis to provide verbal prompts to the talker based on the talker's previous comments. Participants also reported that they do not like hearing their own voice when they played back their message. It is expected that this is a transitional problem; people who often record their voice seem to get accustomed to the sound. However, voice enhancement software with the recorder making personal selections might also act to overcome this problem.

Editing a voice contribution prior to posting required the whole contribution to be re-recorded from the beginning, regardless of how small the error. In contrast, an error in a text entry can be fixed easily. It is reasonable to expect the likelihood of making an error in a voice contribution would increase proportionately with the duration of the recording. Therefore, a long voice contribution would take an increasingly long time to get error free. This would seem to encourage shorter voice postings or suggest that longer voice postings contain a higher percentage of errors. This is reinforced by participants complaining that the postings of

---

[2] With thanks to Phil Marriott and the Wimba suppliers.

their colleagues contained a number of annoying speech errors. This problem seems to be the single biggest one in asynchronous voice communication. It is an area in need of extensive research so as to allow words to be found in a recording and edits to be made easily.

Although the majority of participants preferred text messaging, some did prefer voice to text messages, claiming that it was because it was easier than typing and provided them with a richer means of communication - one that contained verbal cues and an emotional context. Many participants used a mix of voice and text in their messages. Typically, the text was written as a series of dot points and the voice component elaborated on these points. This hybrid might be the appropriate use of the technology.

The people involved in this trial were a group of undergraduates, reasonably competent in text-based communication. They have never been trained in voice-based work in the way actors and media presenters are. They did not have any disabilities like blindness or paralysis. All of them could type on a computer keyboard. Even so, there did seem to be some demand for the present functionality of voice based asynchronous communication - a minority preferred it.

4.1 Hypermedia

When looking around for evidence of the availability and popularity of asynchronous voice technologies, the author found little apart from the limited functionality of dictation and telephone answering machines. Voice based emails were 'out there' but not popular. Fowler (1994) argues that hypertext (including hypermedia) is a hybrid technology having some of the qualities of writing and some of orality. By hypertext (hypermedia), it is meant the hyperlinking of anything that can be imbedded in a web page with anything else. This includes text, pictures, movies and voice files in an interactive manner so that a reader can add a link or response to everybody else's web site at the discretion of the originators. Examples include history sites where web users are invited to add their own contribution and online journals where people are asked to provide comments or links to articles.

Fowler (1994) identifies a list of characteristics for books. These include that they involve a non interactive reader, the text is fixed, permanent and finished, there is a beginning, a middle and an end, it is only the author's voice (autocratic), there are publishing gatekeepers, it has a single path, and encourages conservativism. Hypertext transcends to some extent many of these qualities but it is still centred on text so of limited use to traditional oral cultures. It may, however, be a step in the right direction for applying tacit knowledge for a literate person. Moreover, the hyperlinking and multimedia functionality is attractive to oral peoples provided the output is broadcast like TV and radio.

A technology that has not been mentioned too much is the radio, and its web equivilant; Internet radio (web page as a streaming 'station'). Coupled with software that mixes incoming audio and video streams to the listeners' requirements, this technology can be used to create numerous purpose-built streaming radio stations. These can be used as decentralised, non-hosted radio able to be designed around local groups needs and local issues. However, this needs infrastructure funding and extensive start up training. Hybrid projects (appropriate technologies) such as the use of the telephone to ask a skilled person to look up a web site and then broadcast it on the radio, or the use of broadcast web pages using push technology over CB radio, is another possibility. However, these require a caring, knowledgeable group to fund and appropriately design for those with little understanding of the technical possibilities.

5 A CULTURE AND SOCIAL PERSPECTIVE

Orality also needs to be thought of as a social or cultural diversity issue, if only to inform the design of technology. Western communities, through the availability and accessibility of communication media and their large volume of explicit written forms, are able to broadcast readily and so promote their knowledge at the expense of oral cultures. The potential for communities whose knowledge system is not based on the written form to be silenced in this environment is significant. The Internet is thus a powerful device to ensure paper knowledge gets to shout louder (metaphorically) than knowledge that is based on experience (trial and error). The "West" gets to shout louder than the developing nations (Ashwari, 1999). This reaches a point where Western explicit knowledge assumes itself superior and starts to treat indigenous knowledge as inferior. Without getting into a defence of Western scientific knowledge, there is a lot of assumed "knowledge" that the West uses that is not scientifically based. Christians may think their beliefs more "logical" than indigenous beliefs, the resulting Western values are also believed superior. So, for example, non-western medication is considered not worthy of research funding. One attitude is that other people's knowledge needs to be subsumed into the dominant culture's knowledge, like Christians taking "Pagan" holidays or images, and Western social scientists "explaining" indigenous artefacts using the latest trend of social theory. Note the problems indigenous and ancient peoples are having in reclaiming their cultural stories, artefacts and knowledge. Examples include the Egyptians reclaiming items taken from the Pyramids and Aboriginal Australians claiming ownership rights over Dreamtime stories and their traditional knowledge about fauna and flora, as well as explanations of artefacts held and sold by museums. There is knowledge competition.

The "oral peoples" are struggling to be heard. A community's unique knowledge not only needs to be preserved but also shared and so developed within that community. For communities with an oral tradition, this means more than saving artefacts and writing down their stories. They have a need to keep building their intellectual property by word of mouth. This knowledge sharing method not only passes on knowledge but also provides a social system, including giving authority to elders and providing a supportive means of learning. In times past, the physical distance between communities would protect their knowledge from being absorbed by other "richer" communities. In the aftermath of the communication revolution, this has been problematic. Printing, cheap travel, the control of diseases like malaria, radio, telephone and television have provided the technology infrastructure for communities to share knowledge across vast physical spaces. However, given the Western dominance of communication media in terms of ownership and content, there has not been an equal sharing of knowledge. Media, such as television and now the Internet, using mainly broadcast models of communication, encourage this. The cost of preparing materials, access to adequate technology and a cultural lack of preference to the broadcasting mode of informing blocks many cultures from sharing, let alone developing, their knowledge. This raises the question of how newer, alternative communication technology can be developed, not only to avoid these mistakes of the past, but also to create space for other knowledge transfer modes.

Ife (1999) suggests that responsible community development should "seek to identify the important elements of the local culture, and preserve them". Preservation and development of knowledge diversity and essential tacit knowledge is critical, then, to the continuation of the cultures and communities. Different communities, from artisans to indigenous peoples, have a need to preserve and develop their knowledge in ways relevant to their specific traditions and cultures. This is not merely a "cultural preservation" act but also an economic development one. Trade knowledge builds modern cities and indigenous knowledge has increasing commercial value especially to indigenous communities.

6 A COMMUNITY PROJECT

Researching community issues with orality or asynchronous communication is not as straightforward as the student based experiment outlined earlier. Below are details of a system designed to assist remote primarily oral Aboriginal communities discuss land rights as part of their negotiations with the State Government. The context details impacting on the design serve to illustrate how developing technologies to encourage or assist oral preferences will need to be awake to some very different needs. It is not just a matter of giving people mobile phones.

This project provides the second of two empirical experiences we designed to improve our interpretation of the situation. The first was the student assignment. We did not analyse any data beyond reflect on what the participants said to us, nor did we sample anything, measure anything nor have control groups or repeat experiments. Indeed, with both the students and the Aboriginal elders experience we would consider it unethical to do so. We are not willing to conduct an experiment on the Aboriginals. Our claim to generalisability is of the problem domain or concepts involved rather than in the statistical sampling sense of generalisability. We make no claim that our empirics are generalisable. We believe that after a lot of experience in this situation, reflecting on the perspectives as reported to us by participants, against a conceptual frame of the importance of oral knowledge, we have some suggestions of where those wanted to inquire further may look first.

With this Aboriginal project, the setting was the State Government trying to negotiate with the Aboriginal peoples over land rights following passage of Federal Law that these peoples still had right to pass over much of the State. After some effort to legally identify elders able to talk for certain traditional areas, community meetings were organized. One of the authors was employed as adviser by these communities on interacting with the Government, which included identifying infrastructure needs. This was undertaken first by listening to the general community discussions over 6 days, and then by including an agenda item of whether they would sign off an a application for a Government grant asking for (faxes, phones and computers) oral communication infrastructure. This agenda item was secondary compared the land rights issues. When it was agreed it could be raised, the suggestions made were made to all those involved and in a public place. We allowed plenty of time for informal debate. At the end of the time allocated for discussion a call for a show of hands was made by the coordinator of the elders. Those people who agreed, took the grant application letter back to their local community and signed it if no objections were raised.

Communication between the 20 different peoples that make up South Australia's Aboriginal peoples is not easy and yet there is a need for them to speak as one to make Statewide legislative change. South Australia is approximately the size of Western Europe but with a mixed population of about 1.5 million. The cost of holding a 3-day meeting of these people is about $130,000. The issues are complicated, more so because many do not speak English nor do they have a common language. Further, many are not comfortable with reading large documents. Therefore, little can be achieved in a 3-day meeting. Further, the age and health of the elders complicate matters. Eyesight and hearing is not good, special diets are required and many are diabetic. Added to this, the quality of telecommunication and power supply in many outback areas of South Australia is of limited quality. Making phone calls is a hit and miss affair, repairs slow, many find sending small emails or downloading web pages too erratic and slow to be worthwhile, so video conferencing is not realistic. The most common means of communication is UHF CB, using a backbone of transmitter tower supported by volunteer workers.

An oral tradition is more than just talking instead of writing. It affects who is allowed to do the talking, what can be talked about and how people are to listen. Whole social structures and routines result from the tradition. The elders are respected because they have the knowledge. If it is written down, the young can ignore and fail to protect the elders. Issues can be dealt with only one at a time, with long meetings on one issue preferred to the writing class's style of a several parallel issues being discussed over several weeks. One issue is discussed to some conclusion. Everybody is encouraged to have their say in front of others rather than the writing class's way of doing much of the negotiations on a one to one basis before the main meeting.

Historically, the radio has been very influential in developing political support and in providing education to remote tribes' people all over the world, particularly in the Middle East. It is cheap and popular especially when the instructional material is well mixed with entertainment. This mix, if correctly selected, may also serve to maintain cultural diversity. Programs like the well known English production, "The Archers", were developed in the 1940's by the Department of Agriculture to carry instructional material to farmers during radio "soapy" dramas. Streaming radio (and some TV) solves the problem of supplying radio over very long distances. It also means that a full time commentator presence is not required. A scanning process can be used to select relevant material, both from satellite down loads and from the very numerous Internet stations, a few hours of which can be placed in a "buffeting file". These can be rebroadcast on the Internet (which, in Australia, is now not classified under the Communications Act as broadcasting) and interspersed with local material using a DJ. At the receiving end, the stream would be rebroadcast in UHF and/or FM for a minimal cost.

A large number of people living in remote South Australia use UHF CBs. The public has invested a lot of volunteer work and money onto a 'backbone" of repeater towers. This forms the main emergency service communication systems in the State. It is an easy matter to connect the external microphone of a UHF CB to the sound card of a PC. Using voice activated CBs, it is then possible for someone in a vehicle with a UHF CB to broadcast and be picked up by a CB, which is attached to a PC. Then, using

VOIP, this message can be carried internationally, including back to the streaming radio station. It is thus possible to have a talkback streaming radio station using existing infrastructure for those travelling long distances. While hardly state of the art, it is a solution that integrates the technology quickly and easily, but importantly it is relevant to the community groups.

Using the types of technology suggested above, an unsuccessful attempt was made to raise a Federal grant to set up, record and store these community meetings as the last of the Aborigines who remember the old way near old age. The land title negotiations that were to take place between the Government and Aboriginal communities would be binding for many decades so the intention was to inform future generations how equitable the negotiations were. It was intended to supplement the recordings with background materials, translations and links to official documents into a hypertext (hypermedia) cultural database, which, with interviews with the elders and careful selection and cataloguing of archival materials, would provide an encyclopaedia of the event. Of course, this would be very inadequate in terms of capturing the oral traditions, but some parts could be saved. The idea is not new - many indigenous peoples have now started similar projects. Apart from visual scenes and conversations with people who have the knowledge, archival footage and descriptions of artefacts, maps and stories can be captured. This still undermines the "respect for elders as they have the knowledge" social order that an oral tradition supports and puts a tremendous onus on the administrators as errors in accuracy may do lasting damage. Secret knowledge, such a sexual advice to young women, may also be lost. The thought was that this multimedia database would be more likely used than a book with subsequent generations of Aboriginal peoples.

The Federal Government did not agree that a grant for recording these negotiations as described was warranted – instead they were recorded using the traditional "written" method. That this project was unsuccessful was the motivation for this paper. In the rush and excitement to provide the technology, 'appropriateness', be it in schools or for communities with an oral preference, may be trampled. A compounding issue is that, while many communities want to emphasise their unique culture and differences, this should not be translated into appearing unable or unwilling to be able to master the latest technology. It is being here that a community based asynchronous oral communication system would push the envelop of the communication technology but towards something appropriate for someone apart from the writing classes.

7 CONCLUSION

To recap, this paper is about orality. The authors are concerned that the development of Internet services has been driven by an implicit assumption that information is written. Oral knowledge sharing has been the silent partner. This is thought to be a mistake as a combination of both is thought to be the optimum for innovation and thus social and economic development for both the developed and developing nations. Orality is more situational, decontextualises less and requires different mental skills such as memory.

This study found that the technology requirements to bring orality up to the standards of written e-communication include:

- Improved asynchronous oral functionality, including
  - Improved voice editing facilities,
  - Improved voice based email, including built-in quality microphones in PCs,
  - Visual or voice feedback when recording maybe including Elisa type software,
  - Development of combined dot pointing with speech media, maybe with simple sketching facilities.
- Improved voice conferencing facilities, i.e. less assumption of 'one on one' communication. This should allow several groups of people to meet and talk for up to days at a time, maybe using 'wired' meeting rooms and having access to multimedia facilities to discuss specific cultural artefacts.

The main force working against the suggestions presented in this paper is the status of writing. Literacy is seen as a major development issue so it is possible to imagine the Internet being used to teach reading and writing before being used as an oral communication technology. Writing is still seen by many as 'the truth', the correct or official version of events. Therefore, if a person's words are written down, they are given status and recorded for the future. While these attributes of writing need to be acknowledged, if orality is suppressed then so might innovation.

This written communication on the needs of oral traditions is clearly paradoxical. Oral people should talk for themselves, but if they did, this paper would not be produced. We are sure many in Aboriginal groups we met would say, "Those academics have no right to talk for me, nor write for me, unless I dictate what is said". This we respect and agree with. We do not claim to talk for anyone but ourselves. We believe that the Internet communications technologies should be more inclusive of oral people because we seek knowledge diversity.

REFERENCES

Ashrawi, H. "Democracy, Survival and Co-operation", keynote speech presented to the inaugural *Adelaide Festival of Ideas*, Adelaide, July 1999.
Butler, T. "Making Sense of Knowledge", *ECIS Proceedings*, Vienna, July 2000.
Checkland, P., and Holwell, S. *Information, Systems and IS*, New York: Wiley, 1998.
Chandler, D., 2002. Available at: http://www.aber.ac.uk/media/Documents/litoral/litoral.html
Dewey, J. *How We Think*, Lexington, Mass.: DC Heath, 1910.
Fowler, R. M. "What Hypertext Can Teach Us", *Interpersonal Computing and Technology*, (2:3), 1994, pp. 12-46.
Habermas, J. and McCarthy, T. *Communication and the Evolution of Society*, London: Heinemann, 1979.
Ife, J., *Community Development: Creating Community Alternatives*, Melbourne: Longman, 1999.
Issacs, W. N. "Taking Flight: Dialogue, Collective Thinking and Organizational Learning", *Organizational Dynamics*, (22:2), 1993.
Jones, S. G. *Cybersociety*, California: Sage, 1995.
Kock, N. "Asynchronous and Distributed Process Improvement", *Information Systems Journal*, (11), 2001, pp. 87-110.
Lawson, C. "Towards a Competency Theory of a Region", *Cambridge Journal of Economics*, (23), 1999, pp.151-166.
Olson, D. R. and Torrance N. (Eds). *Literacy and Orality*, Cambridge: Cambridge University Press, 1991.
Ong, W. J. *Orality and Literacy: The Technologizing of the Word*, New York: Methuen, 1982.
Savard, J. "A Theoretical Debate on Social and Political Implications of Internet Implementation", *Wicazo SA Review*, (13:2), Fall 1998, pp. 83-97.
Shirani, A. et al. "Comparison of Two Technologies for Synchronous and Asynchronous Group Communication", *Information & Management*, (36), 1999, pp. 139-150.
Walsham, G. "GIS for District Level Administration in India", *MIS Quarterly*, (23:1), 1999, pp. 39-65.

About the Authors

Mike Metcalfe: associate research professor, University of South Australia, PhD (Adelaide, Australia) 1994, M.Sc. in Business Planning (Salford, UK) 1981. His publications include 4 books plus refereed articles on 'the design of human inquiry', including information strategies, research methods and forecasting; he is 51 yrs, and has worked in the merchant navy, the army reserves, system design, teaching and consulting including two and half years as adviser to the Deputy Premier of South Australia; present research interests include the use of argument and multiple perspectives to define projects; present employment is as thesis advisor for the Information Systems Doctoral School. He can be contacted by e-mail at mike.metcalfe@unisa.edu.au.

Carmen Joham has held a number of management positions where she has designed, programmed and supervised the implementation of information systems for business like General Electric, Mobil and IT Universities. She is also a PhD candidate and Research Associate at the University of South Australia's Information Systems Doctoral School. She was born in Venezuela and holds a Masters in Business Administration from the Adelaide University. She completed both a BA and a BSc in Computer Science in the University of CUMT-Venezuela Her present research interests include information policy systems and strategies, social aspects of information for socio-economic development and globalisation issues. She can be reached by e-mail at carmen.joham@unisa.edu.au or by Telephone: +61 8 8302 0269.

27

CROSS-CULTURAL IS ADOPTION IN MULTINATIONAL CORPORATIONS

A Study of Rationality

Gamila Shoib and Joe Nandhakumar
School of Management, University of Bath, UK

Abstract: A common claim in the literature on Information Systems' implementation in the context of less developed economies or so-called "developing countries" is that the "Western" technology is at odds with the local cultural context, in particular it is believed to mismatch local rationality in the sense of the accepted ways of doing things. In this paper we investigate IS implementation in a company based in a "non-Western" context compared with IS adoption in another company in a "Western" country context. Seen as a particular form of decision-making, the adoption and implementation processes are analysed drawing on the literature on decision-making, rationality in "Western" and "non-Western" contexts. Presenting evidence from these two contexts we argue that multiple forms of rationality exist in any context and that national culture is only one aspect of actors' as well as researchers' sense-making of activities in any given context. Linking the cases back to the literature we reflect on the implications of our findings for cross-cultural research of IT implementation.

Keywords: Rationality, Decision Making, IS Adoption, Cross Cultural Management.

1 INTRODUCTION

Western[1] knowledge and its model of rationality are usually presented in research on developing countries (DCs) as a homogenous body of knowledge, which dominates research and practice (Avgerou, 2000). Recent interpretive studies of Information Systems (e.g. Ciborra, 1999a), have

[1] The term "West" is used, as is commonly done in the "developing countries" literature and discusses at length in Said (1994), to refer to the "first world", namely Westerns and Central Europe as well as the USA.

however challenged the domination of the rational model and its normative status as the accepted form of behaviour within its own Western context. Unlike conventional critiques, discussed later, these interpretive works, having lost the prescriptive normative view of research, allow for alternative perspectives to emerge. Such interpretive works allow researchers, especially in non-Western countries, to be more reflexive about what happens in organisations without the "pressure" of labelling it as "success" or "failure" because of its alleged mismatch with the accepted model. It allows them to move a bit further in the quest to understand rather than to prescribe behaviour.

This paper therefore first investigates the rational model and its critiques. Moving on the extension of the rational model to IS in DCs context is highlighted. This theoretical foundation is followed by a presentation of the empirical work, namely the two cases - Company X and Company Y, representing the two contexts as well as the a) decision making in IS adoption and b) the rational model and technology as its instrument and symbol in action. The final section of the paper discusses the cases in terms of the literature and reflects on the implications for the way IS research, in particular in cross-cultural contexts, is constructed.

1.1 Rationality and Decision-Making in Organisations

In order to understand what it means for an organisation to be "rational", we build on reviews in the organisational behaviour literature of the rational model and its conventional critiques before moving on to present the alternative interpretive perspective. As Miller et al. (1996) argue, the majority of theories on decision-making in the literature adopt the positivist paradigm. In studying an organisation an attempt is typically made to locate its decision-making within one of the three choice paradigms: the (boundedly) rational model (Simon, 1976), the power and politics model (Pfeffer, 1981) or the garbage can model (Cohen et al., 1972).

> *Decision-making, has thus developed from a view of the process as a step by step logical and rational process to a process constrained by cognitive limitations, organisational constraints and finite resources. Considerations of group (ir)rationality as well as unstable and ambiguous conditions have brought in the political and garbage can perspectives (Eisenhardt and Zbaracki, 1992).*

In a summary of themes of newer perspectives on Organizational Behaviour, Weick (1985) recognises the retrospective and political nature of rationality. Success is thus presented as a form of rationalisation/ legitimation. He recommends

> *"Don't treat rationality as a universal prescription. If you live by rationality alone, you lose options (use of intuition, quick response, trial and error) and you lose non-deliberate sources of variety (hunches)" (Weick, 1985: p.132).*

Such a perspective can be seen as part of an emerging tradition reflected in the recent model put forth by Mintzberg and Westley (2001) that seeks to combine elements of two forms of rationality, one that is planned and another that is situated. Mintzberg and Westley's (2001) motivation for proposing this theory is the observation that the rational approach turns out to be uncommon. They recommend that the "thinking first" model of decision-making in organisations be supplemented with two different models "seeing first" (sudden insight) and "doing first" (trial-and-error) embracing intuitive and action-oriented forms of decision-making. Decision-making is done through discussion, collage and improvisation and incorporates emotions, humour, fear and anger. They conclude that there is a need to "*put [thinking] in its place alongside seeing and doing*" and that people in organisations combine all three. It would therefore seem futile to try to label organisations or individuals based on different types of rationality.

Finally, comparing the structure of improvisation with that of decision-making, Ciborra (1999a) concludes that each contains elements of the other. Reconciling the planned nature of decision-making and improvisation with the circumstantial interpretation of the past that gives ultimate meaning to action, The deep nature of decision-making is therefore one of improvisation (p.145). Elsewhere with his colleagues (Ciborra et al., 2000) argue for an alternate view that defies conventional management and IS wisdom.

1.2 Interpretive Perspective on Rationality and Information Systems

The rational choice model extends to the arena of Information Systems. Technology, seen as a tool for change and control (Avgerou, 2000; Ciborra and Hanseth, 2000), has been heavily implicated in the domination of the rational model in organisations. Some works, however, break with this functionalist tradition and are rooted in a more phenomenological, interpretive worldview. They offer alternate views of rationality. Suchman's (1987) work on *plans and situated actions* challenges the assumption of technology (and its use) as rational. Her work shows how the rationality of technology itself is an ascribed property given to it by human actors who label the machine as intelligent given the slightest signs of human intelligence. Her work attempts to subvert the common view of rationality in the rational choice theory of human action. The premise of Suchman's (1987) work is that all purposeful action is situated action. Accountability necessitates rationality and the retrospective use of plans to explain/justify actions. Plans, which hold the uncontested position of *the* correct form of action for a rational actor, are presented as retrospective rhetorical instruments of rationality. Her investigation of the project of building intelligent artefacts contributes to the notion of situated action and the problem of shared understanding.

Finally, Orlikowski (2000), like Suchman (1987) and also Ciborra's work, may be placed in the context of an increasing shift towards situated action and practice, e.g. Crossan et al. (1996) and Orlikowski (2000).

The literature on IS in DCs follows a very similar pattern to the field of IS at large where the mainstream is dominated by the functionalist views and a small stream of interpretive, context-sensitive studies exist. Studies of IS in "non-Western" countries affiliate themselves with, the cross-cultural and comparative management literature or international business literature. However, given technology's implication in the socio-economic development of DCs (Walsham, 2000), studies also appear as part of the development literature (Avgerou and Walsham, 2000). Global Information Systems are also a new theme for researchers on less developed economies. Unlike the situation of the dichotomy of the notion of culture studies on IS in international, cross-cultural and international settings are preoccupied with the notion of culture. The "culture-free" hypothesis is uncommon. The prevailing assumption therefore is that technology is "Western" and the context is culture- and country-specific and is predominantly studied by focusing on the values and beliefs held by individuals and groups within that culture (Abdul-Gader and Al-Bureay, 1993). Unlike management practices, which are, at times, viewed as universal, technology is almost always seen as "Western" (Shoib and Jones forthcoming). This may be again explained by technology's association with socio-economic development and as a symbol of rationality and progress.

As mentioned earlier, some research within the DCs literature is more interpretive and context-sensitive and recognises and celebrates situated local improvised practice in DCs. Most notable are the conference proceedings of IFIP 9.4 in 1998 in Bangkok and in 2000 in Cape Town, namely Avgerou and Walsham (2000) and Sahay (2000) respectively and the work of Appadurai (1996). The papers presented in the edited collections, despite their diverse topics and wide range of country coverage, are congruent with many of our findings.

2 RESEARCH SETTING AND APPROACH

The research approach adopted in this study is interpretivism (Walsham, 1993). Interpretivism assumes that knowledge of the world is a social construction by human actors. Interpretive researchers therefore attempt to study phenomena from the perspective of the actors involved and to understand the meanings that individuals ascribed to them. This study aims to understand decision-making in IS adoption and implementation in a company based in a non-Western context compared with another company in a Western country context.

The main, larger-scale, study was carried out in the subsidiary of a US-based multi-national corporation in Egypt. A comparative study was carried out in the world head quarters (HQ) of a large manufacturing company (Company Y) in USA and offices and plants around the country. Both companies are volume manufacturer of high-value consumer products, which have operations in several countries and each employ several hundred thousand employees world-wide. The companies are highly structured, have a strong management hierarchy with several layers of senior executives.

2.1 Data Collection

The study at Company X involved one of the authors spent 9 months as a participant observer in the IT department in Egypt during 1997-1998. Additional data were gathered over a further period of 15 months (1998-1999) through email communication with members of the organisation, occasional visits and a formal workshop to discuss the findings. At least 88 unstructured and semi-structured interviews were held with all members of the IT and sales organisations to gain an understanding of the decision-making practices which inform IT use in the sales department, especially those relating to a sales force automation (SFA) project, which had taken place in October 1996 and was being upgraded during the field study. The interviews also included so-called practical sessions during which members of the organisation walked the researcher through their daily work and their usage of their laptops. During these sessions with the users copies of user files were taken onto individual floppy disks, both for reference as well as in preparation of the analysis.

Field observations of interactions with other members of the organisation including user support, chats, social events (e.g. lunch, coffee breaks), meetings etc. were logged into a self devised field-tracking-system shortly after they occurred. This fed into a "chronological events list" which served as a filter/index to the wider set of observations and was crucial in the selection of the "significant" situations, which were selected for closer analysis. A field/personal diary was also written periodically to supplement the formal material gathered and provide reflections on the research as a whole. Further electronic documents received via email and pertaining to all aspects of the organisation and its employees were also retained.

At Company Y the data collection was conducted through over 30 unstructured interviews with organizational members and by observation of their organizational practices at the world HQ in USA and offices and plants in the South of USA from 1997 to 1998. Interviewees included senior executives who led the technology project implementing Internet-based information systems from the USA, many IT and functional managers from the USA and UK. Most of the interviews were tape-recorded and transcribed. Documents have also been examined, including promotional documentation, training manuals and Internet-based support documents. Detailed field notes were maintained during the study period to record observations and events during each visit. Additional data were gathered during 1998-2002 through occasional visits and workshops.

2.2 Data Analysis

The material gathered was partly imported into a qualitative data analysis package (NUD*IST) and examined to identify comments and actions of the participants that reflected similar views with respect to IT implementation (Glaser and Strauss, 1967). The iterative examination of interview statements, comments and observed interactions of participants from each company led to the identification of a set of themes that reflected multiple interpretations held by members of each organization with respect to IT adoption. The analysis encouraged these themes to emerge from the data rather than imposing them on the data. In the case descriptions below extracts from the interviews are provided as examples to illustrate our arguments.

3 CASE DESCRIPTIONS AND ANALYSIS

3.1 Company X

The sales department of Company X was based in a modern office block in Cairo. The building also housed the IT department, including the support staff, who ran user training and the help desk. In 1996 the Company X issued its salesmen with laptop computers, so that they could work from home, downsizing its Cairo headquarters and closing its regional depots. As part of the sales force automation (SFA) project, field sales staff (customer business managers, as they were called), were given training on the system (mainly introduction to the laptops and Microsoft Office applications) by an outsourced third party.

According to IT management, the SFA project had little effect on business processes, as the primary objective appears to have been the introduction of the computers themselves. It was left to the sales force to "export" their practices to the system. At a later stage in the project, however, a shipment reporting tool was introduced, which allowed the sales organisation to run queries on it. The interviews revealed that many of the salesmen had great difficulty in using the SFA due to the generic character of the training, the use of an English language interface (when most were Arabic speakers) and decentralisation of report and data generation (previously done by secretaries). As a result, the computers were seen as a "burden" and most salesmen used them as little as possible. IT staff attributed these difficulties to the sales field staff's lack of computer literacy and ways of thinking.

> *"They're not very computer literate that is. Look sales users, sales people in general they do not have strong thinking and problem solving... So they just know enough to keep their work going" - [IT Manager].*

> *"Yes, it's my feeling from the way they operate the machine. They just don't know what's going on. They memorise but they don't understand that is" - [IT Manager].*

Field sales people were seen to be ill educated (with no university degrees), and reliant on memory rather than understanding, especially compared to IT staff who were predominantly Engineering graduates. One expression of this difference was evident in language, where the IT staff tended to be fluent in English and mix Arabic and English in one sentence whereas field sales spoke only Arabic. This was significant, as language and computer knowledge were heavily implicated in performance evaluations and closely linked to promotion:

> *"English has to be important for us ... if I am going to be in charge of a new job or will move up with a higher promotion or so, I have to ... suppose the department head is an Englishmen or Spanish or any other nationality ... how will I deal with him? I will never get promoted unless I am perfect in English." - [Area Manager, Alexandria and Delta region].*

Further informal hierarchies were evident within the sales force between those based in Cairo and those "in the provinces" and within the company as a whole between graduates of the American University in Cairo and graduates of other universities.

The salesmen, on the other hand, considered that the IT staff, working regular office hours in a peaceful, air-conditioned building in Cairo, had little appreciation of the working conditions in which salesmen were expected to use their computers. Spending their days in the heat, dust and noise, travelling long distances over poorly-maintained roads to visit unruly customers in remote locations was very different from the protected "calm" and order of the head office environment where they had been trained:

> *"there is a problem between us. They are people who look at us as negligent with the machines, that we ... do not use them well and consequently you feel when we deal with them that they are fed up of us ... And then at the same time, something could happen incorrectly, unintentionally because I don't understand. If you explained to me correctly from the beginning, I will not do anything wrong" - [Area Manager, Alexandria and Delta region].*

This situation was further exacerbated by the fact that none of the IT staff had completed their orientation session and been to the field with a salesman. These same "ignorant" salesmen, however, present a very different picture if we examine them in interaction with their customers. Here they appear as the rational, technologically capable ones, seeing themselves as superior to their customers because of their (company's) "scientific" ways of doing business, and dismissive of what they see as their customers' backward *baladi*[2] business practices:

> *"Because you are really dealing with a type of people who are not human. You could be dealing with someone who at most can count cases and money. That is all his interaction with the world through these two things only" - [Area Manager, Cairo and Canal region].*

> *"We are not sitting at the corner of an alley ... [The customer] upset me. "If I fill the boot of the car with product" and stuff like that. I told him "No this is in Baladi. No we want numbers" - [Area Manager, Cairo and Canal region].*

They were also amused by their customers' lack of computer knowledge:

[2] An Egyptian term, literally "my country" and associated with *awlad el balad* or "the sons of the country", referring to people who follow and preserve the old Egyptian ways (El Hamamsy 1985, p. 56).

"I once went with it to the field. It was the first time ... for them they were seeing a bag like that with a computer coming out of it ... So while I was standing talking to the customer, another one of our customers was calling him, telling him "Mr X will come to you and is bringing the computer for you with him. And you'll see your numbers and you'll see whatnot, and the screen is like the television". So I went and found him all dressed up and elegant and sitting there. And he brought his brothers and relatives and was sitting waiting" - [Area Manager, Cairo and Canal region].

This salesmen's scorn for their customers, however, also reflected a concern with keeping the upper hand in the relationship:

"So the customer ... when the customer finds an opportunity to feel that I don't understand, he attacks me, because I am always sitting on top of him... So when he finds an opportunity to escape me and feel that I don't understand, ... I am unable to control him. So he must feel continuously that I understand" - [Area Manager, Cairo and Canal region].

In this, the computer could sometimes be an important tool. Even though most salesmen shied away from taking the laptops with them to the customers (perhaps because they were concerned that the customer would sense their incompetence and hesitation in front of English error messages) they regularly took computer print outs (often produced by a more knowledgeable colleague as a favour) using them for "objection handling" and as the last word if a dispute should arise with a customer.

"If you have your facts and figures, pretty much, you're just so powerful. And you're even so much more powerful than the customer himself.. Because you know when he tells you that "Look this is too much for me, I can't take that much", "Why can't you take that much", you know because you've got the computer print out and it shows you the last few months, past six months and the same a year ago and pretty much the numbers can beat him. You know, he can't complain" - [Area Manager, Cairo and Canal region].

But the salesmen also recognised that there were limits to what computers could do for them.

"So there are different circumstances with the customers that the computer cannot understand. Because it is only numbers and information" - [Area Manager, Cairo and Canal region].

Despite the computer's importance as a means of persuasion with the customer, therefore, the key to successful selling was still seen to be practical experience and knowledge of the market or the *suq*, in which the computer's role was restricted to very much a secondary role as the source of shipment data. Thus the salesmen's view of computers was often ambivalent. On the one hand they were a valued tool that supported the salesmen's sense of superiority over their customers, while on the other, they were not just a reminder of their marginality from the "westernised", modern world of the head office, but also a "burden", of limited use in some of the most important aspects of the salesman's work. However, the salesmen coped with their lack of training by relying on an informal network of more computer-literate friends as well as improvising and tinkering while using their laptops.

3.2 Company Y

At Company Y a dedicated IT Division based in the World HQ in the Midwest of USA was responsible for implementing a new internet-based system. This system was implemented to facilitate a global restructuring initiative involving a major plan to merge company's operations carried out separately in various parts of the world into one giant global company. Initially the new IT Division implemented a standardized network (based on TCP/IP) to link-up every computer within Company Y. Many engineers from the HQ Product Development Division who were experimenting with the free web browser made use of the newly implemented network, for example, for publishing technical documents and tracking project status. The IT Division managed the new system centrally, by requiring other divisions to place data content in servers located in the world HQ.

Senior executives from the HQ were able to use browsers for accessing and publishing technical documents and tracking project status. Many managers from Product Development Division claimed that the new systems saved travel costs, as they were able to share engineering drawings and visual images online with team members in other countries.

As these new systems became established in the company, the IT Division began to replace most of the legacy systems such as EDI with Internet systems in order to interact with suppliers. Many senior managers became technological evangelists, promoting the use of systems:

> ".... *part of our mission is to be technology advocates and evangelists; part of our mission is to be educators; part of our mission is to be developers; part of our mission is to be infrastructure developers. We*

> *have, you know, quite a mixed bag of things that we're responsible for" - [Head of IT Division].*

While executives from HQ vigorously promoted Internet-based systems within the assembly plants, the systems staff at the plants considered that the IT Divisions from the HQ had little appreciation of their systems needs. Plant systems managers often joked how ignorant HQ systems staff. Many of the managers from assembly plants felt that the Internet initiative was imposed on them by the head office. Executives at HQ however felt that compared to staff at the HQ the staff at plants were incompetent, middle-aged managers who were not comfortable with advances in technologies.

> *"...we actually have tried a couple of times with true collaboration like news groups and stuff like that [for plant engineers]. It hasn't been as successful - it's a cultural thing. I think there's a tremendous amount of value in a newsgroup ... facilitating a newsgroup, but our culture is just that they just weren't very well accepted. Part of it's a typing things. Part of it's we get a lot of middle aged and above people; some of them aren't comfortable with it. The secretaries tend to use it pretty well and those are good. But, well, we've had at least two attempts to try to get that into the engineering community. It just hasn't been real successful" - [Head of IT Division - Recalling with frustration].*

Although assembly plants were a big challenge for HQ systems managers, they proceeded with their strategy to implement global Internet infrastructures. The IT Division insisted that anyone could use the internet-based system. Plant managers however found it hard to make use of the new systems without any training and they often forgot password and procedures to gain access to HQ systems. They had to rely heavily on their personal assistants for generating printouts of necessary reports from HQ systems that were routinely used in the plant meetings and negotiations:

> *"he [the plant manager] has a new computer on his desk, but very seldom does he do anything other than look at our production coun], there is a programme with production [local] counts on it, he will look at that and see what the hourly rate is coming out at" - [Systems analyst from an assembly plant in the South of USA].*

On the other hand, the plant managers gave a different impression in their dealings with the workers in the plant. For example, the plant managers were seen as technically competent and promoted technological advances in the plant. A senior staff from a plant revealed that he was thrilled to take instructions from his plant manager who did "picture talk" [video conferencing] with him from the HQ. The workers in that plant were impressed with the plant manager's technological accomplishment. On the other hand the plant workers were seen as incompetent by their managers. For plant managers the information generated by these systems was extremely valuable in sustaining their managerial status locally, while on the other hand, this was also reinforcing their inferior status in their relationship with HQ executives.

4 DISCUSSION

Drawing on local descriptions of Egyptian society, e.g. El Hamamsy (1985) and Amin (2001), it is possible to identify two systems of (perceived) rationality: planned and situated. This situated rationality viewed from the point of view of the planned rationality can be seen as irrational, informal, and *ad hoc*. The story told by case X is one of individuals playing multiple roles in multiple spheres of interaction with various interests and degrees of conscious deliberation and knowledgeable action. The dominating theme, however, also described persistently in most discussions of Egyptian society (e.g. El Hamamsy 1985; Amin 2001) is that of "Westernisation" and/or "Egyptianisation". It is a theme that appears to be as formative of individuals' assertion of their identity as it is of society as a whole as well as the research scene at large. In their various roles the actors draw on normative models of behaviour that are most pertinent to their situation. The "Western" model of planned rationality and a local model of situated rationality were identified as relevant to actors' sense-making. This analysis is supported by the discussions of society at large (El Hamamsy, 1985), which differentiate between "Westernised" individuals and "baladi" traditionalist Egyptians. Technology, as a symbol of "Western" rationality, played an important part in the projection of a "rational", "Westernised" image. This was particularly visible in focusing on the salesmen and their interactions in different realms, with the IT staff and with the customers. In their interactions with the IT staff they were the "irrational" incompetent users who were baladi in their ways and lacked problem solving skills and rationality. In their interactions with the customers they drew status from their use of technology (or image of so doing). In other situations they could chose to project an image of an "Egyptian" and disassociate themselves from the "Western" multinational.

If we focus only on the relationship between IT staff and salesmen then it would be easy to conclude that the salesmen were seen as irrational, unplanned and inferior. It would then be easy to see many examples that might seem to confirm the cultural characteristics identified by writers such as Hofstede (1980) and Trompenaars and Hampden-Turner (1998). But if we broaden the scope to include the Helpdesk's managers and the salesmen's relationship with their customers then the picture becomes somewhat different. One of the pair is seen to be "irrational", traditional, deferential and personalised. However, the same person can be simultaneously at both ends of the spectrum, depending on which relationship is being examined. This fits less well with these cultural typologies. Thus with IT staff, salesmen were the irrational traditionalists, while they, in turn, made similar remarks about their customers. IT managers made similar comments about the Helpdesk staff etc. The image of managers presented in the literature may therefore be described as one-dimensional and simplistic.

Local culture, which has been said to emphasise a situated rationality characterised by an appreciation of practical thinking, *bricolage*, improvisation and group loyalty, when viewed without the pressure of the "Western" rational normative model proves itself to be a facilitator and even "saviour" of business when the planned rational model "breaks down". It is therefore possible to argue that success, failure, and breakdowns are all relative constructs rooted in a particular worldview and perception of rationality. When one breaks down, the other, which is concurrently operative, becomes more visible. Researchers presenting and analysing such cases are also implicated in constructing and de-constructing these worldviews.

The story told by case Y is that IT adoption was situated everyday decision-making although there was evidence of rational decision-making, especially in action plans and peoples' accounts of adoption process and rationalizations of them. Many of the decisions were guided by individual's preference and mutual accommodation and negotiation, characterized by opportunism and improvisation (Ciborra's, 1999).

There was also evidence of decisions guided by actions and evolving conditions rather than preferences. Initially there were no intentions to allow users from Product Development division to develop their systems but with the implementation of Internet infrastructure it wasn't possible to stop this from happening. The unintentional nature of many of such decisions seems to imply a "garbage can" (Cohen et al., 1972) and situated form. The managers from HQ IT Divisions reacted to problems and their actions were situated and contextual.

Similar to case X we could make the same argument about the competing planned and situated rational models, which were used by actors as a resource for the purpose of rationalization. HQ executives' view was that the slow adoption of Internet-based IS at assembly plants was a 'cultural thing' and the 'middle-aged and above' plant managers were not comfortable with new IT. Similar to Sales Force in case X, the interactions of the assembly plant managers with the HQ executives and plant workers is a good example of individuals being at both ends of the rationality spectrum and their use of their position to further their situated interests in a particular context.

5 CONCLUSIONS AND REFLECTIONS

This paper has sought to further our understanding of IS implementation in cross-cultural, international settings. By drawing on an in-depth analysis of IS implementation in a company based in a non-Western context compared with another company in a Western country context, the paper argues that multiple forms of rationality exist in any context and that national culture is only one level of actors' as well as researchers' sense-making of activities in any given context.

Our first reflection is on the treatment of technology in the cross-cultural and international context. The literature treats technology as "superior" or "Western" and the context is almost always assumed to be culture- and country-specific. It is predominantly studied by focusing on the values and beliefs held by individuals and groups within that culture (e.g. Abdul-Gader and Al-Bureay, 1993). A more helpful approach would seem to be to avoid treating IT as a tool with unambiguously positive associations and question the prevailing assumption that "traditional" ways are necessarily negative, inferior and irrational and at odds with this allegedly "Western" technology.

Our second reflection is the normative nature of the "success" model used to evaluate IS implementation especially in the context of less developed economies. Case X argued for a situated and relative definition of success and one that does not necessarily see local culture and context as an impediment to implementation. Success of IS implementation in organisations, therefore, needs to be understood as being the product of relationships between social actors who sustain and/or transform them in the course of their interactions. IS implementation success is a relative construct (Avgerou, 2000) and a function of chosen theoretical and analytical perspectives (see e.g. Bada, 2000). Researchers therefore need to be sensitive to the multiple roles and social positions at play in the various contexts as well as the multiple rationalities and alternative perspectives on success and failure available to actors and researchers alike.

Our third reflection is concerned with human agency. Evaluations of local situated culture would seem only able to see it as benchmarked against an idealised, normative model, which, as shown, may be largely more of a political legitimation instrument. It would also seem naïve to suggest that knowledgeable actors are unaffected by the multitude of cultural, and otherwise, influences available to all of us in our everyday life. Rationalities, as generally representative of different ways of doing things, are therefore a resource for actors to draw on in their everyday actions and sense-making of them. There is therefore a need to take account of the reflexivity of actors and their ability to transform or introduce change or act differently. This recognition would allow for the inclusion of "alternate" rationalities, which get drawn on according to the situation in order to advance actors' political interests. Individuals in different contexts thus play multiple roles in multiple spheres of interaction with various interests and degrees of conscious deliberation and knowledgeable action.

The final reflection is on analytical dualisms often adopted in IS research in this context and in general. The findings illustrate a point made by Giddens (1984) with regard to "discursive penetration" (social actors' awareness of their engagement in social reproduction and production) which leads to a "double hermeneutic" where the concepts of the sociological observers are constituted as meaningful by social actors and can become part of the actors' sense-making. The "West/non-West" dichotomy as well as the normative rational model are not theoretical constructs about developed and developing economies and societies or decision-making in organisations, but can also be said to exist in the minds of policy makers and actors and to play a part in their sense-making of their own actions and the assertions of their identities. It is therefore important for researchers to be clear on their own positions with regard to such devices while remaining open to the social actors' own interpretation of such constructs. Researchers therefore need to be clear on their own as well as actors' positions with regard to dualisms at hand. In this context we have identified three such dichotomies and transformed them into dualisms: West/non-West, developed/developing countries, and finally planned vs. situated action.

REFERENCES

Abdul-Gader, A. H. and Al-Bureay, M. A. "An Islamic Perspective to Information Technology: Management Implications", in *Proceedings of the Arab Management Conference*, Bradford, UK, D. T. H. Weir and M. J. Alwani (eds.), 1993.

Amin, G. *Whatever Happened to the Egyptians? Changes in Egyptian Society from 1950 to the Present,* American University in Cairo Press, Cairo, 2001.

Appadurai, A. *Modernity at Large: Cultural Dimensions of Globalization,* University of Minnesota Press, Minneapolis, 1996.

Avgerou, C. "Recognising Alternative Rationalities in the Deployment of Information Systems", *Electronic Journal on Information Systems in Developing Countries,* (3), Paper 7, 2000, pp. 1-17.

Avgerou, C. and Walsham, G. (eds.). *Information Technology in Context: Studies from the Perspective of Developing Countries,* Ashgate: Aldershot, 2000.

Bada, A. O. "'Actually Existing Globalization': A Case Study of IT and Organizational Change in a Nigerian Bank", in *Information Flows, Local Improvisations and Work Practices, IFIP WG 9.4 Conference*, Cape Town, S. Sahay (ed.), 2000.

Cohen, M. D., March, J. G. and Olsen, J. P. "The Garbage-can Model of Organizational Choice", *Administrative Science Quarterly*, (17), March, 1972, pp. 1-25.

Ciborra, C. U. "A Theory of Information Systems Based on Improvisation" in *Rethinking Management Information Systems,*W.L. Currie, and B. Galliers (eds.), Oxford. Oxford University Press, 1999.

Ciborra, C. U. and Hanseth, O. "Introduction: From Control to Drift" in *From Control to Drift: The Dynamics of Corporate Information Infrastructures,* C.U. Ciborra et al. (eds.), Oxford. Oxford University Press, 2000.

Ciborra, C. U., Braa, K., Cordelia, A., Dahlbom, B., Failla, A., Hanseth, O., Hepsø, V., Ljungberg, J., Monteiro, E. and Simon, K. A. *From Control to Drift: The Dynamics of Corporate Information Infrastructures,* Oxford. Oxford University Press, 2000.

Crossan, M. M., White, R. E., Lane, H. W. and Klus, L. "The Improvising Organization: Where Planning Meets Opportunity", *Organizational Dynamics*, (24:4), 1996, pp. 20-35.

Ein-Dor, P., Segev, E., and Orgad, M. "The Effect of National Culture on IS: Implications for International Information Systems", *Journal of Global Information Management,* Winter, 1993, pp. 33-44.

Eisenhardt, K. and Zbaracki, M. J. "Strategic Decision-Making", *Strategic Management Journal*, (13), 1992, pp. 17-37.

El-Hamamsy, L. "The Assertion of Egyptian Identity" in *Arab Society: Social Science Perspectives,* N.S. Hopkins, and S. E., Ibrahim (eds.), Cairo. The American University in Cairo Press, 1985.

Glaser, B. and Strauss, A. L. (1967). *The Discovery of Grounded Theory: Strategies for Qualitative Research,* Chicago: Aldine, 1967.

Hofstede, G. *Culture's Consequences: International Differences in Work-related Values,* London: Sage, 1980.

Khalil, O. and Elkordy, M. "The Relationship of Some Personal and Situational Factors to IS Effectiveness: Empirical Evidence from Egypt", *Journal of Global Information Management,* Spring, 1997, pp. 22-34.

Miller, S. J., Hickson, D. J. and Wilson, D. C. "Decision Making in Organizations", in *Handbook of Organization Studies,* S. R. Clegg, C. Hardy, and W. R. Nord (eds.), London. Sage, 1996.

Mintzberg, H. and Westley, F. "Decision Making: It's Not What You Think". *MIT Sloan Management Review*, (42:3), 2001, pp. 89-93.

Orlikowski, W. J. "Using Technology and Constituting Structures: A Practice Lens for Studying Technology in Organizations", *Organization Science*, (11:4), July-August, 2000, pp. 404-428.

Pfeffer, J. *Power in Organizations*, Marshfield, MA.: Pitman, 1981.

Sahay, S. "Information Flows, Local Improvisations and Work Practices", *IFIP WG 9.4 Conference*, Cape Town, 2000.

Said, E.W. *Culture and Imperialism,* London: Vintage, 1994.

Schutz, A. *The Phenomenology of the Social World,* Evanston, Ill.: Northwestern University Press, 1967.

Shoib and Jones (forthcoming). "Focusing on the Invisible: The Representation of IS in Egypt", *Information Technology and People*.

Simon, H. A. *Administrative Behaviour,* 3rd Edition, New York: Free Press, 1976. (First edition published 1945).

Suchman, L. *Plans and Situated Actions: The Problem of Human Machine Communication,* Cambridge: Cambridge University Press, 1987.

Trompenaars, F. and Hampden-Turner, C. *Riding the Waves of Culture: Understanding Cultural Diversity in Business,* London: Nicholas Brealey Publishing, 1998.

Walsham, G. *Interpreting Information Systems in Organizations*, Chichester: John Wiley and Sons, 1993.

Walsham, G. "Globalization and IT: Agenda for Research" in *IFIP TC8 WG8.2: International Working Conference* Aalborg, Denmark, R. Baskerville, J. Stage, and J. I. DeGross (eds.), Boston: Kluwer, 2000.

Walsham, G. *Making a World of Difference: IT in a Global Context,* Chichester: John Wiley, 2001.

Weick, K. E. "Sources of Order in Underorganized Systems: Themes in Recent Organisational Theory" in *Organization Theory and Inquiry: The Paradigm Revolution,* Lincoln, Y. (ed.), Beverly Hills: Sage, 1985.

About the Authors

Gamila Shoib is a Lecturer in Information Systems at the University of Bath, School of Management, United Kingdom. She was born in Alexandria, Egypt and obtained her PhD in Information Systems from the University of Cambridge, Judge Institute of Management. She was raised in Egypt and currently lives in England. Her research interests focus on social and organisational aspects of the development and use of Information Systems in organizations especially in cross-cultural contexts including less developed economies. Gamila can be reached by e-mail at G.M.Shoib@bath.ac.uk or through her homepage on www.bath.ac.uk /~mnsgmas/.

Joe Nandhakumar is a Senior Lecturer in Information Systems at the University of Bath, School of Management, United Kingdom. He was raised in India and Sri Lanka and currently lives in England. Joe gained his PhD in Information Systems from the University of Cambridge, Department of Engineering. His primary research interest focuses on social and organisational aspects of the development and use of Information Systems. Joe can be reached by e-mail at j.nandhakumar@bath.ac.uk or through his homepage www.bath.ac.uk/~mnsjn/.

Part 6:

Panels

28

PROFESSIONAL SOCIETIES IN INFORMATION SYSTEMS

A Force for Globalisation or Good?

Robert Davison[1], Ernest Jordan[2], Carol Hsu[3] and Chrisanthi Avgerou[4]

[1,3]Dept of Information Systems, City University of Hong Kong, Hong Kong; [2]Macquarie Graduate School of Management, Australia; [4]London School of Economics and Political Science, UK

1 INTRODUCTION

With the increasing importance of both Information Technology and Information Systems (IT & IS) in the economies of the developed and the developing world, professional societies in the IT/IS discipline are in a singularly important position. These societies, which have been established throughout the world, and which represent the interests of large bodies of members, have an important role to play in the development of national and international standards that apply to the conduct of their professional members. Notwithstanding the distribution of the societies, careful scrutiny reveals that while some have a comprehensive set of membership types, regulations, codes and requirements for professional development, others offer a much less rich portfolio of resources and are correspondingly much less formally structured.

Professionalism itself is generally taken to be a 'good thing', whether it applies to IT, medicine, law, accountancy, engineering or any other profession. However, professionalism also involves a considerable degree of standardisation – members of professional societies are expected to conform to established standards expressed in codes of ethics, conduct or practice and may be cautioned or expelled if they fail to so conform. Comparison of standards across countries reveals remarkably little variation in the way these standards are described, although some provide much more detail than others. Exemplar societies on the detailed side include the British Computer Society (BCS) and Association for Computing Machinery (ACM). The relatively recently established Association for Information Systems (AIS) is also moving down this path; it has now set up a standing committee on member misconduct. This committee will start the process of crafting a code of ethics for its members.

The lack of variation in the way standards are presented suggests that professional societies may, intentionally or inadvertently, promote the globalisation of professional norms (i.e. standardisation on a worldwide basis), with all the attendant concerns with which globalisation is often associated, notably absolutism, Westernisation, Americanisation and other characteristics that may detract from practices appropriate to local cultures. This situation is likely to be exacerbated by the pressure on national societies to include regulations deemed appropriate by the global majority, even though these too may not best serve local interests. Examples could range from the protection of intellectual property rights (cf. Steidlmeier, 1993) through to the ethical treatment of employees.

In this panel session, we will investigate the role of professional societies as a force for globalisation, taking a number of perspectives. We believe that this is relevant for the conference, given a) the overall theme – we see the professional societies as key players in the challenging context of globalisation; b) the international nature of the topic and its relevance to both IT in organisations (IFIP WG 8.2) and the social impacts of IT in developing countries (IFIP WG 9.4). We have invited panelists to focus on the role of professional societies from a variety of national/cultural perspectives, according to their expertise. We aim that half the available time for the panel will be taken by the panelists and half will be available for the audience to engage in debate with the panelists and each other.

2 THE PANELISTS AND A SUMMARY OF THEIR POSITIONS

Robert Davison will chair the panel session and discuss the process under way within the AIS' standing committee on member misconduct to craft a code of ethics for the society's members. Key to this development process is the soliciting of values from members as to what is acceptable behaviour and what is not. Society membership now includes people from 50 countries, including many developing countries in Latin America, Africa and Asia, as well as transitional countries in Eastern Europe and elsewhere.

Consequently, there is a unique opportunity for a wide range of people to contribute their views and values towards this code of ethics.

Ernest Jordan notes that the Australian Computer Society (ACS) acts a force for globalisation but that this is not necessarily good. The ACS has a significant role in 'approving' the IT qualifications of potential migrants, thereby facilitating their admission to Australia. This may be a unique position for a society. It means that 'conforming' education and training systems will be much better regarded than non-conforming. As a result aspiring migrants (especially from developing countries) will be 'encouraged' to follow an Australian model (or one that is closely compatible). However, differences between Australian values and those more commonly espoused in developing countries are very apparent.

Carol Hsu will compare and contrast the development of Computer Societies in Hong Kong, Singapore, Taiwan and China. Computer societies in these four countries were all established in the 1960s or early 1970s. However, Hong Kong and Singapore seem to have been more successful in promoting IT professionalism through active support and organisation by their local computer societies. Although Taiwan and China have contributed greatly to computer manufacturing production in the world, the work on establishing IT/IS professionalism seems less successful as compared with Hong Kong and Singapore. Carol will discuss how Taiwan and China can strengthen their IT professional membership with reference to the Hong Kong and Singapore experiences.

Chrisanthi Avgerou will discuss the controversial role of professional 'codes of conduct' in the light of her involvement in IFIP TC9 on the social aspects of computers. She will draw on a comparative study that TC9 conducted in the 1990s that presents and discusses a variety of countries' professional codes and draws some interesting conclusions about the range of issues they cover, the mechanisms of enforcement they foresee, and the extent of their diversity. She will then address the issue of professionalism in relation to the diffusion of standard practices in the context of globalisation.

REFERENCES

Steidlmeier, P. (1993) "The Moral Legitimacy of Intellectual Property Claims: American Business and Developing Country Perspectives", *Journal of Business Ethics*, (12:2), pp. 157-164.

About the Panelists

Robert Davison is an Assistant Professor in the Dept of Information Systems at the City University of Hong Kong. He was born in the UK, lived in various parts of Western and Eastern Europe, and was educated in the UK and Hong Kong, where he received his PhD in Information Systems. He is editor-in-chief of the *Electronic Journal of Information Systems in Developing Countries* and an associate editor of the *Information Systems Journal*. He runs several Web pages for ISWORLD, including: *Professional Ethics* and *IT in Developing Countries*. His recent work has been published in *Communications of the ACM*, *Information Systems Journal*, *IEEE Transactions on Engineering Management*, *Information & Management* and *Information Technology & People*. He has recently completed co-editing a special issue of the *IEEE Transactions on Engineering Management* on Cultural Issues and IT Management. His current work involves an exploration of virtual teams in educational contexts, as well as an ongoing interest in applying Action Research in organizational problem solving. Robert can be contacted by e-mail at: isrobert@is.cityu.edu.hk.

Ernie Jordan is Professor of Management at Macquarie Graduate School of Management in Sydney, Australia where he is the Director of the Electronic Commerce Programs and board member of the Macquarie Risk Competence Centre. Ernie was born and educated in the UK, completed his PhD in Hong Kong, and has spent most of his working life in Australia. His major research interests are in electronic commerce, risk management, the strategic management of IT, with particular emphasis on management support systems, and knowledge management systems. Publications include a range of IFIP conferences as well as *International Journal of Information Management*, *Information Systems Journal*, *Journal of Strategic Information Systems*, *Journal of Global Information Technology Management, and Information Management & Computer Security.* He adopts an Australian community leadership role of academic analysis in the area of business continuity and disaster recovery. Ernest can be contacted by e-mail at: Ernest.Jordan@bigpond.com.

Carol Hsu is a Visiting Assistant Professor in the Dept of Information Systems at the City University of Hong Kong. She also maintains an affiliation as a researcher at the *Computer Security Research Centre* at the London School of Economics and Political Science. She was born in Taiwan and pursued her higher education in the UK, where she recently obtained her PhD in Information Systems. In recent years, her research has been focused on online education and socio-organisational aspects of IS security. Her recent work has been published in *American Conference of Information Systems, and Journal of Information Systems Education.* She is currently involved in a DTI/ESRC project, which examines PKI interoperability through the lens of economics perspective, as well as the implementation of online education on campus. Carol can be contacted by e-mail at: ischsu@is.cityu.edu.hk.

Chrisanthi Avgerou is a Senior Lecturer in Information Systems at the London School of Economics. She was born in Greece and was educated in Greece and the UK, receiving her PhD in Information Systems from the London School of Economics. She chaired the IFIP 9.4 group from 1996 till 2002 and she is currently vice-chair of IFIP TC9. Her main research area is information systems and organizational change, with particular emphasis on the context of developing countries. Among her recent publications are *Information Systems and Global Diversity*, Oxford, OUP (2002), and *Information Systems and the Economics of Innovation*, Cheltenham, Edward Elgar (2003). Chrisanthi can be contacted by e-mail at: C.Avgerou@lse.ac.uk.

29

OPEN SOURCE AND FREE SOFTWARE

Organizational and Societal Implications

Joseph Feller[1], Brian FItzgerald[2], Jan Ljungberg[3] and Magnus Berquist[3]
[1]Business Information Systems, University College Cork, Ireland; [2]Department of Computer Science and Information Systems, University of Limerick, Ireland; [3]Department of Informatics, Gothenburg University, Sweden

Abstract: Although there is evidence of wide-spread organizational and societal adoption of open source and free software (OS/FS) products, processes, philosophy and business models, our understanding of OS/FS in the organizational and societal contexts is still quite limited. In this panel, we seek to stimulate an open and productive conversation by articulating the key research questions which have informed, and emerged from, the study of the socio-cultural, legal, ethical and policy issues associated with OS/FS.

1 PANEL BACKGROUND

Briefly stated, the terms "open source software" and "free software" refer to software products distributed under terms that:

- Allow users to use the software for any purpose.
- Allow users to modify the software (this requires that the source code of the product be made freely available to users).
- Allow users to redistribute (give away or sell) the software in modified or unmodified form.
- Do not require that users pay the author(s) of the software a royalty or fee for engaging in the above activities (although fees may be associated with warranties or related services) (Free Software Foundation, 2003; Open Source Initiative, 2003).

More detailed, formal definitions for the terms are maintained by the Free Software Foundation (FSF) and Open Source Initiative (OSI), however these definitions are substantively identical, and with very few exceptions a software product conforming to one definition will also conform to the other. The decision to use one of these terms rather than the other is ideological, rather than functional; the FSF prefers to use of a term that explicitly refers to freedom, while the OSI believes that the dual meaning of the English word "free" (*gratis* or *libertas*) is confusing and prefers the emphasis on the availability and modifiability of source code (Feller and Fitzgerald, 2002). In Europe, interestingly, the French word *Libre* has been widely adopted to unambiguously capture the connotation intended by the FSF.

Part of the reason for this interest, particularly in the business media, has been the impressive market-penetration of key OS/FS products like the GNU/Linux operating system, the Apache HTTP (Web) Server, BIND (an implementation of the Internet Domain Name System), Sendmail (a mail transport agent), etc. However, from a research point-of-view, the principal attraction is that the terms of distribution associated with OS/FS products are disruptive; they enable new modes of software development and use, stimulate the creation of new software business models and challenge much of the conventional wisdom relied upon by both organizational and societal decision makers and policy creators.

Since the term "open source software" was coined in 1998, mainstream interest in the open source and free software (OS/FS) phenomenon has grown considerably, evidenced by wide-spread media coverage, business and investor interest and a substantial research literature. The early research and analysis on OS/FS was disseminated primarily through the conferences and publications of the ACM, IEEE and related bodies and as such had a "hard" technology focus. However, more recently, a rich multi-disciplinary literature has emerged in mainstream information systems (IS) channels such as AMCIS, ICIS, ECIS and IFIP conferences, and journals like *European Journal of Information Systems* and the *Information Systems Journal.* This latter body of work has more explicitly focused on the social, cultural and organizational aspects of OS/FS, and will be used as the basis of this panel-led discussion.

2 PANEL THEMES

Although the emphasis of the session will be on open discussion, the panelists will briefly present the key findings and critical research questions that have emerged from the literature on OS/FS, with particular focus on the following four themes:

- The world view, ethos and practice of OS/FS development communities, and the implications these have for the organizational and societal adoption of OS/FS development models.
- The economic foundations of OS/FS, and the implications these have for the organizational and societal adoption of OS/FS business models.
- The OS/FS roles of the collaborative user-developer and user-trainer, and the implications these have for the organizational and societal adoption of OS/FS products (vis-à-vis technology acceptance, knowledge management, etc.)
- The philosophical foundations of OS/FS as a public good, and the implications these have for regional policy makers, particularly in developing regions.

REFERENCES

Feller, J., and Fitzgerald, B. *Understanding Open Source Software Development*, London: Addison-Wesley, 2002.

Free Software Foundation. "The Free Software Definition", 2003, Available at http://www.fsf.org/philosophy/free-sw.html, (Accessed at: February 12, 2003).

Open Source Initiative. "The Open Source Definition", 2003, Available at http://www.opensource.org/docs/definition_plain.php, (Accessed at: February 12, 2003).

About the Panelists

Dr. Joseph Feller and **Prof. Brian Fitzgerald** are the lead organizers of the Open Source Software Engineering Workshop Series, hosted by the annual *International Conference on Software Engineering*, and have served as guest editors for OS/FS special issues of the *Information Systems Journal*, *IEE Proceedings – Software* and *Systèmes d'Information et Management* (with Frederic Adam, UCC). They are the authors of *Understanding Open Source Software Development* (Addison-Wesley, 2002) and the editors (with Scott Hissam, SEI, and Karim Lakhani, MIT-Sloan) of *Making Sense of the Bazaar: Perspectives on Open Source and Free Software* (O'Reilly, Forthcoming). Their work on OS/FS has been published in a variety of international journals and conference proceedings. Joseph Feller is a College Lecturer in Business Information Systems at the University College Cork, Ireland, and can be contacted at jfeller@afis.ucc.ie. Brian Fitzgerald is Frederick A Krehbiel II Chair in Innovation in Global Business & Technology at University of Limerick, Ireland, and can be contacted at brian.fitzgerald@ul.ie.

Prof. Jan Ljungberg and **Dr. Magnus Bergquist** are doing research on open source from a social and organizational point of view, e.g. incitement structures for cooperation and knowledge sharing. They have published their results in the *Information Systems Journal* and *European Journal of Information Systems*. Jan Ljungberg is a professor at the Department of Informatics at Gothenburg University and research leader of the knowledge management group at Viktoria Institute. He can be reached at janl@viktoria.se. Magnus Berquist is associate professor in cultural anthroplogy and a senior faculty member of the Department of Informatics, Gothenburg University. He can be reached at magnus@viktoria.se.

Index of Contributors

GPSR Compliance
The European Union's (EU) General Product Safety Regulation (GPSR) is a set of rules that requires consumer products to be safe and our obligations to ensure this.

If you have any concerns about our products, you can contact us on

ProductSafety@springernature.com

In case Publisher is established outside the EU, the EU authorized representative is:

Springer Nature Customer Service Center GmbH
Europaplatz 3
69115 Heidelberg, Germany

www.ingramcontent.com/pod-product-compliance
Lightning Source LLC
LaVergne TN
LVHW011859200926
848484LV00002B/487

9781402074882